*F. Fasching, S. Halama,
and S. Selberherr (eds.)*

Technology CAD
Systems

Springer-Verlag Wien New York

Dipl.-Ing. Franz Fasching
Dipl.-Ing. Stefan Halama
Univ.-Prof. Dipl.-Ing. Dr. Siegfried Selberherr
Institut für Mikroelektronik
Technische Universität Wien, Austria

© 1993 Springer-Verlag/Wien
Softcover reprint of the hardcover 1st edition 1993

Typesetting: Camera ready by authors

Printed on acid-free paper

With 199 Figures

ISBN-13:978-3-7091-9317-4   e-ISBN-13:978-3-7091-9315-0
DOI: 10.1007/978-3-7091-9315-0

Preface

We would like to present the proceedings of the first "Workshop on Technology CAD Systems" (TCS) which was held at the Technical University of Vienna, Austria, on September 6th, 1993, in conjunction with the "Fifth International Conference on Simulation of Semiconductor Devices and Processes" (SISDEP 93).

This workshop is a first attempt to account for the increasing interest of the semiconductor industry, universities and vendors in the design, development and application of Technology CAD Systems. The continuously growing demands on numerical process and device simulation result in large computer programs with hundred thousands lines of code, becoming more difficult to manage and maintain by the programmer, and uncomfortable and complicated to use by the process and device engineer. The recently emerging Technology CAD Systems try to manage this increasing complexity both from the programmers' and users' point of view. The TCS Workshop brings together TCAD system developers and users from all over the world, ranging from industrial in-house developers over university researchers to professional TCAD vendors.

The organizers of the TCS Workshop have prepared a program with fourteen outstanding invited papers for oral presentation. Their respective origins reflect the international nature of the conference: 8 from the USA, 4 from Europe and 2 from Japan. The TCS Workshop is organized in five paper sessions covering the following topics: the US industrial perspective, the US university perspective, the European perspective, the Japanese perspective and the vendor perspective.

The proceedings are printed from direct lithographs of the authors' manuscripts. The editors are not responsible for any inaccuracies, comments or opinions expressed in the papers.

We would like to express our sincere appreciation to the authors for their high quality contributions and for their cooperation and effort.

Franz Fasching
Stefan Halama
Siegfried Selberherr

Institute for Microelectronics
Vienna, September 1993

Supporting Organizations

Austria Mikro Systeme International AG
Bundesministerium für Wissenschaft und Forschung
Der Bürgermeister der Stadt Wien
Digital Equipment Corporation, Austria
Die Erste Österreichische Spar-Casse – Bank
Erwin-Schrödinger Gesellschaft für Mikrowissenschaften
Forschungsförderungsfonds für die gewerbliche Wirtschaft
IEEE Austria Section
IEEE Electron Devices Society
IEEE Region 8
Der Landeshauptmann für Niederösterreich
Kammer der gewerblichen Wirtschaft für Wien
Siemens AG Österreich
Siemens AG, Zentralabteilung Forschung und Entwicklung
Siemens Entwicklungszentrum für Mikroelektronik Ges.m.b.H.
Technische Universität Wien
Textilmaschinenfabrik Dr.Ernst Fehrer AG
Österreichische Elektrizitätswirtschafts-AG
Vereinigung Österreichischer Industrieller

Table of Contents

Technology CAD at AT&T

P. Lloyd, C.C. McAndrew, M.J. McLennan, S. Nassif, K. Singhal, Ku. Singhal,

P.M. Zeitzoff, M.N. Darwish, K. Haruta, J.L. Lentz, H. Vuong, M.R. Pinto[†],

C.S. Rafferty[†], and I.C. Kizilyalli[‡]

AT&T Bell Laboratories,
1247 South Cedar Crest Boulevard, Allentown, PA 18103, USA
[†]AT&T Bell Laboratories, 600 Mountain Avenue, Murray Hill, NJ 07974, USA
[‡]AT&T Bell Laboratories, 555 Union Boulevard, Allentown, PA 18103, USA

Abstract

Technology computer-aided design (TCAD) is essential to the design of modern
integrated circuit fabrication processes. TCAD tools must not only model real processes
accurately, to allow predictive simulation during technology research and development,
but must work together as an integrated system to allow efficient exploration of
technology options. Sensitivity and statistical analyses using an integrated TCAD system
provide rapid technology characterization, including the examination of process
extremes, before fabrication. This predictive capability reduces the technology design
interval, and enables the design of optimized, manufacturable designs. This paper
describes the integrated TCAD system in use at AT&T.

1. Introduction

Efforts to reduce the dimensions and increase the performance of integrated circuits (ICs)
are becoming more challenging than in the past. Small structures are more difficult to
manufacture than the large structures of past technologies, and are more sensitive to
process variations. Their electrical characteristics are often distorted by new physical
effects, making it more difficult to use them as conventional devices. Small oxide
thicknesses, channel lengths and junction depths require small supply voltages, forcing
digital circuits to operate in more of an analog regime. At the same time, circuits of
greater complexity must be produced in shorter development cycles, forcing designers to
rely more on computer-aided design (CAD) tools to develop and verify their designs.

Tools supporting this Technology CAD (TCAD) at AT&T cover a wide range of
disciplines that includes process and device simulation, compact model parameter
extraction, interconnect simulation, and circuit simulation. As the demand for these tools
has grown, the tools themselves have matured into a system providing an end-to-end
simulation capability for process to cell level design [1]. This system is always in a state
of continuous improvement.

Accuracy of the physical models within the tools is paramount to the predictive capability of the overall system. Simulated results must be carefully verified against experimental data, and models must be improved when discrepancies arise. When individual tools produce accurate results, the entire system of tools can be used to study high level design issues. Examples include optimization of transistor characteristics and of manufacturability, and studies of the effects of statistical process variations on yield and circuit performance. A tight link to manufacturing data allows on-going verification of our TCAD tools. Figure 1 gives a high level block diagram of our system, showing how our individual tool set fits into our overall TCAD process.

In recent years, the UNIX® environment has provided the glue needed to bind our tools into an integrated system [1]. Task threads were described in shell scripts, and programs (written, for example, in *awk* [2]) were used to filter and translate data being passed from one tool to another. As software systems have continued to grow, new software technologies have emerged to help us manage the increased complexity. Object oriented programming, for instance, has provided us with a new way to organize data and the procedures that operate on them, in contrast to earlier systems, that had different file formats for individual tools and a network of conversion programs to link one tool to another. For example, in process simulation, object oriented programming has significantly simplified tool development, by providing a simple and unified access mechanism to objects that represent a device structure. Increased complexity in the control of the tools is managed by using an extension language in place of the input deck parsers of earlier tools. Finally, the actions among groups of tools are coordinated by messages passed via intertool communication facilities.

Besides point tools and software systems for integrating them, people and communications amongst them are essential to define tasks. More important, people are necessary to detect weaknesses in tools and method.

The organization of this paper is as follows. In section 2, we describe the capabilities of our point tools. These tools have been used in an end-to-end system for some time. The integration within this system is being improved using several emerging software technologies, as described in section 3. In section 4, we present several applications of the system, each one emphasizing a different aspect of TCAD simulation. In the last section we present concluding remarks and some directions for the future.

2. TCAD Point Tool Capabilities

TCAD activities at AT&T involve tools for process and device simulation, interconnect simulation, measurements, compact modeling, parameter extraction, and circuit simulation, and an optimizer that can work with all these tools. In this section we describe the tools used for each of these tasks, and the capabilities of each point tool.

2.1 Process Simulation

Process simulation is used during technology development to refine a process recipe, and during technology characterization to model the input structure for device simulation.

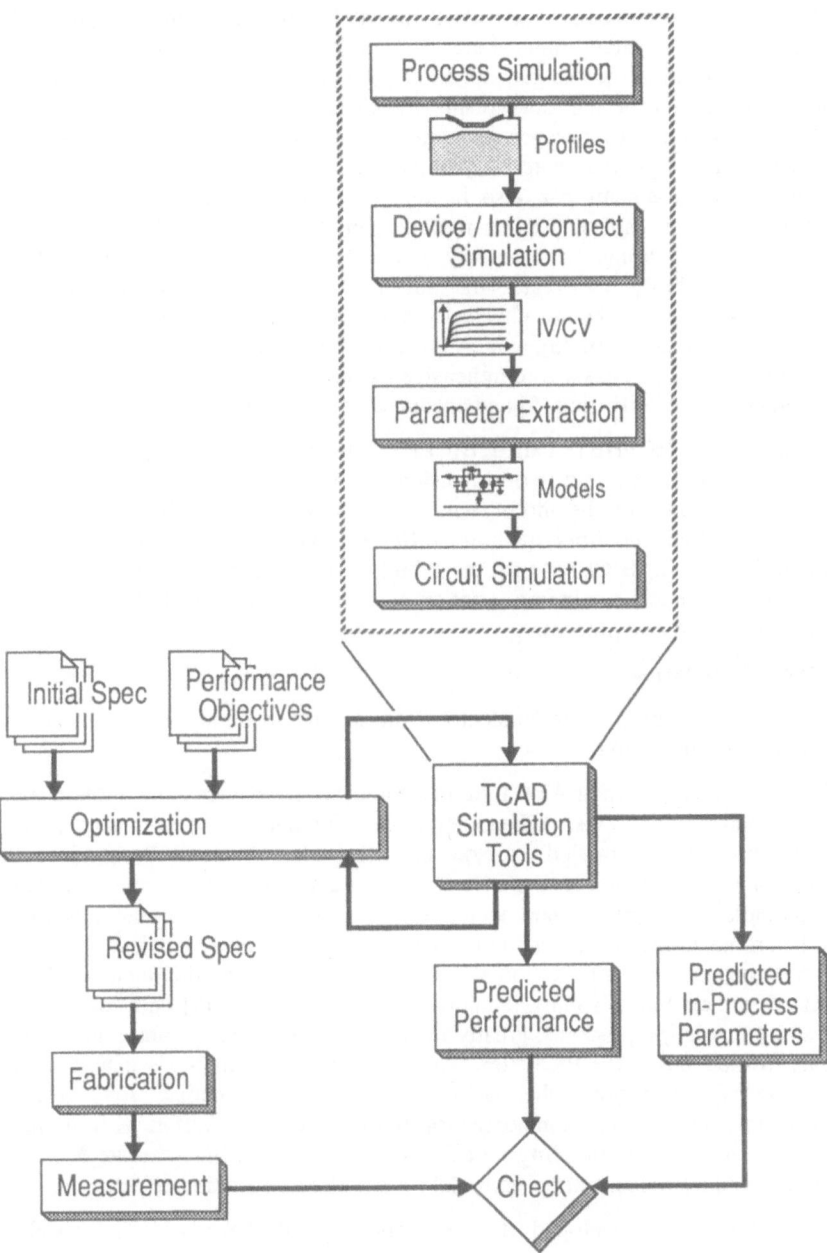

Figure 1. The AT&T Integrated TCAD System

PROPHET [3,4] was developed at AT&T Bell Laboratories for process simulation in one, two and three spatial dimensions. It simulates all the basic processing steps including predeposition, deposition, epitaxial growth, etching, oxidation, diffusion and ion implantation; and it has state-of-the-art physical models that are calibrated to measurements obtained from reported literature and from AT&T fabrication facilities. All model coefficients and material parameters are kept in an extensive database. Experimental measurements can also be integrated into this database, so that SIMS profiles, for instance, can be used to represent implant statistics. PROPHET exploits object-based programming, so it is easy to add new simulation modules to the core solver. This capability accelerates the development of new physical models. For example, PROPHET has several physically based models and one empirical model for transient enhanced diffusion [3], which is important for low-temperature and/or short anneals. PROPHET also has a comprehensive oxidation model that includes effects such as oxidation enhanced diffusion [5] and enhanced oxidation in HCl ambients [6].

A forerunner for PROPHET, called BICEPS [7], was also developed at AT&T for process simulation in one and two dimensions. It simulates the same basic processing steps, but lacks some of the more sophisticated models, found in PROPHET. Like PROPHET, BICEPS monitors its finite difference grid at each fabrication step and expands it as needed. BICEPS also has a built-in `bias` command that allows simple electrical measurements (including junction capacitances and threshold voltages) to be simulated in the midst of processing.

2.2   Device Simulation

Device simulation is used to model electrical measurements of single devices, largely for technology characterization.

PADRE [8] was developed at AT&T Bell Laboratories for device simulation in one, two and three spatial dimensions. Like other device simulators, PADRE does a full two carrier solution of the coupled Poisson and continuity equations, to produce device characteristics for steady state, transient and small signal conditions. In addition, PADRE handles arbitrarily shaped device structures, heterostructure materials, including abrupt heterojunctions, and solves both electron and hole energy balance equations to model short channel and hot electron effects. The numerical solution methods developed for PADRE are extremely robust. Adaptive gridding facilities [9] both refine and relax the mesh as simulation proceeds, greatly reducing numerical errors and putting all mesh elements to their best use. Predictor/corrector continuation methods enable it to trace negative resistance regions that arise in $I-V$ characteristics, for instance, in complementary metal-oxide-semiconductor (CMOS) circuits suffering from latch-up. PADRE also has verified mobility models that are derived from accurate Monte Carlo simulations, and a single event upset model for simulating radiation effects.

MEDUSA [10] was developed at the University of Aachen as a mixed mode device/circuit simulator with physically based models for both metal-oxide-semiconductor field effect transistors (MOSFETs) and bipolar junction transistors (BJTs). Because it can solve device level (drift-diffusion) equations along with circuit equations, changes to doping profiles and device dimensions can be studied directly in circuits like inverters and memory cells. The MEDUSA MOSFET model is based on several simplifying assumptions that make it fast and efficient, though less general than PADRE. Because of its modest computation requirements, MEDUSA is used for routine technology characterization where its assumptions are valid and where large data sets are required for parameter extraction.

We also use Monte Carlo simulations to analyze MOSFET and BJT reliability, and to fit transport coefficients in the energy balance models of PADRE. BEBOP [11], developed in collaboration with the University of Bologna, is our first generation Monte Carlo simulator. BEBOP can solve for both electron and hole transport self-consistently with the Poisson equation, for an isotropic band structure with arbitrary (nonparabolic) $E(k)$. More recently, we have developed a general full-band code that includes models of additional physical mechanisms, such as Zener tunneling and carrier transport through oxides. Through algorithmic improvements, such as phase-space discretizations using refinable simplexes [12], our full-band code is comparable in speed to analytic band codes, yet gives rigorously correct results for high fields and hot carriers. An analysis of bipolar degradation has recently been reported using this capability [13].

2.3   Interconnect Simulation

Propagation delay associated with interconnect is becoming a significant factor limiting the performance of ICs in advanced technologies. Accurate characterization of interconnect is thus necessary for circuit design.

RESCAL [14] is a two-dimensional Laplace equation solver that calculates the capacitance matrix for one or more conductors embedded in dielectric materials. Both conductors and dielectrics can have arbitrary shapes, and RESCAL can handle infinite and semi-infinite boundaries. TLP [15] is a two-dimensional solver that computes the transmission line parameters for rectangular conductors embedded in planar dielectric slabs. TLP uses a frequency domain moment method, based on ellipsoidal harmonics. TCLP is a version of TLP that includes an extension language both to enable users to program high level tasks easily and to provide a graphical user interface (GUI). QP is a three-dimensional solver, also based on frequency domain moments, that computes capacitances for conductors composed of metal parallelepipeds embedded in planar dielectric structures.

EASI is an above-silicon simulator that generates the three-dimensional metal, polysilicon and insulator structure of the IC interconnect from a process recipe and its associated mask information. It uses empirical models for etching and deposition, and is therefore faster though less accurate than process simulators with more complex models.

A program TWINE runs the interconnect simulators for various combinations of runner placements, dimensions and spacings. Results are fed to GOALIE, a program that extracts circuit models specified in terms of conductor dimensions and spacings between conductors that come directly from layout.

Thus, given a process recipe and mask information, we can simulate the topography of the interconnect using EASI, compute the interconnect capacitances using RESCAL, TCLP and QP, and extract compact models for use in circuit simulation.

2.4   Measurements

Measurement is the final arbiter for any semiconductor simulation system. Our TCAD system includes several types of measurement systems. A variety of measurements, including SIMS, RBS and SEM/TEM, are used for on-going verification of process simulation models. A measurement laboratory that includes wafer probes, thermal chucks, and instruments to measure DC currents, AC impedance, s-parameters, noise, and so on, is used to verify our TCAD characterizations. Manufacturing line test measurement systems also provide data that are compared to playbacks of the compact models used for circuit design, for on-going verification of the process extremes predicted by our TCAD system.

Many of the software concepts and practices that underly our integrated TCAD system are also used in our measurement laboratory control systems. The tester and device descriptions and the measurement modules for each instrument and each type of measurement are manipulated as objects that support inheritance. The control software is written in an extension language, so it can be integrated tightly with a GUI. A test wafer, for example, is represented by a wafer diagram in the GUI, see figure 2. Clicking on any site in that diagram causes the automatic prober to move to that site and do a measurement. The data object created by this procedure can then be accessed by other task modules, for instance a parameter extractor.

Archiving of data is important. It provides a reference database of information spanning many manufacturing technologies. We keep such an archive of both measured and simulated data. This database is continually used for verification and on-going development of our process and device simulators and of our compact semiconductor device models used for circuit simulation.

2.5   Circuit Level Simulation and Models

The highest level in our TCAD system is circuit simulation. IC designers at AT&T currently use ADVICE [16] for this task, although a new circuit simulator CELERITY is being developed and will be deployed soon. CELERITY uses advanced numerical techniques for fast, robust analyses, and uses an extension language to provide flexible control over solution algorithms.

The main link between circuit level simulation and lower level TCAD tasks is through the compact models used in circuit simulators to characterize the behavior of individual circuit components. These models are fitted to the data produced from process and device simulation, providing circuit designers with a CAD environment that accurately characterizes the manufacturing lines that will make their circuits.

BJTs are modeled in ADVICE with variants of the Gummel-Poon BJT model [17], including an extended model [18] that includes quasi-saturation effects. Three and four terminal BJTs can be modeled, as can heterojunction bipolar transistors (HBTs) and self-heating effects [19].

ADVICE contains series of MOSFET models that trace the evolution of compact models at AT&T, CSIM [20], ASIM [21], ASIM3 and AUSSIM (see Figure 3). CSIM has been successfully used for digital design for almost two decades. However, with decreasing

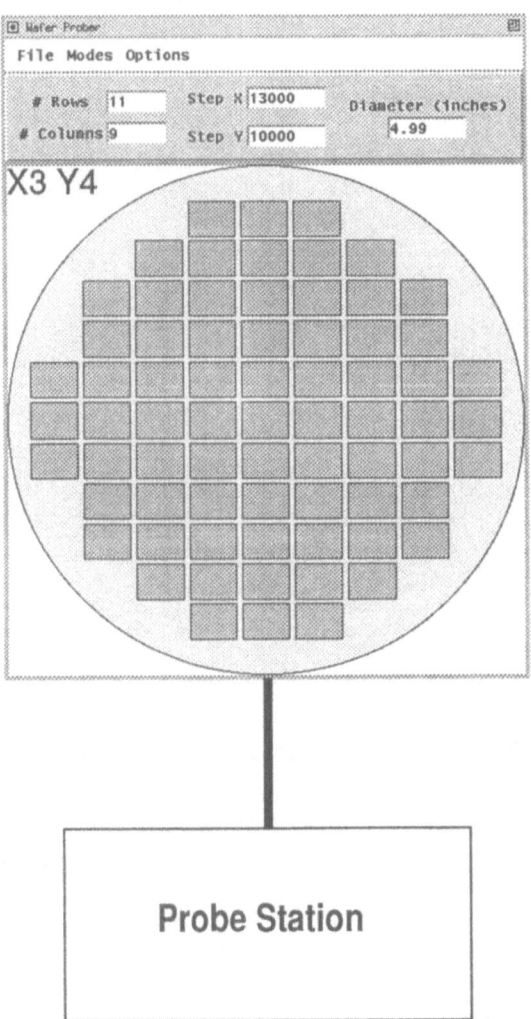

Figure 2. Graphical Interface for Measurement Equipment

device dimensions, decreasing supply voltages, and increasing circuit speeds, it is no longer accurate enough for design in advanced technologies that need greater analog modeling accuracy. To meet these needs, a charge-based model ASIM was developed, that models subthreshold conduction. However, ASIM is a regional model, and so has kinks and glitches at (and is inaccurate near) the boundaries between below and above threshold regions, and between triode and saturation regions. ASIM is also a source-referenced model, and so has asymmetric capacitance coefficients. To overcome these problems, ASIM3 was developed. ASIM3 is a charge-based, single-piece, bulk-referenced model, that is accurate over bias, geometry and temperature. AUSSIM is a simplified form of ASIM3 that is significantly faster to compute, but is still much more

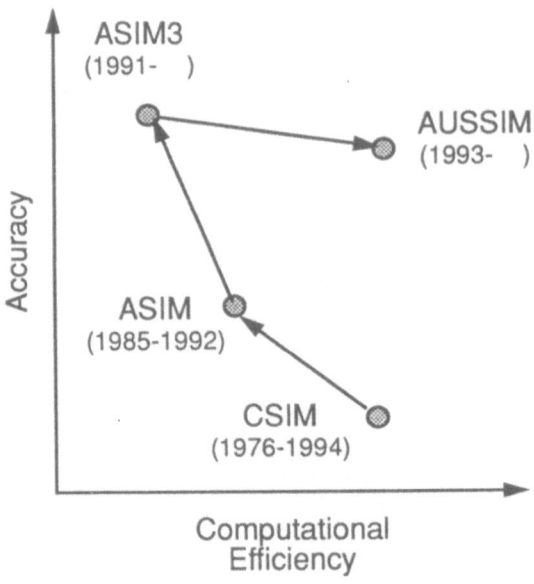

Figure 3. The Evolution of Compact MOSFET Models as AT&T

accurate than CSIM. Both ASIM3 and AUSSIM satisfy the "requirements for good analog models," and pass the smoothness and noise benchmark tests, presented in [22], and they also both pass some additional MOSFET model tests that we have developed.

Smooth, C_∞–continuous models, such as ASIM3 and AUSSIM, have many advantages over regional models (see [23] and references therein). They are more accurate near region boundaries, give better DC convergence, allow longer time steps in transient analyses, are easier to extract parameters for, give more accurate distortion analyses, and allow the use of advanced numerical techniques such as homotopy and high order integration methods for transient analysis.

Symbolic formulation of a compact semiconductor device model is only a small part of the process of developing a model useful for IC design. The model must be implemented in one or more simulators, including digital timing simulators, and model parameters must be extracted for a range of target manufacturing technologies. These processes are long and tedious, and prone to error. To help in these tasks we have developed the CAMELOT system, that is described in more detail below.

2.6   Optimization and High Level Control

Having accurate simulation tools in an integrated environment leads to opportunities in analysis and optimization. For instance, it is natural to ask "What are the process settings that give optimum device characteristics for my design application?" or even "What process settings result in the best yield for this design?". To answer these types of manufacturability questions, we have developed CENTER [24,25] for analysis, optimization and high level control tasks.

A typical goal of a nominal technology optimization is to maximize drive current while keeping threshold voltage and leakage current within specified bounds, by adjusting processing parameters such as implant doses and energies and anneal or drive-in times. CENTER can do this because it is capable of optimizing multiple objectives subject to specified constraints.

In manufacturability applications, CENTER can make use of process control databases to predict statistical variations in device and circuit performances. CENTER's optimization capabilities can then be used to reduce circuit sensitivities and hence enhance manufacturing yield.

Our integrated TCAD system allows robust design threads to be built, for example by coupling process and device simulators. TCAD tools configured into a design thread are interfaced to CENTER either via a two-way pipe or via shell scripts and data files. These uniform communication interfaces make it simple to configure tools to work with CENTER, and our integrated TCAD system makes execution of the design threads robust. Both of these factors significantly simplify the application of optimizers and other high level control tasks to complex design threads.

CENTER can also be used to help improve the physical models in other point tools. For example, the oxidation models used for process simulation have in the past been derived from experiments done at high temperatures. With the trend to low temperature processing, the predictions of the high temperature models may become inadequate. CENTER has been used to optimize the fit of the oxidation models to both high and low temperature oxidation data, the latter from recent measurements. This has significantly improved the accuracy of our process simulators.

Because no single optimization method works best on all design problems, CENTER has six optimization algorithms including simulated annealing, constrained sequential quadratic programming, and a nonlinear least squares minimizer. In addition, CENTER contains modules for other high level tasks including sensitivity analysis, worst case analysis, and mathematical model building. CENTER also provides a simple, consistent way to allow new numerical analysis modules to be applied to design problems.

3.   TCAD Framework Facilities

At AT&T, TCAD tools have been used in an integrated approach to modeling since the mid 1980's [26,27]. These tools are organized under a common access program *tcad* that manages security, tracks usage patterns, provides on-line documentation, and supports communication with tool developers through news messages and electronic mail. In the past, *tcad* and the MECCA characterization system [27] have relied on the UNIX® operating system to organize individual point tools into an end-to-end system. Simulation tasks, for instance, were controlled by shell scripts. Data files were transformed from one format to another using tools such as *awk*, *sed* and *m4*. Recently, new software technologies have emerged. They tighten the integration of the overall system by improving communication between the tools and giving users better facilities to customize an integrated system to meet their needs.

Object-oriented programming [27] is one such technology that we have leveraged to improve the exchange of data between simulators. For many years, data have been stored as textual descriptions in ASCII files. Indeed, standards have even emerged to define such file formats [28,29]. For simple data types, such as $I-V$ characteristics or one-dimensional doping profiles, this treatment is certainly adequate. For intricate two- and three-dimensional semiconductor structures, however, it begins to break down. Such structures have complex relationships between data that are difficult to manage. Object-oriented programming offers a natural solution to this problem, and has therefore become the basis for new standards being developed, for instance, by the TCAD Subcommittee of the CAD Framework Initiative (CFI) [30].

For semiconductor process and device modeling, object classes are created to represent material regions (including points, edges, faces and solids) and the finite element solutions that are contained within them (including doping profiles, electron density, etc.). All objects are tagged with a unique identifier and moved to an object-oriented data repository for persistent storage. TCAD tools act as producers and consumers of such objects by retrieving objects, doing simulations, and producing new objects in the repository.

Since the object-oriented paradigm encourages data encapsulation and the construction of well defined interfaces, the resulting code is modular and easy to maintain. We have adopted C++ as our high level programming language since it integrates easily with existing simulators written in C and FORTRAN.

New technologies are also emerging to provide better coordination among tools. Traditional UNIX® sockets have become the building blocks for higher level communication schemes developed by standards organizations such as the CFI [31]. Known generically as Intertool Communication (ITC), these low-bandwidth links are used to pass simple status messages or references to data. Tools can register to receive certain types of messages, and can post messages globally or send messages to specific tools. When higher bandwidth is required, for instance in passing semiconductor data between tools, shared memory [32] can be used. We generally rely on sockets using simple ITC protocols, and sometimes use named pipes [33] to avoid having to manage temporary files.

The development of GUIs and extension languages has significantly improved the interfaces to our tools. GUIs provide easy access to on-line help and to plotting facilities, and they aid both the novice and the expert user. Extension languages such as Scheme [34] and TCL [35], act as a powerful command language in which the user can express new algorithms, or test new modules for diffusion or mobility models, for example. The use of a common extension language gives a uniform user interface for all tools. Given programmatic access to a tool, users can write their own routines to manipulate inputs, post-process outputs, and customize the tool operation.

One such tool that makes use of a GUI is the interconnect capacitance simulator TCLP, see figure 4. It provides a GUI in which the user can "sketch" the problem domain and interactively query capacitance values from the drawing. It also provides a means of associating dimensions on the drawing with variables that can be manipulated in the

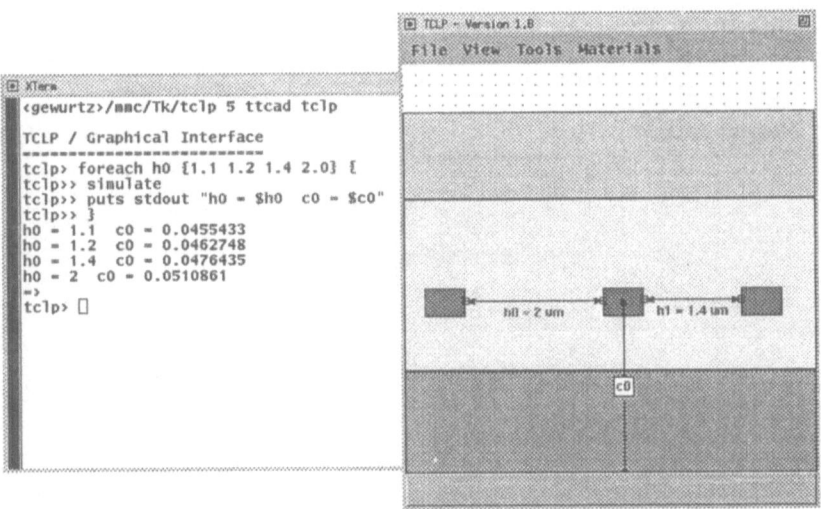

Figure 4. TCLP Interconnect Capacitance Simulator

extension language. After marking the distance between conductors with a dimension
h0, for instance, the user can write an extension language statement to vary h0 over a
range of different lengths and calculate the resulting capacitance values. These values
can be passed to other extension language routines for post-processing, to do, for
instance, statistical analysis or optimization. In the UNIX® paradigm, these
calculation/analysis functions would have been done by separate programs sharing
intermediate data via ASCII files. In the new paradigm, all computation tasks are tightly
integrated in a single program, and the tasks themselves are more easily customized by
the end user to suit a specific application.

4. Application Examples

In this section we present examples of applications of the use of our integrated TCAD
system for technology optimization, for statistical and sensitivity analysis, for technology
characterization, and to aid compact modeling activities.

4.1 Technology Optimization

We have used CENTER to optimize several manufacturing technologies, most recently
an AT&T CMOS technology. The goal was to study the modification of an existing
technology, to be optimized for low supply voltage ($V_{dd} = 1.2$V) operation. Figure 5
shows a block diagram of the TCAD tool system configured for this application.

CENTER adjusts specified *design variables* to optimize defined *design performances*.
Design performances include quantities to be minimized or maximized, called *design
objectives*, as well as *design constraints* to be satisfied.

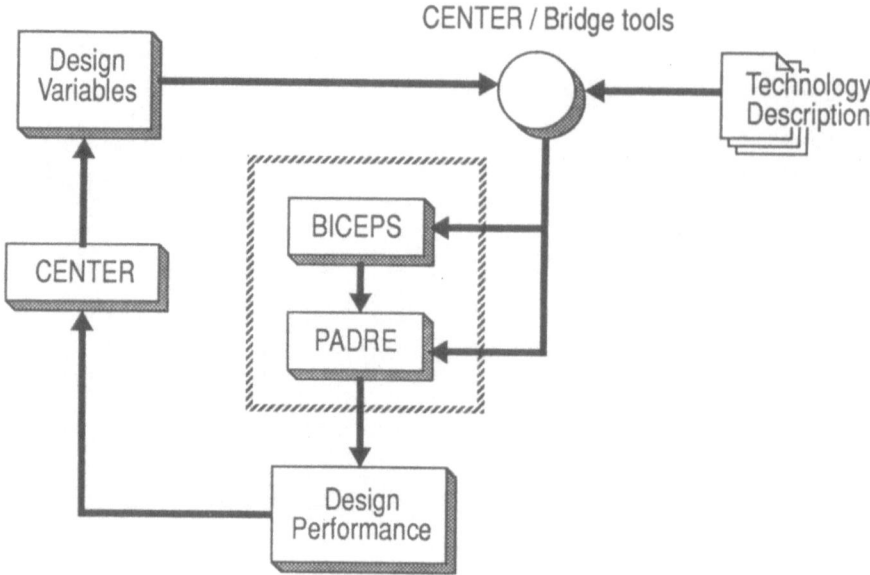

Figure 5. TCAD System Configured for Technology Optimization

For both n– and p–channel transistors, the design objectives were to maximize the drive current I_{on} (I_d at $V_{gs} = V_{ds} = 1.2$V and $V_{bs} = 0.0$V) and minimize the body effect (change in threshold voltage V_{th} as V_{bs} changes from 0.0 to -1.2V). The design constraints were that V_{th} be within the range 0.4–0.5V, leakage current I_{off} (I_d at $V_{gs} = V_{bs} = 0.0$V and $V_{ds} = 1.2$V) be less than 0.15nA/µm, and $\partial V_{th}/\partial L_m$ (the rate of change of V_{th} with masked channel length L_m, at the minimum L_m) be less than 1.2V/µm. The design variables to be adjusted to optimize the technology were the minimum L_m, and the doses and energies of the V_{th} adjust and lightly doped drain (LDD) implants, for both n– and p–channel transistors. The initial values of these design variables were those of the original technology.

Table 1 shows the results of the optimization. Clearly, the drive current has been significantly increased, while meeting the specified design constraints. The cost was a slight increase in the body effect.

4.2 Statistical and Sensitivity Analysis

Recently, our integrated TCAD system was used to study the effect of process variations at the circuit level for a BiCMOS technology [37]. Specifically, the propagation delay τ_d of a BiCMOS inverter was studied with respect to six process parameters X_i identified as being the most critical for BJT performance.

Figure 6 shows a block diagram of the TCAD tool system configured for this application. Doping profiles for the BJTs were simulated using BICEPS, and were used to build a quasi two-dimensional, physical BJT model in MEDUSA. Since the focus of the study

Design Variable	Design Objective/Constraint	Specification	Optimization Result
	I_{on}	maximize	up 150%
	$\Delta V_{th}/\Delta V_{sb}$	minimize	up 10%
	V_{th}	$0.4 < V_{th} < 0.5V$	satisfied
	I_{off}	$<0.15nA/\mu m$	satisfied
	$\partial V_{th}/\partial L_m$	$\leq 1.2V/\mu m$	satisfied
minimum L_m			down 15%
V_{th} adjust dose			down 64%
V_{th} adjust energy			up 18%
LDD dose			down 74%
LDD energy			up 134%

Table 1. Results of Low Voltage CMOS Technology Optimization

was on statistical variations affecting BJT performance, the MOSFETs in the BiCMOS inverter (see figure 7a) were simulated using compact models with fixed parameters. Only the physical BJT model was varied statistically from sample to sample. Using the mixed mode device/circuit simulation capability of MEDUSA, the transfer characteristic for each statistical sample was computed directly and analyzed to extract the propagation delay τ_d. Note that without a mixed mode simulator, an intermediate step of computing BJT device characteristics and extracting compact model parameters would have been necessary, adding considerable time and complexity to the problem.

Random values for the six process parameters X_i were generated from a uniform distribution using the Latin Hyper-Cube (Monte Carlo) method. Using this method, a total of sixty samples was adequate to represent the range of process fluctuations.

Table 2 summarizes the correlations of τ_d to the process parameters X_i. Correlations for other BJT performance parameters are reported as well. The dominant influence on propagation delay τ_d is the epi-layer thickness X_1, with a correlation coefficient $r = 0.97$. This is expected, since the load capacitor discharges through the collector of transistor Q_2 during the transient cycle, making the circuit performance sensitive to the collector resistance. Reducing the epi-layer thickness decreases the collector resistance, and thus reduces the propagation delay.

This result was verified by experiment, by producing a series of wafers with different nominal epi-layer thicknesses and measuring the propagation delay through a chain of inverters. Because of differences in the way τ_d was determined, it is difficult to compare the absolute values of the simulation and measurement results. From simulation, τ_d was computed directly as the delay for a single inverter, with a load of 3pF, driven with a slow (5ns rise-time) ramp, whereas τ_d from measurement was computed as the period of a ring oscillator, loaded with both 5pF and 10pF per stage, divided by the number of stages. Nevertheless, figures 7b and 7c show that the strong correlation between τ_d and the epi-thickness is evident in both simulated and measured results. For the epi-growth

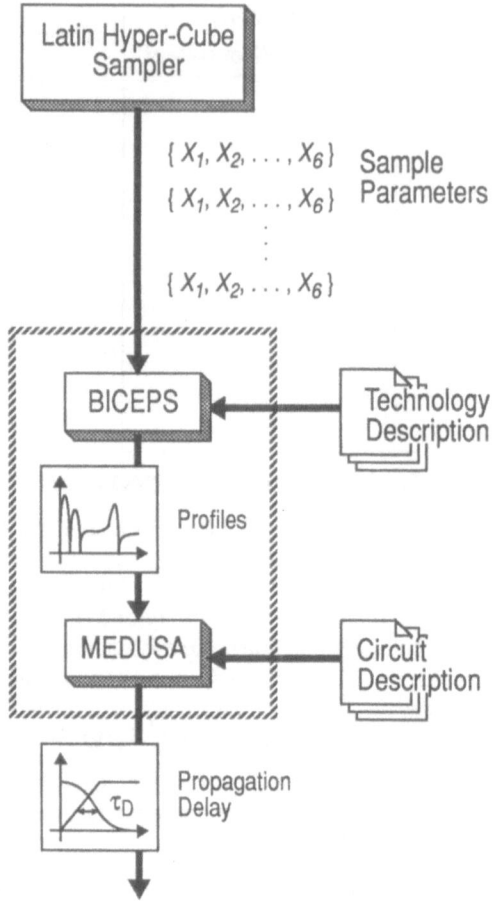

Figure 6. TCAD System Configured for Statistical Analysis

process used in this study, the epi-thickness varied across each wafer, increasing by roughly 10% near the edge of a wafer. A corresponding increase of nearly 10% in the propagation delay was found in inverters near the edge of the wafer.

4.3   Technology Characterization

Technology characterization involves a set of parameter extractions that cover a range of temperatures and device geometries and account for nominal and worst-case performances that arise from process variations. A characterization starts with a specified process sequence. Process simulators are used to generate the doping profiles for the structures being characterized. Profiles for the nominal and extreme cases are obtained by varying the implant doses, drive-in times, processing temperatures, and so on, according to the variations expected on the fabrication line. Extreme case files include not only fast and slow cases for digital design, but also various mismatch cases

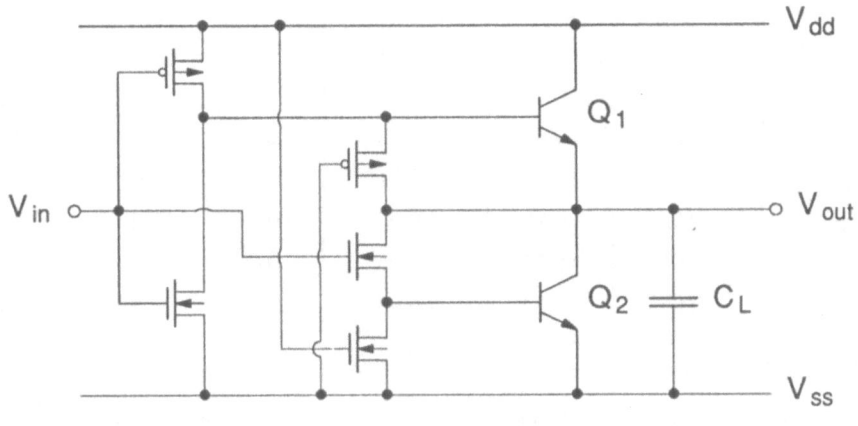

(a) BiCMOS Inverter Circuit

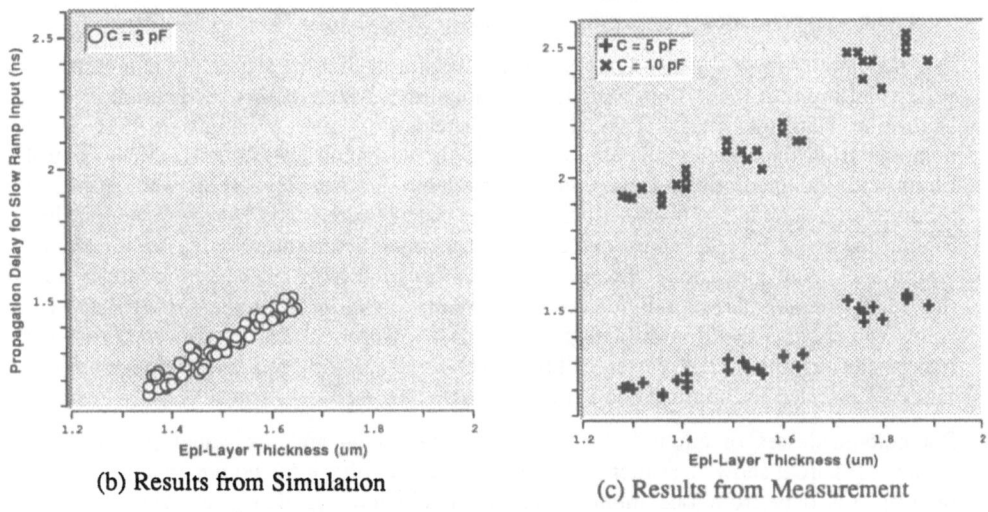

(b) Results from Simulation

(c) Results from Measurement

Figure 7. BiCMOS Inverter and Scatter Plots of τ_d Versus Epi-Layer Thickness X_1

Parameter	Epi Thickness X_1	Emitter Dose X_2	Base Oxide Etch X_3	Tub Dose X_4	S/D Anneal X_5	RTA X_6
τ_d	0.97	−0.05	0.28	−0.01	0.23	−0.05
f_t	−0.26	0.22	−0.91	0.02	0.15	0.11
β	−0.04	0.10	−0.96	−0.04	0.14	0.14
w_b	0.05	−0.22	0.95	−0.07	−0.11	−0.11
ρ_b	−0.05	0.22	−0.93	−0.04	0.18	0.15

Table 2. Correlation Coefficients Between BJT Electrical and Process Parameters

required for aggressive linear design. The resulting profiles, along with geometry specifications, are used as input to device simulators that generate DC, $C-V$ and transient characteristics. These simulations are repeated for the various geometries, temperatures, and extreme processing cases. Parameter extraction tools are then used to create model parameter files used in circuit simulations. This approach gives predictive process characterizations, and enables silicon designers to evaluate a technology before it is completed. Early test wafers are then analyzed, and used as part of the process of on-going verification and improvement of our TCAD system. This allows out tools to be used with some confidence to predict the behavior of next generation technologies.

Our simulation system is also coupled to manufacturing line measurement databases, to verify and monitor the system's predictive capability. Much of this verification is done using the statistical and graphical tools provided by the S language [37]. The manufacturing databases track production variations, which are plotted against control limits and the predictions from compact models. Statistical analyses and graphical reports are generated to ensure conformance to specifications, and to reduce design risk. Figure 8 shows a typical report, that summarizes process variations in I_{on} for a nominal length n–channel transistor. The left segment shows the trend over a six month period, the middle section shows individual measurements made during the last month, and the right segment is a $Q-Q$ plot of data for the last month. Shaded regions represent the manufacturing limits, and the solid horizontal lines are the predictions from compact models generated for nominal and extreme cases.

For certain classes of circuits, the matching between n– and p–channel transistors is critical. Figure 9 shows a scatter plot of the threshold voltage for both n– and p–channel devices, for data from one month from the manufacturing database. The unshaded region, inside the box, represents 2σ limits for the technology. The six sided polygon, drawn with a solid line, is the convex hull formed from the predicted compact model parameter files for the extreme processing cases, which are the vertices of the polygon. To maximize yield, the process is controlled so that the number of sample points falling within the convex hull is maximized.

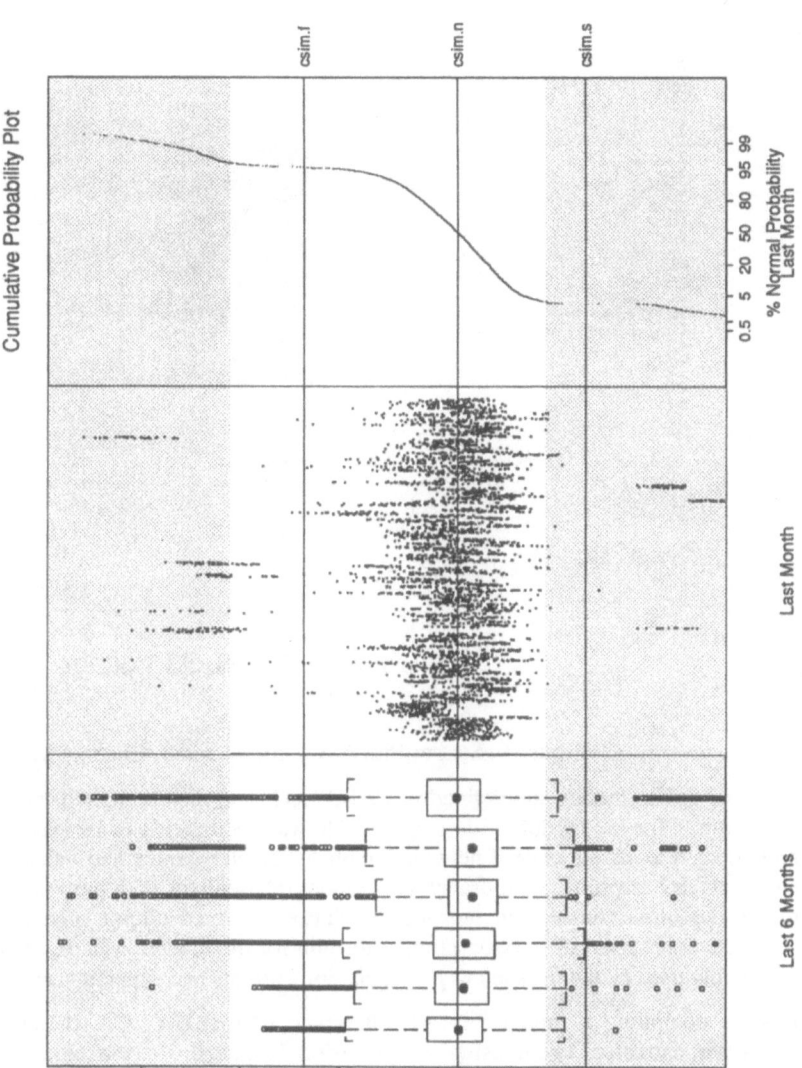

Figure 8. Graphical Summary of Recent Measurements in the Manufacturing Database

4.4 Compact Modeling Aids

As noted above, there is a significant effort, and a significant hazard for error, involved in hand coding a compact device model in a circuit simulator. To automate this task, and guarantee its correctness, we have developed the CAMELOT system.

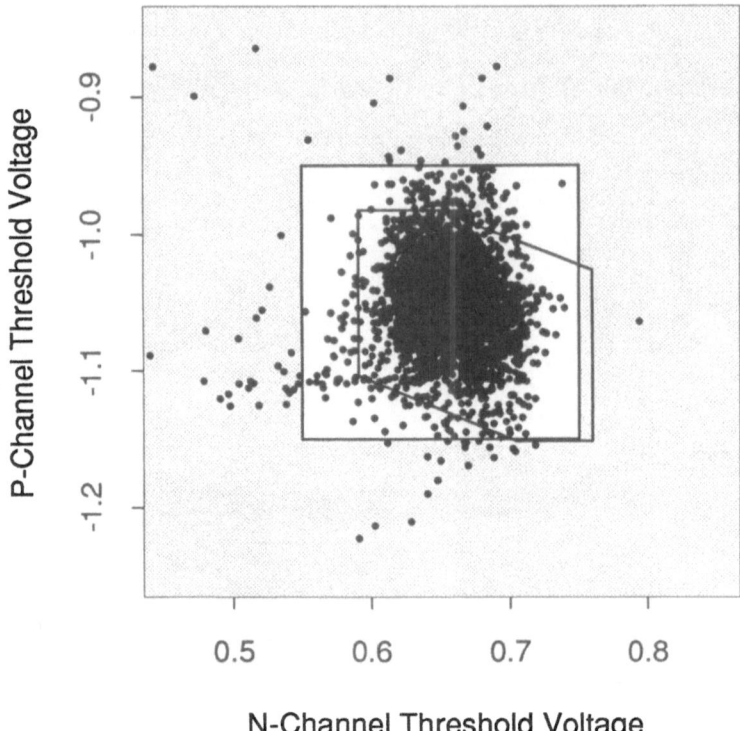

Figure 9. Scatter Plot of V_{th} for $n-$ and $p-$Channel MOSFETs

Compact device models are defined in CAMELOT, and then compiled into target circuit simulators. The model definition language allows specification of all aspects of a device model, such as its equivalent network graph, parameter names and values, temperature and geometry mappings, and branch constituent relations for both current and charge. A model definition can also include noise relations, printed output specifications, and a complete description of the extraction algorithm for the model. Macros are available for commonly used elements, such as $p - n$ junction currents and capacitances,

Besides compiling a model into ADVICE and CELERITY, CAMELOT generates a parameter extractor, again using an extension language. In the parameter extractor, models, data, parameters and optimizers are defined as objects, and an extraction algorithm, coded in an extension language, is defined and implemented as a model object being applied to subsets of the data objects and parameter objects, see figure 10. The model objects are created directly from the model definition, and besides calculating device terminal currents they also compute conductances and capacitances as seen from the terminals of the device. This allows both DC and AC data, e.g. measured or simulated capacitances and/or s –parameters, to be fitted during an extraction. The data objects used within CAMELOT can come directly from either measurements or from

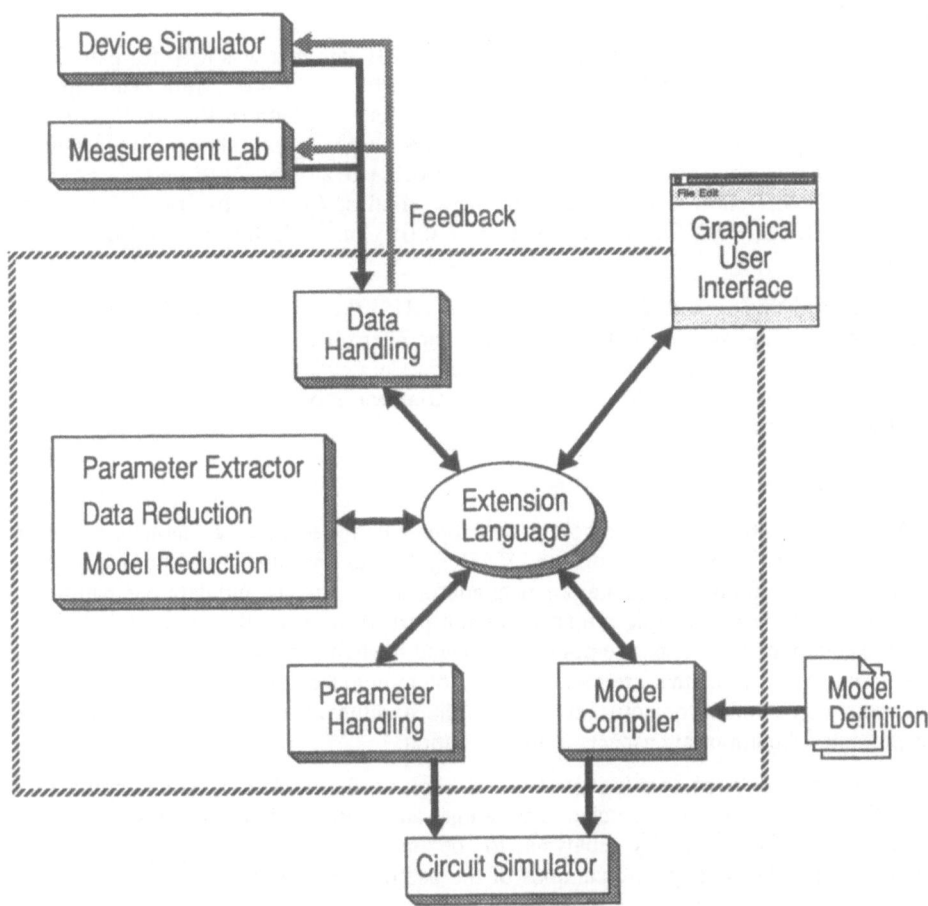

Figure 10. Architecture of the CAMELOT System

device simulations.

Using CAMELOT, all information about one compact device model, or a hierarchy of models such as three and four terminal $n-$ and $p-$channel MOSFETs, is stored in one place, rather than being scattered throughout many subroutines in program code for a circuit simulator. We find this to be convenient for the conceptual development of a model. There are also several other advantages to using a model compiler and a single high level model definition. First, there is no duplication of any information. Duplication of information can cause errors in models that are implemented in separate pieces, because of inconsistencies that arise when updates to a model are not propagated to all pieces of the model. Second, derivatives and small-signal (linearized or AC) models are guaranteed to be correct by construction. Third, codes for different derived functions, such as circuit simulation, parameter extraction, and stand alone evaluation, are guaranteed to be consistent with each other.

Figure 11 shows playbacks of ASIM3, against PADRE data that was used for extraction in CAMELOT. The data is for an n–channel MOSFET of $L_m = 0.25\,\mu$m. The top left plot of figure 11 shows threshold characteristics, I_d as a function of V_{gs} for $V_{ds} = 0.1$V and $V_{bs} = 0.0, -0.6, -1.2$ and -2.4V. The top right plot shows subthreshold characteristics, I_d as a function of V_{gs} for $V_{ds} = 0.1, 1.0, 2.0$ and 3.0V and $V_{bs} = 0.0$V. The bottom plots show triode-saturation characteristics, I_d as a function of V_{ds} for $V_{gs} = 0.5, 1.0, 1.5, 2.0, 2.5$ and 3.0V and $V_{bs} = 0.0$V, on both linear and logarithmic scales.

We have used this system for the ASIM3 and AUSSIM MOSFET models, for many flavors of BJT models, for diode models, and for 3-terminal resistor models. We have found that the system greatly simplifies compact model development and implementation in ADVICE and CELERITY. All our new compact models are defined using the CAMELOT system.

5.   Conclusion

TCAD is an essential part of technology development and technology characterization at AT&T. Critical to the success of our TCAD system is on-going tool improvement to ensure accuracy, robustness, and a practical approach to software and data compatibility. We also recognize that people are an important part of the overall system. Our TCAD system is evolving from separate point tools bound in shell scripts, that communicate via files, to integrated tools that are modules in a common extension language environment, that operate on common objects that represent the structures being simulated. This paper has presented the important aspects of this evolution.

Interoperability and compatibility of point tools and simulation data significantly simplifies the job of process characterization, and allows high level tasks such as optimization and sensitivity analysis to be used effectively during technology development. We have given examples of the use of our TCAD system for high level tasks.

Acknowledgements

We would like to thank R. V. Booth, W. T. Cochran, W. M. Coughran Jr., E. Grosse, H. K. Gummel, G. A. Howlett, P. A. Layman, S. Liu, L. W. Nagel, R. K. Smith, M. J. Thoma, G. Zaneski, and many other colleagues at AT&T Bell Laboratories for contributions to this work. The use of the AT&T Bell Laboratories DDL facility in Allentown is gratefully acknowledged.

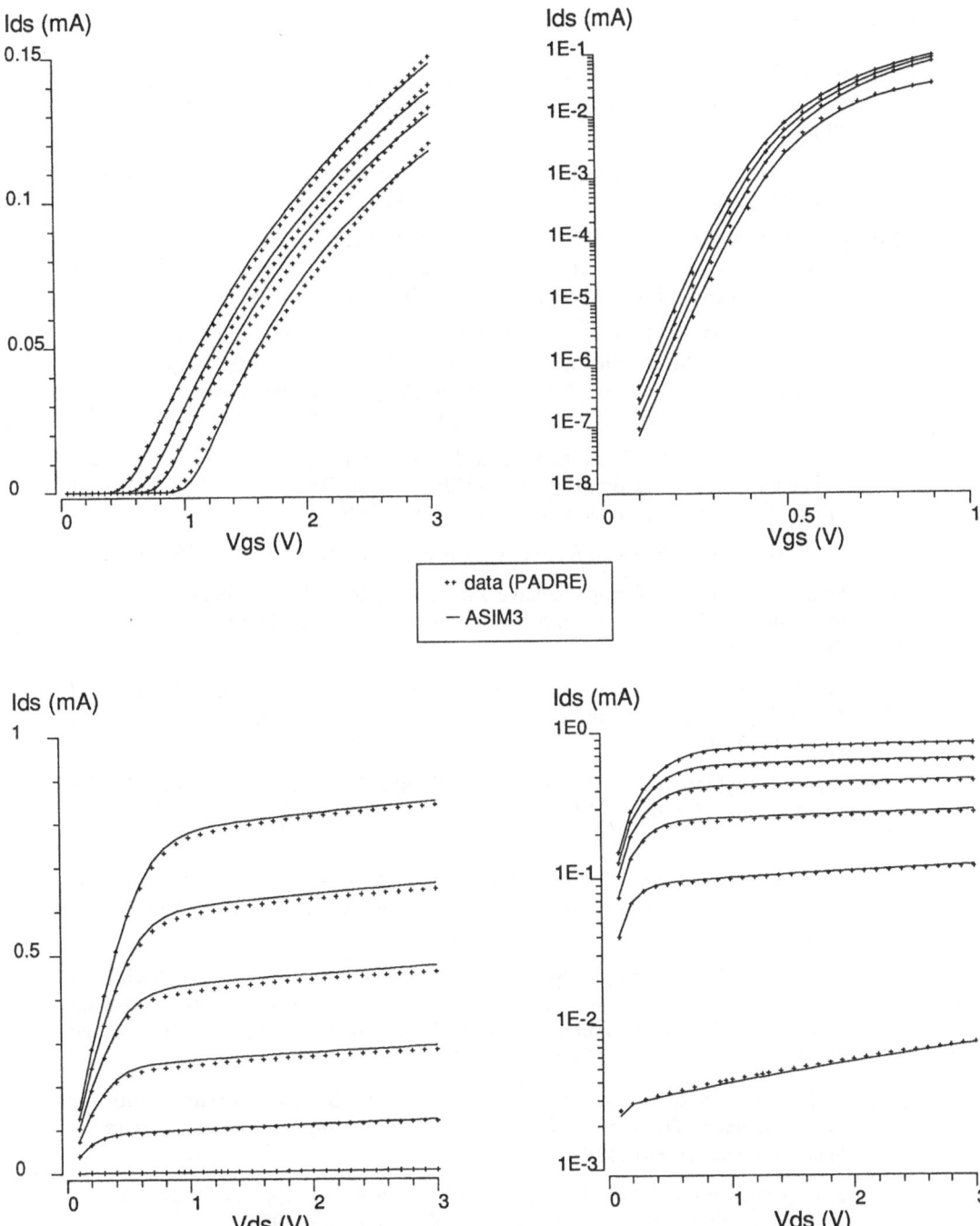

Figure 11. Data Fits of the ASIM3 Compact Model, $L_m = 0.25\mu m$

References

[1]   P. Lloyd, H. K. Dirks, E. J. Prendergast, and K. Singhal, "Technology CAD for Competitive Products," *IEEE Trans. Computer-Aided Design*, vol. 9, no. 11, pp. 1209-1216, Nov. 1990.

[2]   A. V. Aho, B. W. Kernighan, and P. J. Weinberger, *The AWK Programming Language*, Reading, MA: Addison-Wesley, 1988.

[3]   C. S. Rafferty, M. D. Giles, H.-H. Vuong, S. A. Eshraghi, M. R. Pinto and S. J. Hillenius, "Anomalous short-channel body coefficients due to transient enhanced diffusion," Proc. VLSI Process/Device Modeling Workshop (VPAD), May 1993.

[4]   M. R. Pinto, D. M. Boulin, C. S. Rafferty, R. K. Smith, W. M. Coughran, I. C. Kizilyalli, and M. J. Thoma, "Three-dimensional characterization of bipolar transistors in a submicron BiCMOS technology using integrated process and device simulation," IEDM Technical Digest, pp. 923-926, Dec. 1992.

[5]   A. M. Lin, D. A. Antoniadis, and R. W. Dutton, "The Oxidation Rate Dependence of Oxidation-Enhanced Diffusion of Boron and Phosphorus in Silicon," *J. Electrochem. Soc.*, vol. 128, p. 1131, 1981.

[6]   D. W. Hess and B. E. Deal, *J. Electrochem. Soc.*, vol. 124, no. 5, p. 735, 1977.

[7]   Penumalli, B.R., "A Comprehensive Two-Dimensional VLSI Process Simulation Program, BICEPS," *IEEE Trans. Electron Devices*, vol. ED-30, no. 9, pp. 986-992, Sep. 1983.

[8]   M. R. Pinto, "Simulation of ULSI Device Effects," in *1991 ULSI Science and Technology*, J. Andrews and G. Cellar eds., Electrochem. Soc. Proc., vol. 91-11, pp. 43-51, 1991.

[9]   W. M. Coughran Jr., M. R. Pinto, R. K. Smith, "Adaptive Grid Generation for VLSI Device Simulation," *IEEE Trans. Computer-Aided Design*, vol. 10, no. 10, p. 1259, Oct. 1991.

[10] W. L. Engl, R. Laur, and H. K. Dirks, "MEDUSA–A Simulator for Modular Circuits," *IEEE Trans. Computer-Aided Design*, vol. 1, no. 2, pp. 85-93, Apr. 1982.

[11] F. Venturi, R. K. Smith, E. Sangiorgi, M. R. Pinto, and B. Riccò, "A General Purpose Device Simulator Coupling Poisson and Monte Carlo Transport with Applications to Deep Submicron MOSFETs," *IEEE Trans. Computer-Aided Design*, vol. 8, no. 4, pp. 360-369, Apr. 1989.

[12] J. D. Bude and R. K. Smith, "Phase Space Simplex Monte Carlo for Semiconductor Transport," *Proc. Eighth Intl. Conf. on Hot Carriers in Semiconductors*, Oxford, 1993.

[13] J. D. Bude and I. C. Kizilyalli, "New Mechanism for Bipolar Degradation in Submicron BiCMOS," *Proc. Symp. on VLSI Technology*, 1993.

[14] B. R. Chawla and H. K. Gummel, "A Boundary Technique for Calculation of Distributed Resistances," *IEEE Trans. Electron Devices*, vol. ED-17, no. 10, Oct. 1970.

[15] T. A. Lenahan, "Calculation of Transmission-Line Parameters for 2D IC Interconnects," AT&T Technical Memorandum 52174-120988-01, 1988.

[16] L. W. Nagel, "ADVICE for Circuit Simulation," *Proc. ISCAS*, Apr. 1980.

[17] H. K. Gummel and H. C. Poon, "An Integral Charge-Control Model for Bipolar Transistors," *Bell Syst. Tech. J.*, vol. 49, no. 5, pp. 827-852, May 1970.

[18] G. M. Kull, L. W. Nagel, S.-W. Lee, P. Lloyd, E. J. Prendergast, and H. Dirks, "A Unified Circuit Model for Bipolar Transistors Including Quasi-Saturation Effects," *IEEE Trans. Electron Dev.*, vol. 32, no. 6, pp. 1103-1113, Jun. 1985.

[19] C. C. McAndrew, "A Complete and Consistent Electrical/Thermal HBT Model," *Proc. IEEE BCTM*, pp. 200-203, Oct. 1992.

[20] S. Liu and L. W. Nagel, "Small-Signal MOSFET Models for Analog Circuit Design," *IEEE J. Solid-State Circuits*, vol. SC-17, no. 6, pp. 983-998, Dec. 1982.

[21] S.-W. Lee and R. C. Rennick, "A Compact IGFET MODEL–ASIM," *IEEE Trans. Computer-Aided Design*, vol. 7, no. 9, pp. 952-975, Sep. 1988.

[22] Y. Tsividis and K. Suyama, "MOSFET Modeling for Analog Circuit CAD: Problems and Prospects," *Proc. CICC*, pp. 14.1.1-14.1.6, May 1993.

[23] C. C. McAndrew, B. K. Bhattacharyya, and O. Wing, "A Single Piece, C_∞ Continuous MOSFET Model Including Subthreshold Conduction," *IEEE Electron Device Lett.*, vol. 12, no. 10, pp. 565-567, Oct. 1991.

[24] K. Singhal, C. C. McAndrew, S. R. Nassif, and V. Visvanathan, "The CENTER Design Optimization System," *AT&T Technical Journal*, vol. 68, no. 3, pp. 77-92, May/June 1989.

[25] H. K. Dirks, R. Erwe, J. L. Lentz, C. C. McAndrew, S. R. Nassif, E. J. Prendergast, and K. Singhal, "The Modeling and Optimization of GaAs HFET Structures," *Proc. NASECODE VI*, Dublin, Ireland, pp. 28-39, July 1989.

[26] P. Lloyd, "Application of Numerical Simulation in Modeling of IC Device Structures," in *Proc. NASECODE III*, Galway, 1983.

[27] E. J. Prendergast, "An Integrated Approach to Modeling," in *Proc. NASECODE IV* D b in, 1985.

[28] G. Booch, *Object-Oriented Design with Applications*, Benjamin-Cummings, 1991.

[29] S. G. Duvall, "An Interchange Format for Process and Device Simulation," *IEEE Trans. Computer-Aided Design*, vol. 7, no. 7, pp. 741-754, 1988.

[30] F. Fasching, C. Fischer, S. Selberherr, H. Stippel, W. Tuppa, and H. Read, "A PIF implementation for TCAD purposes," *Simulation of Semiconductor Devices*

and Processes (SISDEP), vol. 4, pp. 477-482, Sept. 1991.

[31]   A. Wong, W. Dietrich, and M. Karasick eds., "Semiconductor Wafer Representation Architecture, Version 1.0," CFI TCAD Framework Group, *CAD Framework Initiative, Inc.*, Document TCAD-91-G-1, June 1992.

[32]   D. Hare and K. DeVilbiss eds., "Inter-Tool Communication Architecture," CFI Inter-Tool Communication TSC, *CAD Framework Initiative, Inc.*, Document 55, June 1991.

[33]   W. R. Stevens, *UNIX® Network Programming*, Englewood Cliffs, NJ: Prentice-Hall, pp. 153-169, 1990.

[34]   *Op. cit.*, pp. 110-115.

[35]   G. Springer and D. P. Friedman, *Scheme and the Art of Programming*, Cambridge, MA: MIT Press, 1989.

[36]   J. K. Ousterhout, "TCL: An Embeddable Command Language," *Proc. 1990 Winter USENIX Conference*, pp 133-146, 1990.

[37]   I. C. Kizilyalli, T. E. Ham, K. Singhal, J. W. Kearney, W. Lin, and M. J. Thoma, "Predictive Worst-Case Statistical Modeling of 0.8µm BICMOS Bipolar Transistors: A Methodology Based on Process and Mixed Device/Circuit Level Simulators," *IEEE Trans. Electron Dev.*, vol. 40, no. 5, pp. 966-973, May 1993.

[38]   R. A. Becker, J. M. Chambers, and A. R. Wilks, *The New S Language*, Pacific Grove, CA: Wadsworth, 1988.

Technology CAD at IBM

R.W. Knepper, J.B. Johnson[†], S. Furkay[†], J. Slinkman[†], X. Tian[†], E.M. Buturla[†],

R. Young[‡], G. Fiorenza[‡], R. Logan[‡], Y.S. Huang[‡], R.R. O'Brien[‡], C.S. Murthy[‡],

P.C. Murley[‡], J. Peng[+], H.H.K. Tang[‡], G.R. Srinivasan[‡], M.M. Pelella[‡],

D.A. Sunderland[‡], J. Mandelman[‡], D. Lieber[*], E. Farrell[*], and M. Kurasic[*]

IBM Semiconductor Research and Development Center,
East Fishkill Facility, Hopewell Junction, NY 12533, USA
[†]IBM Technology Products Division, Burlington, VT, USA
[‡]IBM SRDC, Hopewell Junction, NY, USA
[*]IBM T.J. Watson Res. Lab., Yorktown Heights, NY, USA
[+]AMC, Santa Barbara, CA, USA

Abstract

The IBM suite of TCAD tools for semiconductor process and device modeling is described. The series includes FEDSS for process modeling, FIELDAY for device modeling, FOXi/FIERCE for resistance and capacitance calculation, EXCALIBR and MGP for compact device model generation, and SEMM for soft-error failure probability prediction. Comprising the VATS series of tools, the programs interact through a common database representation, are accessible via a graphics-user-interface WIZARD, and provide for inputs, outputs, and meshing through a number of pre- and post-processor programs.

1. Introduction

As the cost of developing new semiconductor technology at ever higher bit/gate densities continues to grow, the value of using accurate TCAD simulation tools for design and development becomes more and more of a necessity to compete in today's business. The ability to tradeoff wafer starts in an advanced piloting facility for simulation analysis and optimization utilizing a "virtual fab" S/W tool set is a clear economical asset for any semiconductor development company. Consequently, development of more sophisticated, accurate, physics-based, and easy-to-use device and process modeling tools will receive continuing attention over the coming years.

The cost of maintaining and paying for one's own internal modeling tool development effort, however, has caused many semiconductor development companies to consider replacing some or all of their internal tool development effort with the purchase of vendor modeling tools. While some (noteably larger) companies have insisted on maintaining their own internal modeling tool development organization, others have elected to depend totally on the tools offered by the TCAD vendors and have consequently reduced their modeling staffs to a bare minimal support function. Others are seeking to combine the best of their internally developed tool suite with "robust", "proven" tools provided by the vendors, hoping to achieve a certain synergy as well as savings through this approach.

In the following sections we describe IBM's internally developed suite of TCAD modeling tools and show several applications of the use of these tools. Beginning with the methodology of modeling within IBM, we will show the similarities and differences in the approaches taken for both FET and bipolar modeling. After describing features and applications of the tools, we discuss the programs being developed for interfacing the simulators, as well as for obtaining output graphics. The intermediate data representation, as well as the installation platform(s), will be discussed briefly.

2. S/W Tool Set in Use at IBM

At IBM an internally-developed process and device modeling set of TCAD tools [1 – 16] has been in place and seen extensive use for the past three or four generations of semiconductor technology development. Numerical simulation tools were originally developed independently at several locations throughout the company with the specific needs of either bipolar or FET device development in mind. For example, in the original 2-dimensional finite element process simulator developed for bipolar technology (SAFEPRO), particular attention was placed in the algorithm and methodology for modeling the two species interaction between arsenic and boron in order to correctly model the effect of the steep arsenic emitter profile on the more mobile but highly sensitive boron base profile [1,2]. SAFEPRO was later "converted" into a 2D finite difference simulator FINDPRO with enhanced diffusion physics for point defects and their effect on the impurity diffusion [6,7]. This S/W tool development for the bipolar device technology was done at the East Fishkill site and was invaluable in the design and optimization of the recent high performance (20 GHz) ATX4 double polysilicon bipolar technology [17].

On the other hand, the 2D finite element process simulator FEDSS originally developed for the FET technology business, took a somewhat different approach. For impurity diffusion the simulator used an algorithm based on the assumption of the adequacy of single species diffusion, and concentrated instead on developing an early tool for modeling oxidation, etch, and deposition. Both FEDSS and SAFEPRO/FINDPRO received various enhancements and were used extensively in process simulation for technology design, development, and optimization in their respective intended arenas throughout the 1980's and early 90's in IBM research and development at the Yorktown, Burlington, and East Fishkill locations.

The development of numerical device simulation S/W tools also occurred separately at two locations devoted somewhat independently to either bipolar or FET development. At the IBM Burlington, VT (FET) location the 2D/3D finite element FIELDAY program was written and used extensively for FET device modeling and analysis. Likewise, a 2D finite difference program 2DP was developed at the East Fishkill (bipolar) location and also heavily used for bipolar device studies, design, and model generation. Both programs solve the same semiconductor equations, i.e. the carrier continuity equations and Poisson's equation, and assuming that the same mobility, ionization, and band-gap narrowing formulations are used, both programs give the same results. 2DP used Boltzman statistics whereas FIELDAY was further enhanced to include Fermi-Dirac statistics, as well as a number of other physics-based additions including the Hydrodynamic energy-balance equations and, more recently, the lattice energy equation.

In 1991 the Semiconductor Research and Development Center (SRDC) was formed, combining the TCAD S/W development efforts at both the East Fishkill and Burlington locations into one project. At that time the decision was made to continue the development of the FIELDAY program and discontinue further development of 2DP. Likewise, when both FEDSS and FINDPRO were examined, it was decided to combine the best of both programs and apply these to the further development of the FEDSS process simulator. Internal development is therefore continuing on the FEDSS and FIELDAY programs for process and device modeling within the IBM SRDC/TP organization(s).

Figure 1 is a block diagram illustration of the IBM Technology Computer-Aided Design (TCAD) suite of tools (coined VATS for VLSI Analysis Tools Series), including simulators, pre- and post-processors, meshing programs, and a graphics-user-interface, along with the intermediate database representation.

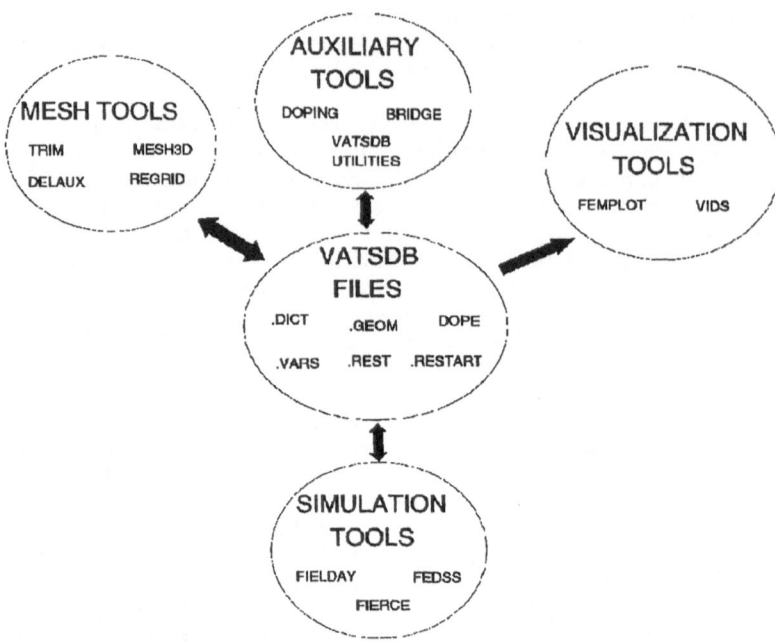

Figure 1. Block diagram illustrating relationship of VATS (VLSI Analysis Tools Series) simulation programs and pre/post-processors to database representation.

These tools are supported and in use today within the IBM SRDC, Technology Products, and Research organizations at the Fishkill, Burlington, and Yorktown locations, as well as at other sites involved in semiconductor design/development. Internally developed simulation tools in use today for semiconductor process and device modeling and described in this paper are the following:

- FIELDAY (2D/3D Device Simulation)
- FEDSS (2D Process Simulation)
- FOXi/FIERCE (2D/3D Capacitance, Resistance Simulation)
- DMG Program (Bipolar Model Generator)
- EXCALIBR (Extraction Program for CMOS Model Generation)
- SEMM (Soft Error Modeling of Alpha and Cosmic Particles)

These are used in connection with auxiliary pre- and post-processor programs for gridding, profile inputs, outputs, and a graphics user interface:

- WIZARD (Graphics User Interface)

- TRIM (mesh generation using ASCII control file)
- DELAUX (interactive mesh generation)
- REGRID (mesh refinement and optimization)
- MESH3d (extrudes a 2D mesh into 3D)
- DOPING (generates 2D doping profiles)
- BRIDGE (interpolates doping unto the mesh)
- FEMPLOT (graPHIGs-based plotting program)
- VIDS (3D Graphics Plotting for Output and Visualization)

Communication and interaction between the above IBM TCAD tools is made possible by use of a common intermediate data representation (VATS/DB) for data storage. VATS/DB consists of six files having the same basename but different suffixes -

.dict - index file to all the data
.dope - contains doping distribution
.geom - contains mesh and contacts
.rest - contains nodal results
.vars - snapshot of the FIELDAY input file
.restart - restart information for FEDSS

The above programs are all currently accessible through the WIZARD Graphics User Interface with the exception of DMG (device model generator) and SEMM (soft error rate modeling).

The tools are installed and maintained on the IBM RS/6000 workstation platform running the AIX operating system, as well as MVS/ESA for mainframe application. In addition to the Fortran compiler, use of the tool set also requires the "C" compiler for dynamic memory allocation (feature of FIELDAY), graPHIGs for plotting and meshing, and DATA EXPLORER for output visualization capability utilizing VIDS. Partial support is also provided at this time for the new AIX/ESA mainframe operating system.

In addition to the above suite of internally developed tools, a number of university and vendor (process, device, and extraction) modeling programs are also available, such as SUPREM3, SUPREM4, etc., most of which are also accessible via the WIZARD GUI. The above list has included only a subset of the S/W tools normally defined as TCAD tools. For the purposes of this paper we exclude other important technology simulation tools such as lithography modeling S/W, stress simulation programs, circuit analysis tools, placement and wiring systems, and tool and equipment modeling programs.

After discussing the overall modeling methodology, we will describe the features and demonstrate the application of several of the key simulation programs listed above.

3. Modeling Methodology

As described above, bipolar and FET modeling tool development has evolved somewhat independently within IBM. In the bipolar arena a methodology, illustrated by Figure 2a, was conceived and implemented [1 – 7] beginning with a process description and ending up with a lumped equivalent circuit model for circuit analysis through the steps of 2D process modeling, 2D device simulation, and device model generation. The programs have been linked through a common intermediate data representation so that flexibility was obtained in allowing the use of either FEDSS or FINDPRO for the 2D process simulation, FIELDAY or 2DP for the 2D device simulation, DMG for compact circuit model generation, and the IBM ASTAP (now ASX) program for the circuit analysis. Recently, the DMG program concept has been extended to allow the generation of SPICE bipolar models which are desired for external OEM customers utilizing IBM advanced bipolar technology.

In the simulation of FET devices and processes a slightly modified approach has been in use, as described in Figure 2b.

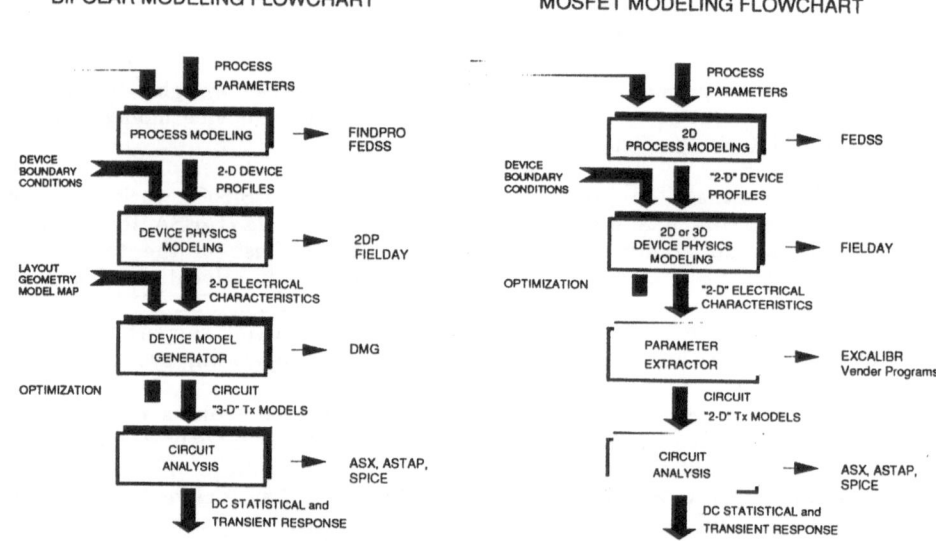

Figure 2. TCAD Modeling Methodology: (a) Bipolar and (b) MOSFET Technology Simulations.

As in the bipolar case FEDSS and FIELDAY are used for process and device simulation, but at this point due to the more 2D nature of the FET device, parameter extraction techniques can be used to construct an accurate source-drain current function, which then is used to build a complete FET compact device model with the addition of appropriate capacitances, resistances, and parasitics for use in either ASTAP (ASX) or SPICE circuit analysis. This alleviates any necessity for constructing a fully 3D distributed model as an interim step, as is done in the case of the bipolar modeling methodology in 2a.

As shown in Figure 2b, the use of FEDSS and FIELDAY for technology development often ends with the device simulation results themselves, which are then studied and iterated upon for the purpose of tuning and optimizing the technology. FEDSS is available only as a 2D simulator, while FIELDAY is used for both 2D and 3D simulation in the design and optimization of FET devices. Device and process studies and design optimization clearly represent the majority of the usage for these two tools in the area of FET technology development. Although the FIELDAY results are often used to develop equivalent circuit ASTAP (or SPICE) models, as is the case for bipolar modeling, measured data is usually preferred for extracting model parameters when available.

4. FEDSS Process Simulator

The 2D semiconductor process simulator FEDSS (Finite Element Diffusion Simulation System) contains the following features and capability:

- Diffusion
- Oxidation
- Ion Implantation
- Pre-deposition and evaporation
- Epitaxy
- Material deposition
- Etching

FEDSS was initially written in the early and mid 1980's at the IBM Burlington location by Salsburg, Hansen, Borucki, et.al. [11, 12]. The simulator was first created as a diffusion program for FET diffusion analysis and was shortly thereafter modified to include oxidation, implant, etch, and deposition. Diffusion models were originally written as single species models for arsenic, boron, and phosphorous which were thereafter enhanced to include two species interactions via a four (diffusion) coefficient coupling matrix for each pair of impurity species [12]. Additional species including Sb, and In have more recently been added to the program capability [18]. FEDSS includes four arsenic clustering models, in order to correctly compute the active dopant in an NPN transistor emitter, for example.

The oxidation model implemented in FEDSS is based on a two stage approach involving first oxidation utilizing a physical viscous oxidation model published originally by Chin, et.al. [19] in the Stanford program SOAP, which is then followed by redistribution of dopants [12]. This oxidation algorithm has been found to predict physically correct shapes for a variety of oxidation processes such as the LOCOS bird's beak isolation process, STI (shallow trench isolation), sidewall oxidation processes, and DRAM trench oxidation, as determined from high resolution SEM analysis.

FEDSS contains both analytic models and Monte Carlo simulation for ion implantation in order to cover a wide range of physical situations. The analytic models include Gaussian, joined half-Gaussian, Pearson IV, and dual-Pearson distributions [12, 18]. The models are based on tables with statistical parameters for arsenic, boron, phosphorous, BF2, antimony, and indium which are interpolated over energy and dose. A Monte Carlo implantation capability suitable for amorphous targets, based upon the TRIM program [20], is available to use when the effects of implanting through (up to three) multiple layers must be modeled accurately.

The results of an application of the FEDSS program for the simulation of diffusion and oxide deposition are illustrated in Figures 3-6. The simulation was done to study latchup in a CMOS process and was used as input to FIELDAY for predicting the latchup behavior [21]. Figure 3 shows the overall latchup structure assumed for the study with the doping contours superimposed.

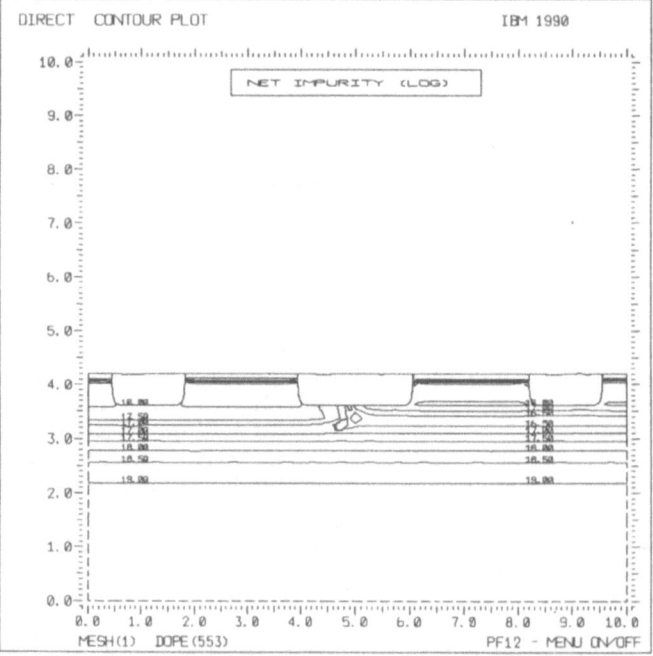

Figure 3. FEDSS simulation of CMOS well profiles for input to FIELDAY latchup simulation. Note the NWELL on the left, the PWELL on the right, and the shallow trench isolation above the junction of the wells.

The N-Well is on the left and the P-Well on the right, with the two wells separated on the surface by shallow trench oxide isolation (STI). A P+ pocket was diffused in the N-Well and an N+ pocket in the P-Well in order to complete the full latchup structure. The presence of a heavily doped P+ substrate which outdiffuses toward the wells can also be seen in the figure. Figure 4 is a closeup view of a region of STI with the N-Well/P-Well junction clearly visible.

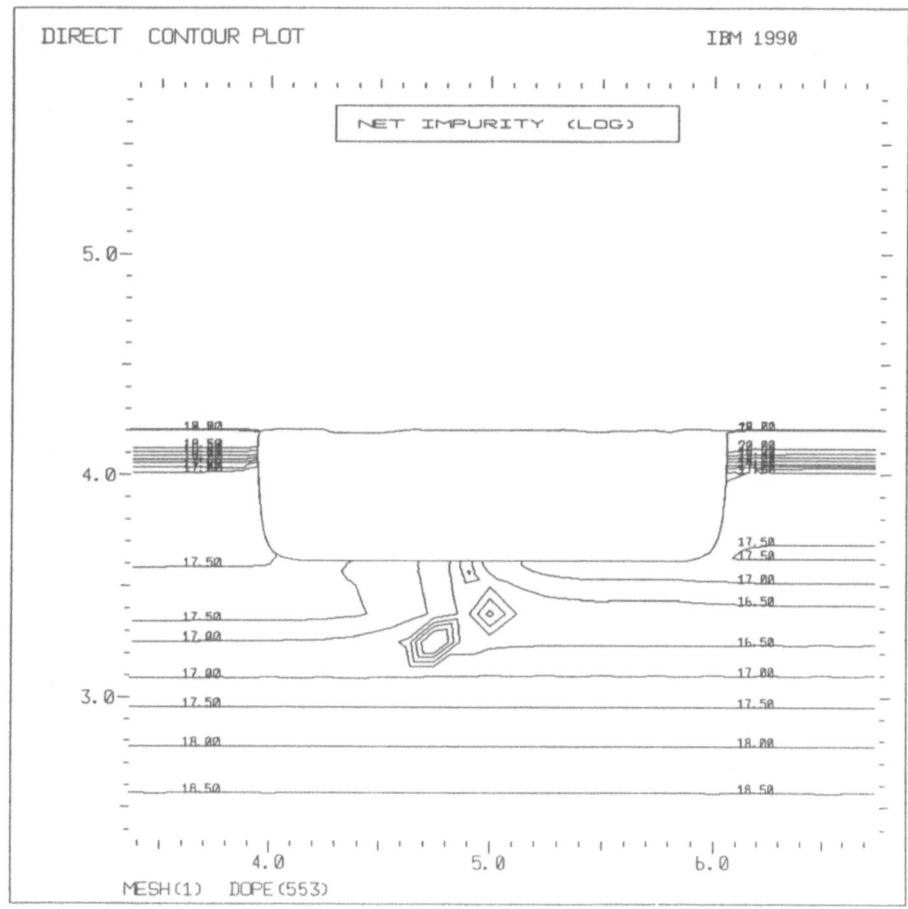

Figure 4. Closeup view of the N-Well/P-Well Junction and STI region of Fig. 3.

A closeup of the corner of the oxide-filled STI region is seen in Figure 5, showing the finite element mesh.

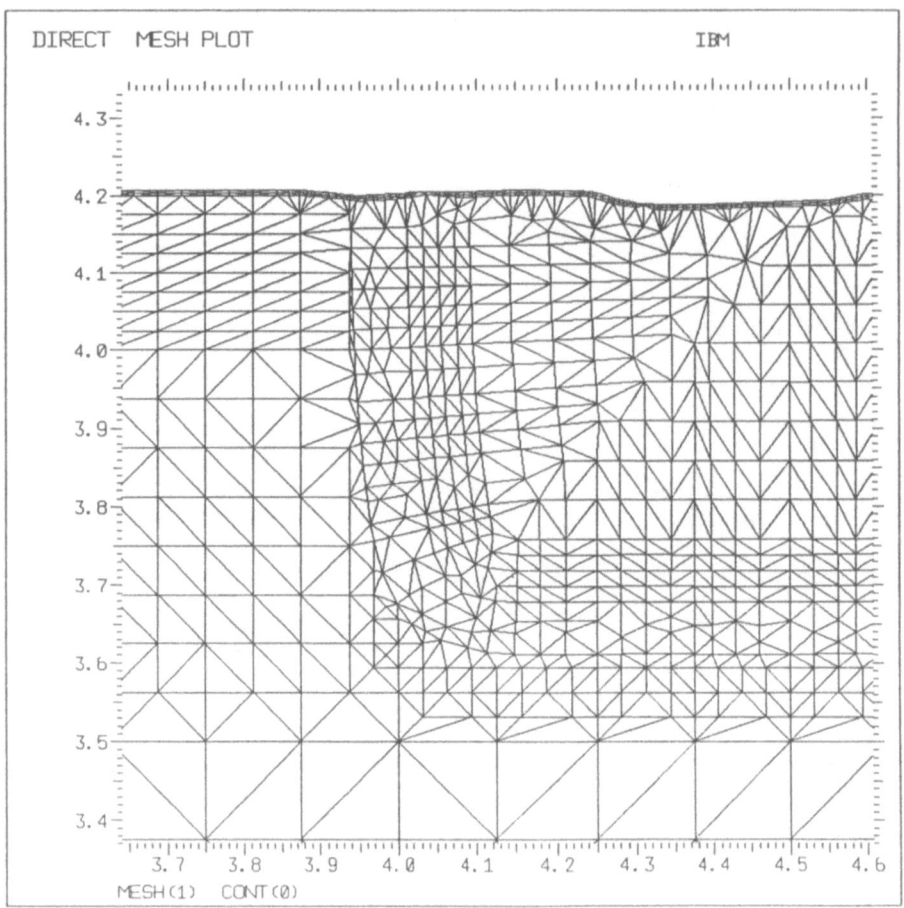

Figure 5. View of STI corner region from the latchup structure of Fig. 3 showing the mesh structure. The STI region is filled with oxide.

Seen in Figure 6 is a vertical profile cut through the P-Well region. One can see the N+ source/drain pocket at a junction depth of 0.2 μm, the p-well profile and the heavily-doped P+ substrate.

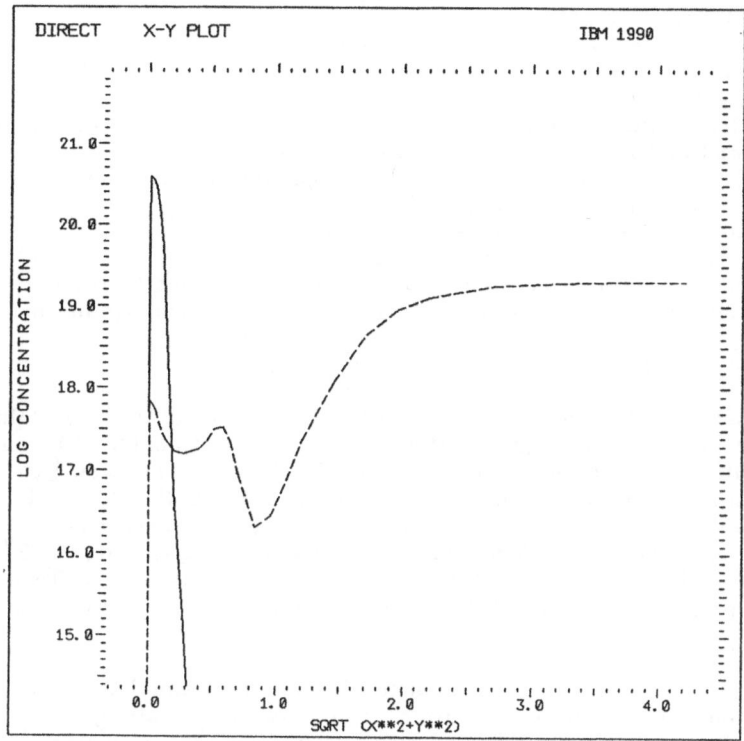

Figure 6. Vertical profile through the P-Well region of Figure 3 showing the N+ source/drain pocket diffusion, the P-Well profile, and the heavily-doped P+ substrate.

5. FIELDAY Device Modeling Program

FIELDAY is a 2D/3D finite element device simulation program having the following features:

- Full 3D capability (right prismatic and tetrahedral elements)
- Hydrodynamic model for holes and electrons
- Fermi-Dirac carrier statistics
- Incomplete ionization

- Heat equation
- Generalized velocity limited contacts
- Voltage and current boundary conditions
- Attached circuit elements (R, C, G)
- A.C. small signal analysis
- Steady state and transient simulation
- Heavy doping effects
- Bulk & interface trapped charge in insulator material
- Fully-coupled (DD) and partially-coupled (Hydro) solution strategies
- Direct and iterative matrix solution methods
- FET and bipolar physical models and parameter extraction
- Heterostructure capability
- Gallium Arsenide simulation capability
- Alpha and cosmic particle impact
- Voltage continuation method for negative resistance IV trace
- Band-to-band tunneling and other field-enhanced leakage mech
- Impact ionization and recombination mechanisms
- Isolated silicon regions

The original FIELDAY program was written in the late 1970's by E. Buturla, P. Cottrell and co-workers at the IBM Burlington location [8 – 10] and was later rewritten by J. Johnson and S. Furkay at the same location [14]. The program has seen pervasive use throughout IBM for both FET and bipolar device simulation primarily for studying and optimizing device behavior and also in the construction of device models for circuit analysis [15, 16, 22]. The program is enhanced and re-released on a regular basis by the Technology Modeling project in East Fishkill and the Technology Simulation department in Burlington.

FIELDAY can be run either as a drift-diffusion (DD) simulator, solving the electron and hole continuity equations and Poisson's equation, or it can be run as a complete Hydrodynamic simulation program, in which case the electron and hole energy balance equations are added to the solution algorithm (but decoupled from the DD solution). A recent enhancement allows the user the option of including lattice energy (self heating) in the simulation. The DD, Hydro, and Lattice Energy options are all fully 3D in their implementation, allowing the user the option to run the simulation either in 2D or 3D mode. The program contains numerous mobility models for both FET and bipolar application, from which the user may select a model of choice.

The program can be linked to FEDSS to receive its starting 2D doping profiles, or alternately, the program DOPING can be used to generate 2D (or 3D) starting profiles either analytically or from experimental data, applying these profiles on the starting mesh. Mesh generation is done with one of the tools TRIM, DELAUX, and/or MESH3D and can be refined with REGRID. Outputs are viewed via FEMPLOT and/or VIDS. The program can be accessed and run through the graphics user interface WIZARD, as explained in a later section.

Additional features soon to be available in FIELDAY are the following:

- Fowler-Nordhiem gate oxide tunneling
- Automatic voltage ramp-up for breakdown simulation
- Addition of extended precision arithmetic for Hydro
- Explicit position-dependent minority carrier lifetimes

Three examples will be shown to demonstrate the capabilities of the FIELDAY program - analysis of the CMOS latchup example shown for FEDSS above, simulation of a 0.1 μm L_{eff} N-FET using Hydro FIELDAY to obtain carrier temperature near the drain region, and a 3D example previously published for a rounded gate DRAM trench-bounded FET device [22].

CMOS Latchup Example:

Figure 7 shows a cross-section drawing of a CMOS inverter circuit with parasitic bipolar NPN and PNP transistors shown schematically.

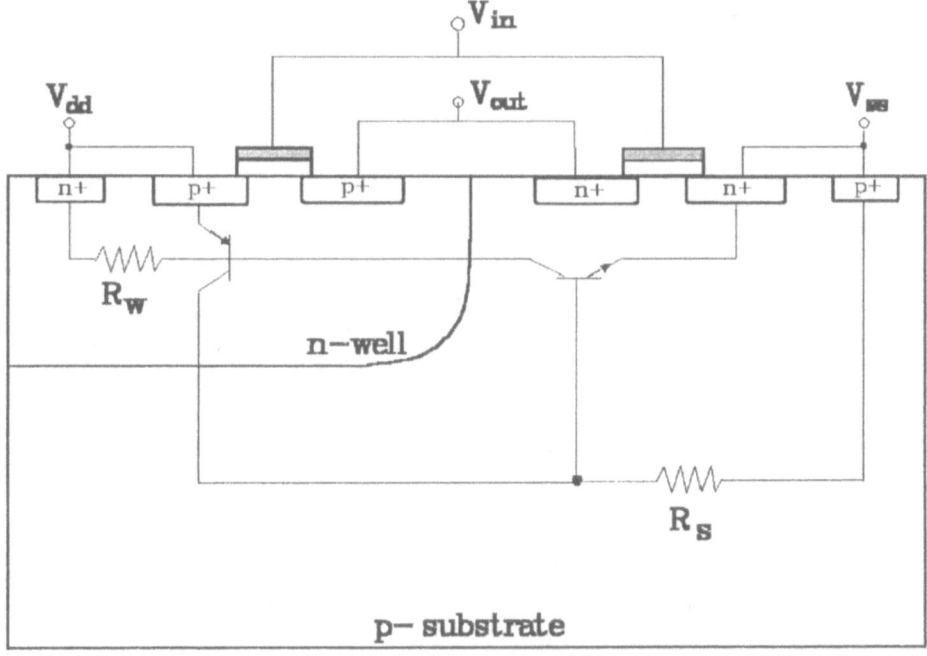

Figure 7. CMOS inverter cross section schematic showing parasitic vertical PNP and lateral NPN devices exhibiting latchup configuration.

These parasitic "latchup" transistors occur naturally in the structure due to the source/drain, N-Well, and/or P-Well junctions. It is well known that such a structure is susceptible to latchup if the product of the transistor betas is greater than one. Spreading resistance components from the Well-biasing (guard ring) junctions to the $P+/N+$ source/drain latchup junctions is represented by R_w for the N-Well and R_s for the P-Well. Obviously, R_w and R_s are highly dependent on the placement and layout of the circuit and guard ring diffusions.

The circuit actually simulated in FIELDAY, however, is shown in Figure 8 with the gates and drain junctions eliminated for simplicity, since these elements are not germane to the latchup simulation.

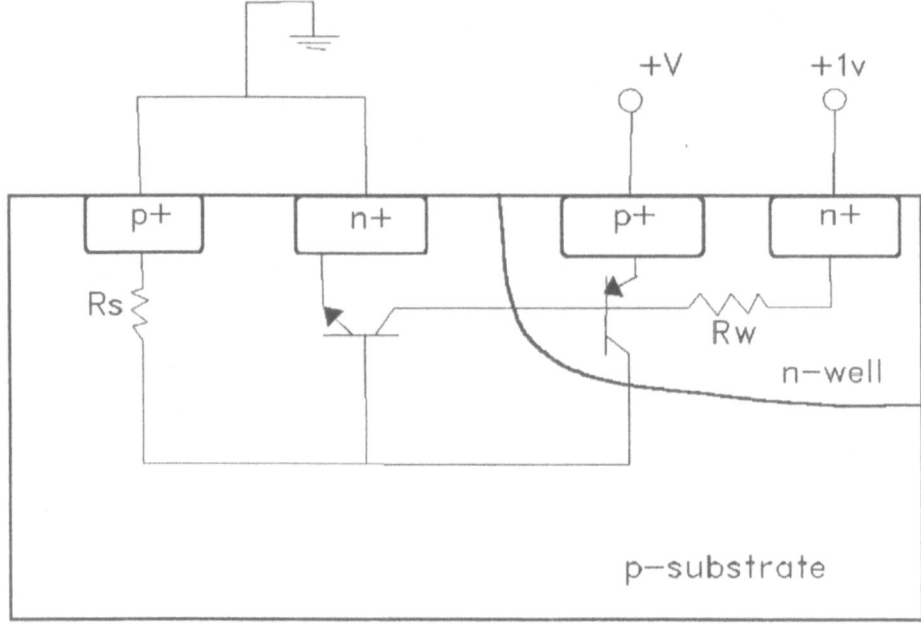

Figure 8. Simulation structure schematic used as basis for FIELDAY CMOS latchup study. R_s and R_w represent spreading resistance in the P-Well and N-Well, respectively.

With the P+ guard ring and N+ source grounded, 1 volt applied to the N+ guard ring, and a voltage ramp applied to the P+ source junction, the two resistors and voltage ramp are varied while latchup is studied. FIELDAY simulation results of the dc characteristics of the two parasitic bipolar devices are plotted in Figure 9.

CMOS latchup structure: parasitic device behavior

Current (micro_amps/micron)

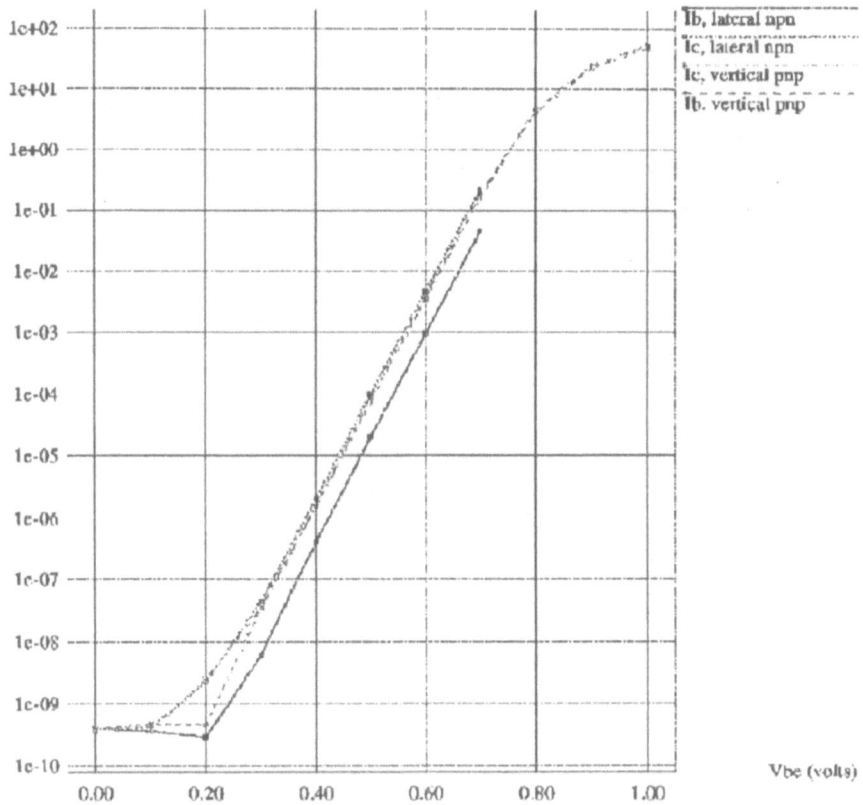

Figure 9. Static IV (Gummel) plots for the parasitic lateral NPN and vertical PNP latchup devices from the structure of Fig. 3.

The NPN transistor has a current gain (beta) of about 5-6 while the PNP gain is near unity.

The results of the latchup overshoot simulation (current-voltage characteristic) are seen in Figure 10.

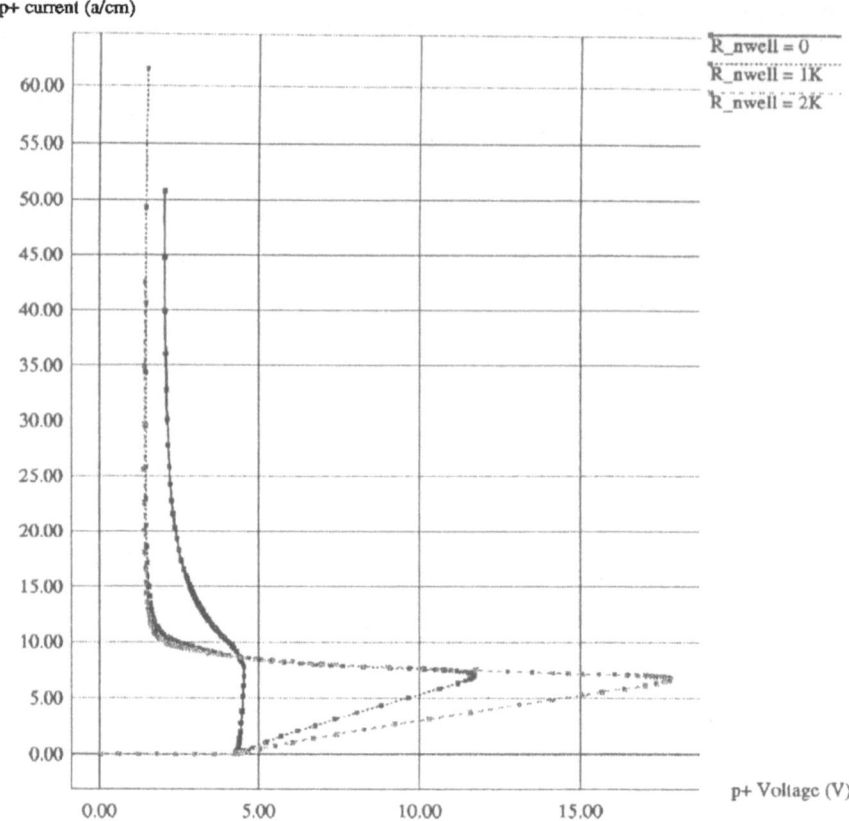

Figure 10. P+ pocket current versus applied voltage for the latchup structure of Figs. 3 & 8 showing the snap-back characteristic traced out with FIELDAY's voltage continuation method.

The snap-back indicating a latchup occurrence is obvious in the figure. With R_s set at 10 ohms as R_w is varied from 2K ohms to 0, the latchup threshold voltage reduces from 18 volts to around 4 volts, as seen in the plot. The curves are shown for room temperature.

Short Channel N-FET Hydro-FIELDAY Simulation Example:

Figure 11 shows the generic structure of the 0.1 μm N-FET device mesh obtained using TRIM for initial mesh construction and REGRID to optimize the mesh.

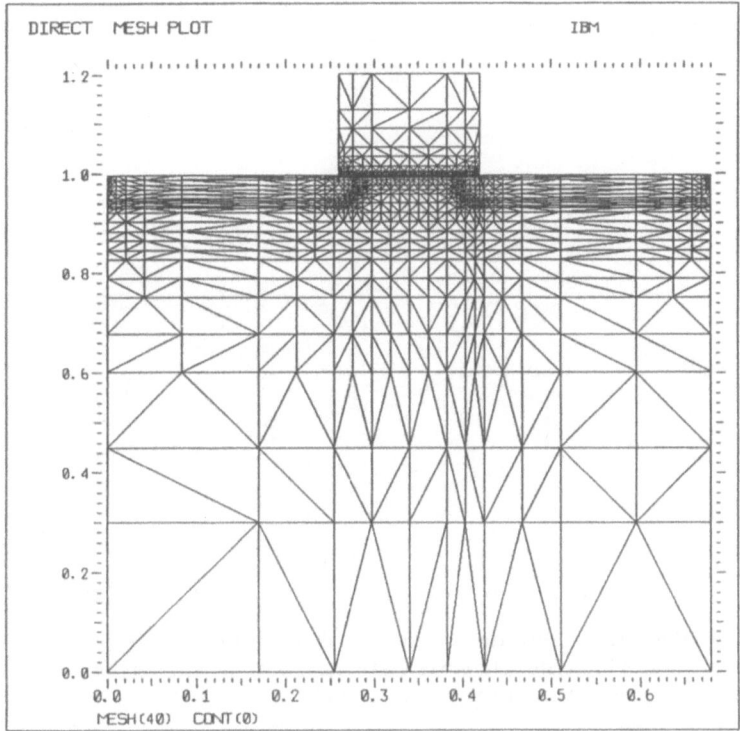

Figure 11. Device and mesh structure for FIELDAY simulation of 0.1 μm L_{eff} N-FET Device.

The gate physical length is seen to be about 0.16 μm giving an L_{eff} of about 0.1 μm after accounting for the source and drain lateral diffusion. A closeup of voltage contours near the drain is shown in Figure 12.

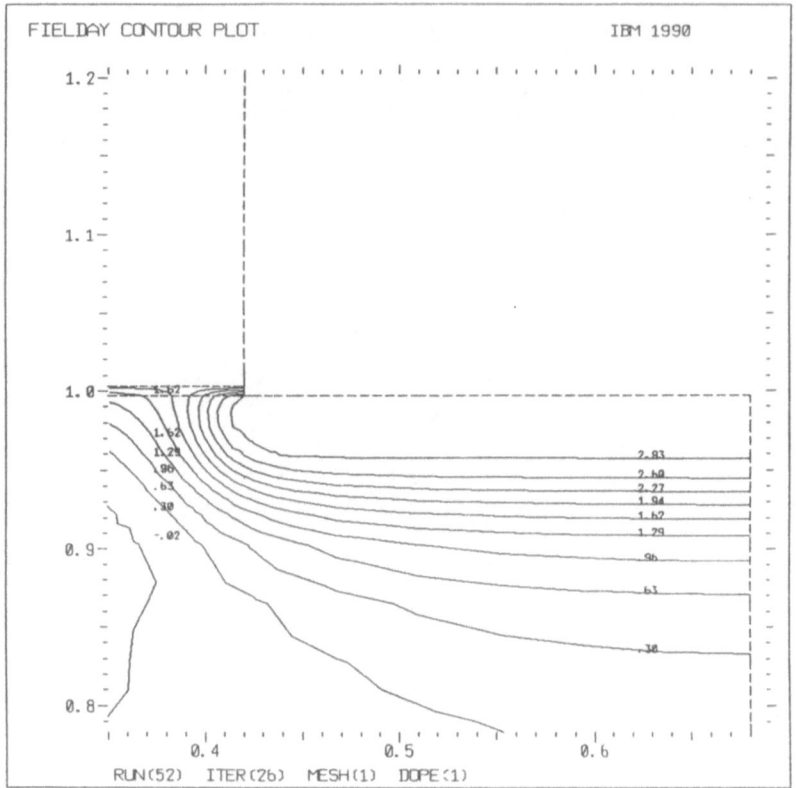

Figure 12. Potential contours near the drain junction for the 0.1 μm FET example of Fig. 11. The physical gate position is shown by vertical dashed lines. The gate oxide thickness is about 60A.

The electric field is maximum very near to the drain junction and near to the surface, as indicated by the spacing of the contours. The potential contours are seen to extend into the gate oxide. The simulation was done for 2.5 volts applied to the drain with 0.8 volt on the gate.

In Figure 13 the electron temperature contours are plotted for the region near the drain. Maximum temperature is something above 3000 K. Hole temperature is near 300 K in this simulation and is not of interest.

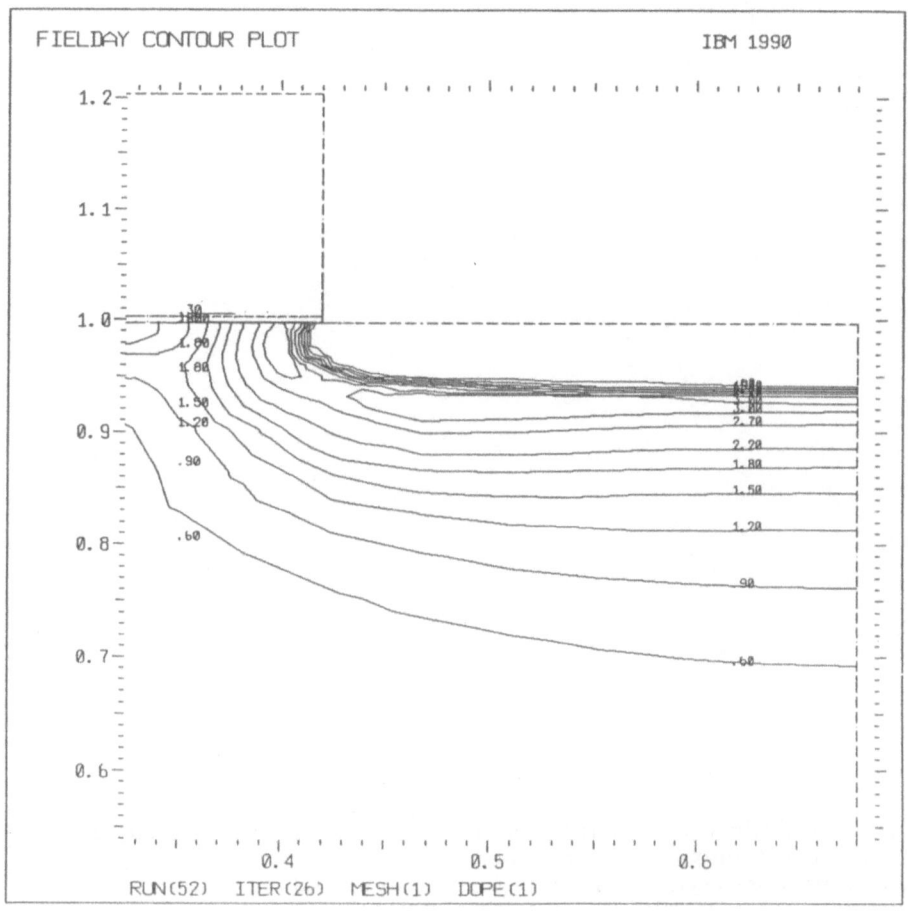

Figure 13. Electron temperature contours near the drain junction for the 0.1 μm FET example of Fig. 11. The contours are in units of 1000 K.

The drain-source characteristics (I_d vs V_{ds}) are shown in Figure 14 where V_{gs} varies from 0.4 to 0.8 volt.

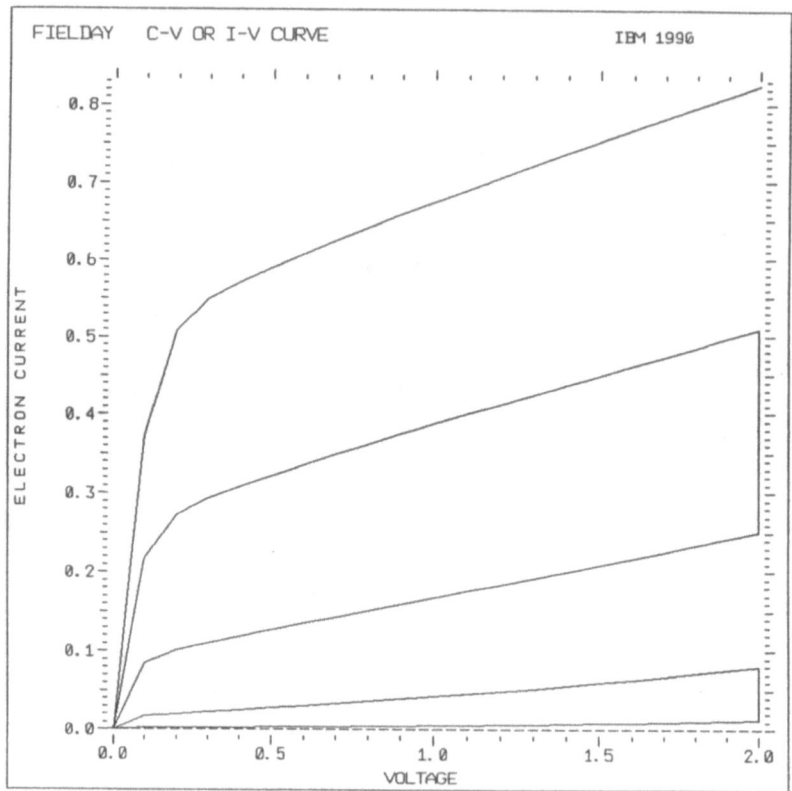

Figure 14. Drain-source characteristics for the 0.1 μm FET example of Fig. 11. Drain current units are in amps/cm and drain-to- source voltage in volts. Gate voltage is varying from 0.4 to 0.8 volt in 0.1 volt increments.

One observes a rather severe effect of channel length modulation by the drain junction in the region of current saturation due to the 2D influence of the drain field on the channel. (The simulation is simply an example with no attempt to design or optimize the device.)

    3D FIELDAY Simulation Example:

For certain TCAD device simulation problems 3D modeling is essential if one is to obtain a complete understanding of the issues of optimization and design. Figure 15 gives an example of such a problem where 3D simulation with the FIELDAY program was required.

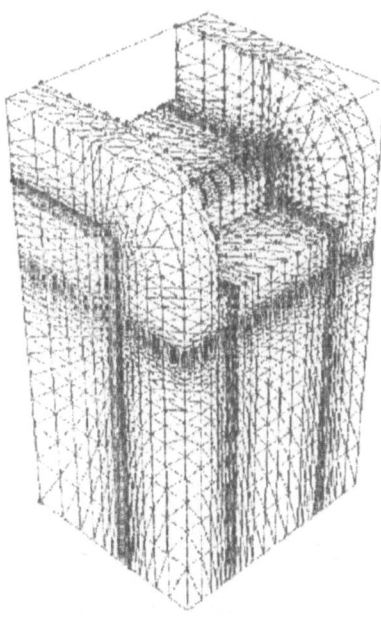

Figure 15. Finite element mesh structure for a 3D trench-bounded FET model (per Ref. 22).

The figure shows the finite element mesh for a trench-bounded MOSFET model [22], as might be found in a DRAM cell having a deep trench storage capacitor. FEDSS-generated doping distributions were mapped into the 3D FIELDAY device structure for this simulation. In this case the authors of [22] were interested in studying the tradeoff(s) between peak gate dielectric electric field (as it relates to device reliability) and the device electrical characteristics as a function of corner radius of curvature. Use of 3D modeling was essential for accurate prediction of electrical behavior due to the submicron features of the structure. Problems of this nature may typically require 25,000 and even up to 100,000 nodes. Such problems have been run on an IBM RS/6000 workstation with run times on the order of 15-30 minutes per dc bias point.

The predicted FET transfer characteristics for the device of Figure 15 are shown in Figure 16 for the case of 3.3 volts applied from drain to source.

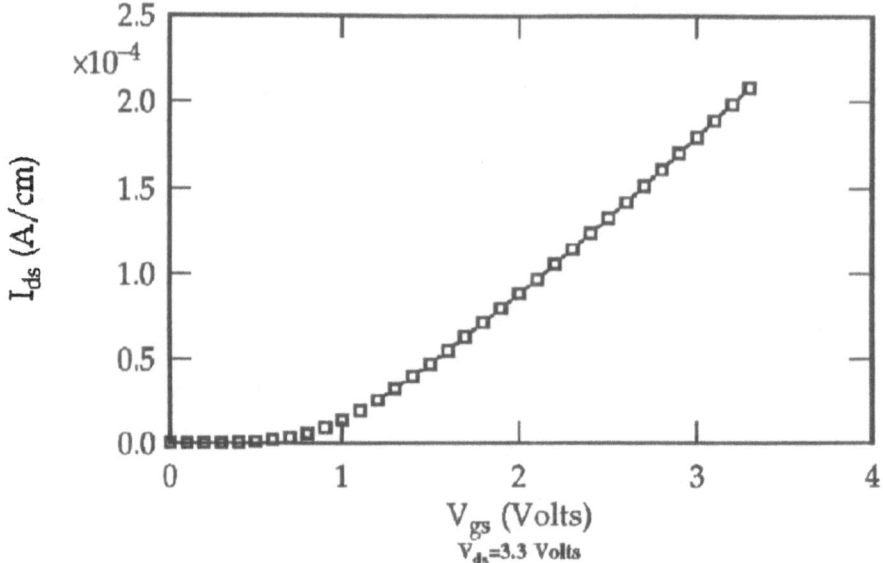

Figure 16. FIELDAY-predicted transfer characteristics for the device of Fig. 15.

6. FOXi/FIERCE Program for Capacitance/Resistance Modeling

The FOXi/FIERCE program is actually comprised of two programs, as the name suggests. FOXi (FIERCE object-oriented X-windows Interface) is an interactive UNIX tool for generating 3D objects, attributes, and finite element meshes. The program was written by D. Lieber and M. Kurasic at the IBM T.J. Watson Research Lab. It is used as an input front-end to FIERCE for building up 3D stuctures upon which FIERCE performs resistance and/or capacitance analysis. Objects in FOXi can be bricks, cylinders, cones, etc. Attributes are contacts, permittivities, and conductivities associated with the materials. FOXi is constructed around a graphic representation utilizing six window panes for control - "resources, view, tree, model, material, and condition".

FIERCE (FInite Element Resistance and Capacitance program) is a 2D/3D simulation program for calculating resistance and capacitance for multiconductor, multidielectric structures. It was initially written by P. Cottrell and E. Buturla in IBM Burlington, VT [13], and recently rewritten by J. Johnson and X. Tian at the same location. FIERCE reads in the structure, mesh, and attribute files from the FOXi 3D solid model and then computes a capacitance or resistance matrix as an output data file.

An example of the use of FOXi/FIERCE is shown in Figure 17 which contains a view of a typical FOXi panel showing the construction of a metalization BEOL (back-end-of-line) structure comprised of word lines and bit lines separated by an insulating dielectric.

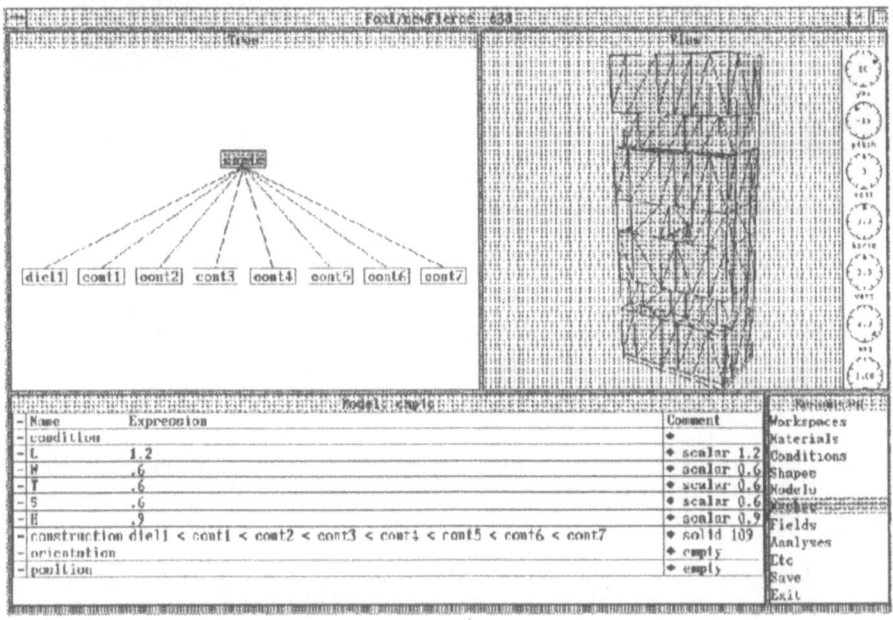

Figure 17. FOXi panel showing construction of a 3D BEOL (back end of line) structure. The structure shows 3 levels of wiring and stud interconnect separated by dielectric isolation.

The upper left window entitled "Tree" shows the subblocks used to build up the solid model. The "View" window shows a perspective of the 3D finite element mesh for the structure. The user has complete flexibility in manipulating and rotating the structure shown in View by using the mouse to point to and adjust the circular, clock-like "control wheels" shown at the far right-hand-side. Two examples of solid structures are given in Figures 18 and 19. Figure 18 illustrates a 3D meshed structure comprised of two metal lines with an interconnecting stud via which is of use in simulating via resistance. Figure 19 shows a closeup of the BEOL metal line structure given in the View window of the FOXi panel in Fig. 17.

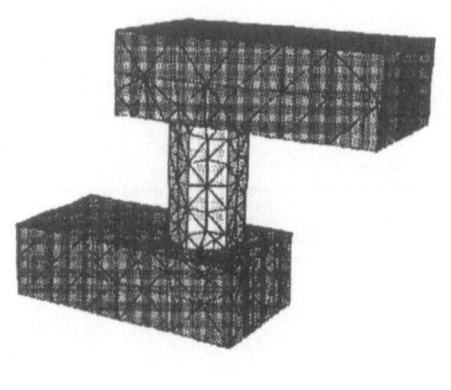

Figure 18. FOXi 3D meshed structure of   Figure 19. Closeup of 3D interlevel wiring
2 wires on adjacent wiring levels intercon-   structure shown in the panel of Fig. 17.
nected by a stud via.

This structure could be used to calculate the interlevel wiring capacitance
between bit lines and word lines on a prospective DRAM memory chip.

7. Parameter Extraction and Circuit Model Generation

The model generation procedure for obtaining MOSFET equivalent circuit
models for circuit analysis in ASTAP/ASX or SPICE involves typically parameter extraction techniques using as input either 2D FIELDAY simulation
results or measured data, as was explained in Figure 2b. This can be accomplished since the MOSFET is still primarily a 2D device (excluding devices of
very small width dimension where fringing fields and/or gate oxide tapering
may necessitate 3D FIELDAY simulation). Consequently, the model generation procedure for MOSFET devices is actually a simpler task than that for
bipolar devices and does not normally require constructing an elaborate 3D
distributed model as part of the procedure.

EXCALIBR (EXtractor/CALIBRator) is a parameter extraction device modeling program developed recently at IBM by D. Sunderland [23] for accurately determining the source-drain current of a MOSFET equivalent circuit model. Using a method termed "matchpoint fitting", the program can quickly and efficiently obtain a best fit to a set of dc MOSFET characteristics utilizing either measured data or 2D FIELDAY calculations as input. The method entails identifying a unique measurement condition (matchpoint) for each adjustable parameter in the model, utilizing an iterative procedure for minimizing the error at each point. The program has been typically used to calibrate the parameters of ASTAP/ASX models and SPICE Level 3 models for both 0.5 μm and 0.25 μm MOSFET devices. EXCALIBR is accessible through the WIZARD graphics user interface.

As discussed briefly above, the DMG (Device Model Generator) program developed for obtaining bipolar equivalent circuit (compact) models utilizes a more complex procedure of first generating a 3D distributed model and then extracting from this a lumped equivalent circuit model suitable for circuit analysis in ASTAP/ASX or SPICE. The procedure was first developed and used throughout the early-to-mid 1980's as the MGP program [4, 5, 7] and was then further enhanced and renamed the DMG program for utilization to SiGe heterojunction NPN transistors as well as VPNP transistors and complex integrated memory cell devices [24].

The structure of DMG, described in Figure 20, is comprised of two subprograms (DMG-A and DMG-B).

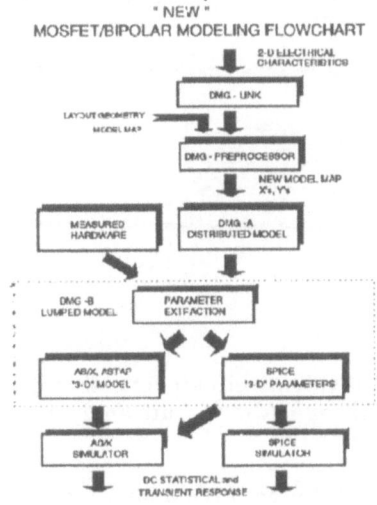

Figure 20. Expanded flowchart describing structure of DMG program.

DMG-A builds a somewhat complex 3D distributed model comprised of
filamentary elements each derived from the 2D FIELDAY results, linked with
current dependent intrinsic or extrinsic base resistance elements, and assembled according to a "model map" provided by the geometrical layout. DMG-B,
then, uses a parameter extraction procedure to fit the elements of the lumped
equivalent circuit model, shown in Figure 21, to the fully distributed 3D
model's terminal characteristics, calling and running the ASTAP program to
accomplish this task.

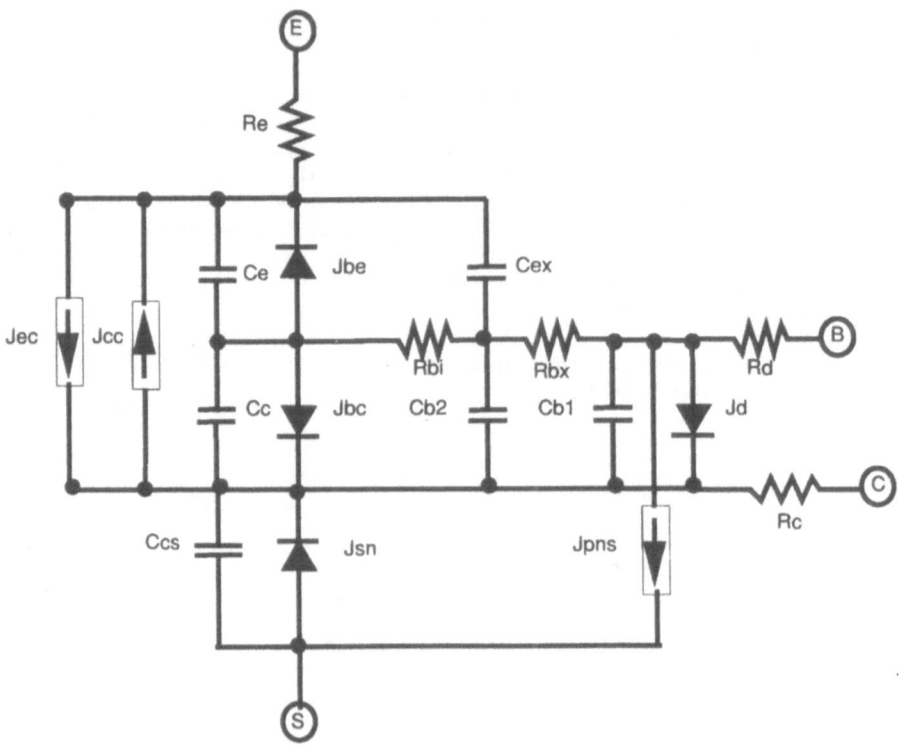

Figure 21. Compact equivalent circuit model topology of DMG-generated
models utilized for ASTAP and ASX circuit analysis.

This approach was found to be invaluable for obtaining accurate bipolar
models since the ability to model the base resistance correctly and to include
the effect of the collector current in the sidewall region of advanced

polysilicon transistors necessitated constructing an effective 3D representation of the device before extracting a compact model for circuit analysis. As seen in Figure 20, the procedure allows DMG-B to be fed with measured hardware data, if such is available, as an alternate to linking in the results of 2D FIELDAY simulation as input. DMG has been recently enhanced to allow the generation of bipolar SPICE models as well as the traditional ASTAP/ASX models [25].

Figures 22-25 show the results of the use of the DMG program for modeling and comparing an NPN SiGe epitaxial-base HBT (heterojunction bipolar transistor) with an advanced double polysilicon NPN BJT (bipolar junction transistor) [24, 26, 27]. Since the structures of the two devices are basically different in their approaches for constructing the base region, as shown in Figure 22, a modification of the original NPN (double poly) model generator program was completed in order to model the epi-base structure.

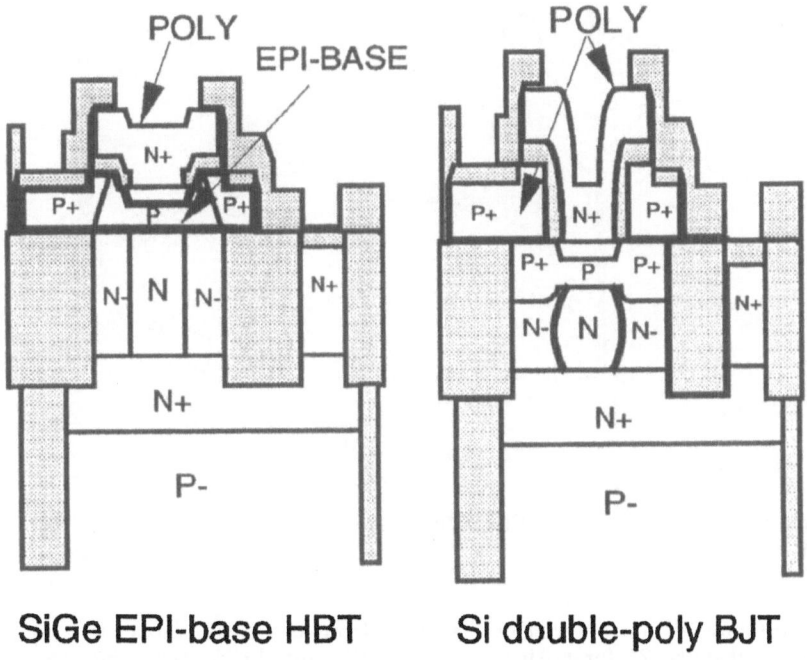

Figure 22. Schematic cross-sections of advanced bipolar transistors modeled with the DMG program [24].

Figure 23 shows the starting SIMS profiles and the simulated (FINDPRO) profiles for both devices including the arsenic emitter and implanted self-aligned
collector, the boron base dopant, and (for the HBT) the Ge graded-base profile.
The effect of base dose variation was studied for the I/I base case and is
shown in the figure.

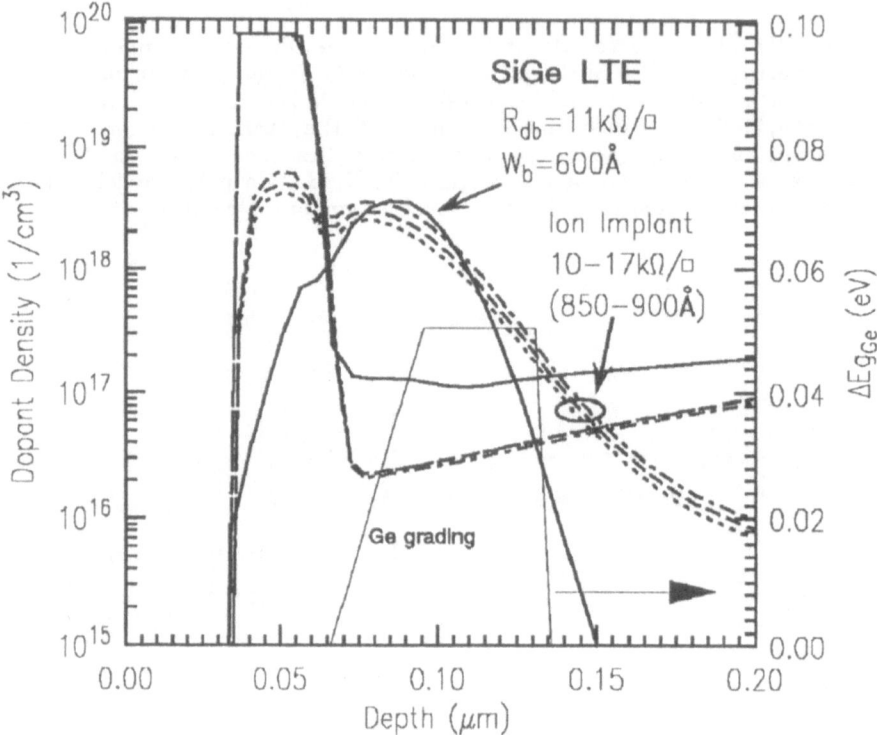

Figure 23. SIMS calibrated FINDPRO simulations of the npn intrinsic vertical
profiles [24].

The dc Gummel characteristics resulting from the DMG modeling program are
shown in Figure 24 for both the SiGe HBT and the Si BJT and are compared
with measured data (for the HBT).

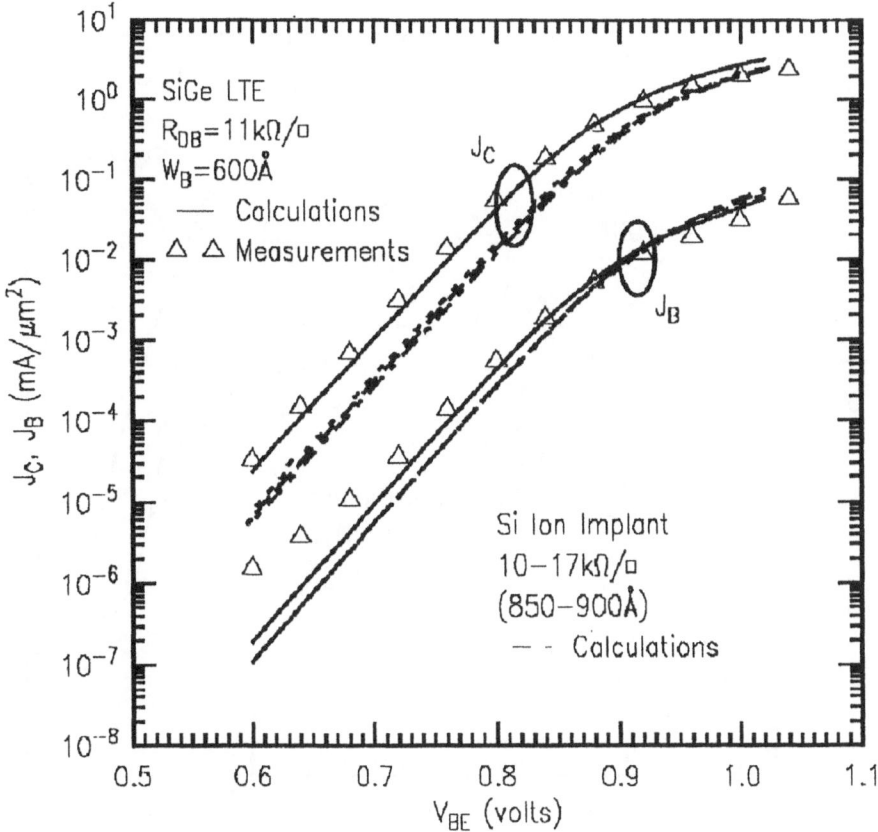

Figure 24. DMG model simulation results of dc IV characteristics versus experimental data for SiGe LTE-base and Si I/I-base NPN transistors [24].

The collector current enhancement for the SiGe HBT over the Si BJT is seen to be about a factor of 4 or 5 and is predicted successfully by the simulation results as shown in the figure. The small variation of collector current with base dose is also shown for the Si BJT device. The small difference in base current between the HBT and the BJT transistors is probably due to a small difference in the emitter polysilicon interface surface recombination velocity measured in the devices and assumed in the simulation. Finally, the predicted f_t high frequency cutoff versus collector current is shown in Figure 25 and compared with experimental measurements (for the SiGe device), showing good agreement with a peak f_t of about 55 GHz.

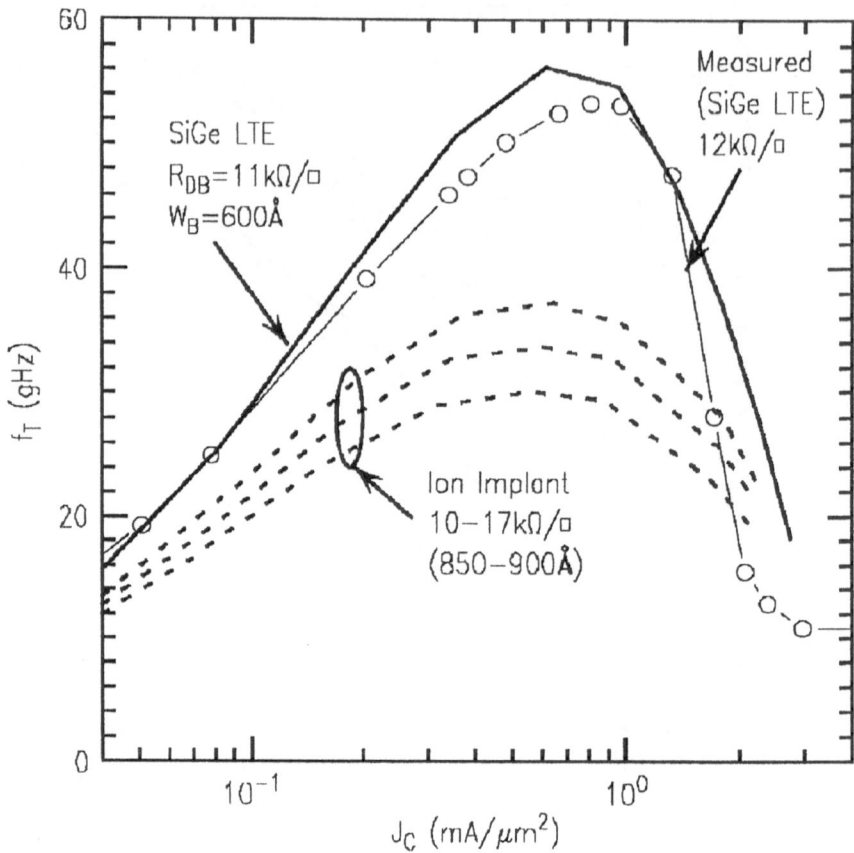

Figure 25. Cutoff frequency (f_t)simulation results versus measured data for SiGe LTE-base and Si I/I-base NPN transistors [24].

8. Simulation of the Probability of Soft Errors in Semiconductor Integrated Circuits

The effect of alpha and cosmic particles in inducing soft error failures in silicon integrated circuit chips has been known and understood for some time [28, 29] and is of obvious concern to circuit designers for reliability implications. SEMM (Soft Error Modeling Methodology) is a Monte Carlo program which has been developed within the IBM Technology Products Division at the East Fishkill location for predicting the soft-error-induced failure rate of both bipolar and MOSFET silicon chips [30 − 33]. The program uses as input chip layout information, vertical device profiles, the measured alpha particle flux from the chip BEOL metallurgy and package, and cosmic history files for the particular global location where the chip is going to be in use. The program has been successfully used in IBM over the past several mainframe generations to understand observed soft failures due to both normal alpha and cosmic sources and abnormal contamination sources, as well as to optimize and qualify chip designs so as to minimize SER failure probability.

The SEMM program methodology is described on the flowchart in Figure 26.

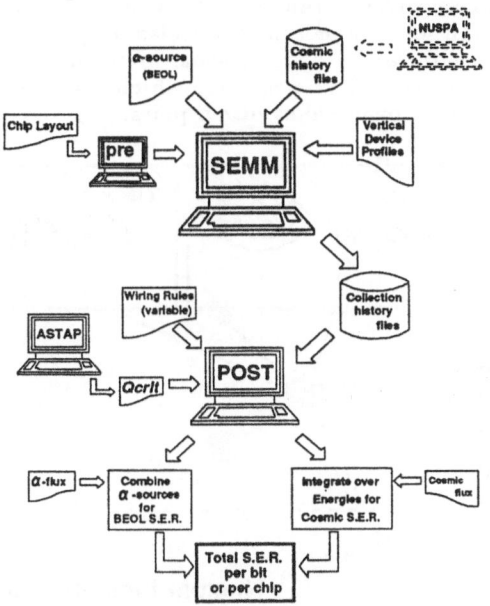

Figure 26. Flow diagram describing the SEMM program methodology for simulation of soft errors in silicon due to alpha and cosmic-generated particles [30 − 33].

The main feature of the program deals with the generation and collection of electron-hole pairs produced by high energy particles traversing the chip in question. The particle, which could either be an alpha particle or a recoiling nucleus, can have an ionizing track length extending from a fraction of a micron to many microns in the chip, and generates in its wake e-h pairs at the rate of one pair per 3.6 eV loss in energy, a typical track perhaps representing a million e-h pairs worth of charge. The SEMM program effectively allows these charges to diffuse into or away from a junction, be swept into a junction by the so-called electric field funneling effect [29] or simply to recombine. The program then accumulates the amount of charge picked up by any junctions of concern, determining if the charge adds up to more than a user-defined so-called critical charge "Q_{crit}". Both the amount of charge and the actual waveform of the pulse are utilized in determining probability of circuit failure.

Both BEOL alpha particles and cosmic particle radiation sources are included and impinge on the chip with appropriate statistical distributions of energy, location, and angle, dependent on the source. The reactions to cosmic particles such as a high energy proton or neutron are determined from cosmic history files which have previously been set up by the NUSPA (NUclear SPAllation) program to cover a range of 20 to 900 Mev [34]. As shown in Figure 27, these high energy cosmic particles may randomly collide with silicon nuclei in the chip and fragment some of them, thus producing daughters of alpha and other secondary particles, which may then travel in various directions producing their own wakes of e-h pairs.

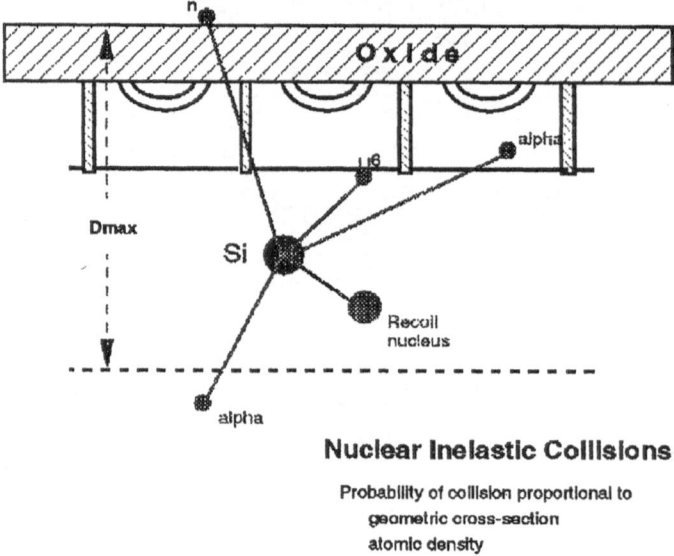

Nuclear Inelastic Collisions

Probability of collision proportional to
geometric cross-section
atomic density
Dmax

Figure 27. Pictorial description of point burst event occurring when a high energy proton or neutron strikes a (recoiling) nucleus in a semiconductor chip causing the generation of secondary particles [30 – 34].

BEOL radiation sources (metallization, C4 Pb-Sn solder balls, and ceramic package) tend to dominate the SER on a chip with low critical charge while cosmic radiation tends to be more dominant in those cases having high critical charge.

SEMM is usually run in cooperation with the circuit designer who provides inputs for fixed versus variable Qcrit and even for statistical Qcrit determination, if so desired. This part of the procedure is handled as a post-processor, as shown in Figure 26. The results are finally scaled according to alpha and cosmic flux and integrated over energy to obtain a final SER per circuit or per chip. The number of random events required in a SEMM run for accurate prediction depends inversely on the chip failure rate, so runs of several CPU hours on an IBM RS/6000 workstation or IBM 3090 mainframe are not uncommon in order to get a statistically significant number of failures. SEMM is not accessable from the WIZARD GUI at the present time.

9. WIZARD Graphics User Interface for Easy Control of Set-up, Input, Output, and Meshing

WIZARD (Window Interface for eaZy Application, Research, and Development) is an IBM internally-developed GUI for ease-of-use in the control of input, output, meshing, and interfacing the IBM TCAD S/W tool set. WIZARD was initially developed by X. Tian at the IBM Burlington, VT Technology Products facility and has been recently enhanced by R. Young and G. Fiorenza at the East Fishkill TP location. WIZARD was developed by utilizing the TCLTK package from UC Berkeley [35, 36] for the purpose of providing a window environment for the user to invoke any of the programs in the VATS suite of tools. It is finding utility in helping the user set up and run a FIELDAY job, for example, by using TRIM, REGRID, DELAUX, and MESH3D to generate and optimize a 2D or 3D mesh, providing input profiles either through DOPING, BRIDGE or a previous FEDSS simulation, browsing and visualizing an output with DBbrowser, FEMPLOT and/or VIDS, and linking the results to a circuit model generator or parameter extraction program such as DMG, EXCALIBR, or other extraction tool for generating a SPICE model.

WIZARD contains the following features:

- Database management
- Input setup interface (for FEDSS, FIELDAY, FOXi/FIERCE, etc.)
- Start and run device or process simulator
- Results checking capability
- Mesh refinement and control
- Doping to a specified mesh
- Simulation library
- SIMS library
- Process macro facility

New features recently added and/or planned for WIZARD are

- Implementation of an OOP (Object Oriented Prog.) framework
- OOP Namelist Manager
- OOP File Manager

- OOP Help Manager
- OOP Simulation Library Manager
- Choice of color scheme and key environment paths
- Additional program interfaces for university and/or
    vendor tools (SUPREM3, SUPREM4, MINIMOS, etc.)
- Incorporate demo quickstart examples
- On-line manuals

An example of the use of WIZARD is illustrated in Figure 28 which shows two panels from a FIELDAY job setup. Fig. 28-a shows a portion of the entry window for WIZARD. Selection of any of the buttons along the top of the window brings up a pull-down-menu with the available programs for that item. Fig. 28-b shows the details of the FIELDAY input setup panel.

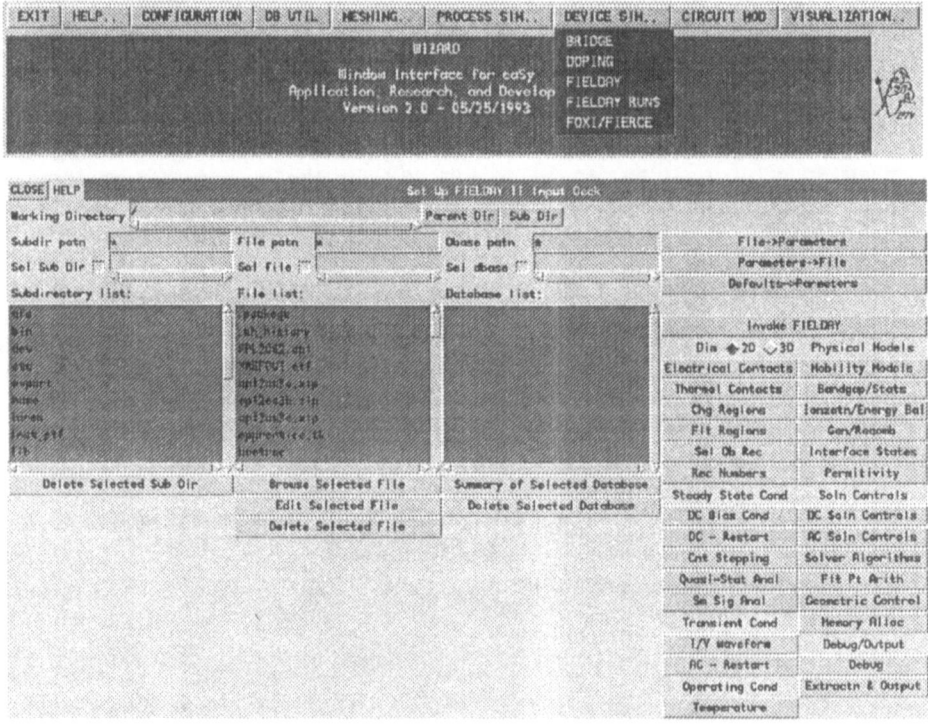

Figure 28. WIZARD (Window Interface for eaZy Application, Research, and Development) panels. (a.) Main entry panel showing user selected options for meshing, process simulation, device simulation, etc. (b.) WIZARD input deck setup screen for FIELDAY.

By using the right-hand buttons, the user may select from various mobility models or other physical models, select either 2D or 3D solution, select the steady state and/or transient input conditions, the solution controls, output, debugging, etc. Also available for browsing, editing, deleting, etc. are the lists of subdirectories, files, and databases in the users library. Similar panels are accessible for setting up and running the other TCAD device and process modeling programs listed in the entry window.

10. Output Plots and Visualization Graphics

The VATS suite of TCAD tools has historically utilized the FEMPLOT program for obtaining output plots and resultant characteristics. Recently, however, a new program for output visualization, coined VIDS for Visual Interpretation of Device Simulations, has been under development by E. Farrell at the T.J. Watson Research Labs. VIDS is an interactive visualization tool for 2D and 3D graphics which is based upon the IBM Data Explorer program product. It contains several display modes for scalar and vector fields, multiple windows with IV plots and data profiles, display of material types/mesh/contact nodes, options to display measured IV data or SIMS data, customized titles and color maps, and on-line helps for each panel. VIDS is accessable from the WIZARD GUI.

11. Summary

The IBM suite of TCAD tools for semiconductor process and device modeling has been described. These include FEDSS - a finite element 2D process modeling program, FIELDAY - a finite element 2D/3D device simulation program, FOXi/FIERCE - a dual program for constructing 3D structures for calculation of resistance and/or capacitance, EXCALIBR and DMG - programs for parameter extraction and generation of compact device models for circuit simulation, and SEMM - a Monte Carlo program for predicting soft-error failure probability in a silicon chip. The key features of these programs have been described and examples of their application have been shown.

The majority of these TCAD programs comprise the VATS (VLSI Analysis Tools Series) technology modeling software tool set. The VATS tools are installed and operational on the IBM RS/6000 workstation using AIX and the IBM 3090 mainframe with the MVS operating system. The tools are accessible with the WIZARD graphics-user-interface and interact through a common database representation format. A series of pre-processor and post-processor programs are used for mesh generation and refinement (TRIM, DELAUX, REGRID, and MESH3D), inputting doping contours (BRIDGE and DOPING), and output/visualization (FEMPLOT and VIDS).

Acknowledgements

The authors would like to acknowledge the present and prior contributions of
L.F. Wagner, F. Morehead, M.J. Saccamango, P.T. Nguyen, C.M. Hsieh, S.N.
Mohammad, K. Souissi, D. Chidambarrao, D. Cole, O. Bula, K. Varahramyan,
J. Egley, A. Nandedkar, B. Pejcinovic, E. Rorris, R.F. Lever, P. Peressini, and
S. Kapur in developing and maintaining the TCAD tool set described above.

References

1. R.W. Knepper, S.P. Gaur, F.Y. Chang, and G.R. Srinivasan "Advanced
   bipolar transistor modeling: Process and device simulation tools for
   today's technology", IBM J. R & D, Vol. 29, No. 3, pp. 218-228, May 1985.
2. R.R. O'Brien, C.M. Hsieh, J.S. Moore, R.F. Lever, P.C. Murley K. W.
   Brannon, G. R. Srinivasan, and R. W. Knepper, "Two- dimensional process
   modeling: A description of the SAFEPRO program", IBM J. R & D, Vol.
   29, No. 3, pp. 229-241, May 1985.
3. S.P. Gaur, P.A. Habitz, Y.J. Park, R.K. Cook, Y.S. Huang, and L.F.
   Wagner, "Two-Dimensional Device Simulation Program: 2DP", IBM J R &
   D, Vol. 29, No. 3, pp. 242-251, May 1985.
4. F.Y. Chang and L.F. Wagner, "The generation of three-dimensional bipolar
   transistor models for circuit analysis", IBM J R & D, Vol. 29, No. 3, pp.
   252-262, May 1985.
5. R.W. Knepper, "Problems in High Performance Bipolar Device Modeling",
   Proc of the 1987 IEEE Bipolar Circuits and Technology Meeting, pp. 1-4,
   September 1987.
6. E. Rorris, R.R. O'Brien, F.F. Morehead, R.F. Lever, J.P. Peng, and G.R.
   Srinivasan, "A New Approach to the Simulation of the Coupled Point-
   Defects and Impurity Diffusion", IEEE Trans CAD, vol. 9, pp. 1114-1122,
   October 1990.
7. R.W. Knepper, "Modeling Advanced Bipolar Devices for High Performance
   Applications", 1990 IEEE IEDM Technical Digest, pp. 177-180, December
   1990.
8. E.M. Buturla and P.E. Cottrell, "Two-Dimensional Finite Element Analysis
   of Semiconductor Transport Phenomena", Int. Conf. on Numerical Methods
   in Electric and Magnetic Field Problems, Santa Margherita, Liguria, Italy,
   June 1976.
9. E.M. Buturla, P.E. Cottrell, B.M. Grossman, and K.A. Salsburg "Finite-
   Element Analysis of Semiconductor Devices: The FIELDAY Program",
   IBM J. R & D, vol. 25, pp. 218-231, 1981.
10. K.A. Salsburg, P. E. Cottrell, and E. M. Buturla, "FIELDAY Finite
    Element Device Analysis", Proc of the NATO Advanced Study Institute
    Course on Process and Device Simulation, Urbino, Italy, July 1982.
11. K.A. Salsburg and H.H. Hansen, "FEDSS-Finite Element Diffusion Simu-
    lation System", IEEE Trans. Electron Devices ED-30, p. 1004, Sept. 1983.

12. L. Borucki, H. Hansen, and K. Varahramyan, "FEDSS-A 2D semiconductor fabrication process simulator", IBM J of R & D, Vol. 29, No. 3, pp. 263-276, May 1985.
13. P.E. Cottrell and E.M. Buturla, "VLSI Wiring Capacitance" IBM J. R & D, vol. 29, no. 3, pp. 277-288, May 1985.
14. E. Buturla, J.B. Johnson, and S. Furkay, "A New 3D Device Simulation Formulation", NASECODE VI, Dublin, Ireland, July 1989.
15. D.C. Cole, E.M. Buturla, S.S. Furkay, et.al., "The Use of Simulation in Semiconductor Technology Development", Solid-State Electronics, vol. 33, no. 6, pp. 591-623, 1990.
16. S. Furkay, et. al., "Process and Device Simulation of a BICMOS Technology"; E. Buturla and S. Furkay, "DRAM Design with an Integrated Process and Device Simulation System"; E. Buturla and S. Furkay, "3D Device Modeling Short Course", 1992 NASECODE Conference, Vienna, Austria, May 1992.
17. K.H. Brown, D. A.Grose, R.C. Lange, T.H. Ning, P.A. Totta "Advancing the state of the art in high-performance logic and array technology", IBM J of R & D: Vol. 36, No. 5, pp. 821-828, Sept. 1992.
18. R.R. O'Brien, J. Peng, and C.S. Murthy, Unpublished work.
19. D. Chin, S.Y. Oh, and R.W. Dutton, "A General Solution Method for Two-Dimensional Nonplanar Oxidation", IEEE Trans. Electron Devices ED-30, pp. 993 ff, 1983.
20. J.F. Ziegler, J.P. Biersack, and U. Littmark, "The Stopping and Range of Ions in Solids", vol. 1, Pergamon Press, New York, 1984.
21. J. Slinkman and S. Furkay, Unpublished results.
22. W. Lee, S.E. Laux, M.V. Fischetti, G. Baccarani, A. Gnudi, J. Mandelman, J. M. C. Stork, and E. F. Crabbe, "Numerical modeling of advanced semiconductor devices", IBM J. R & D, vol. 36, no. 2, pp. 208-232, March 1992.
23. D.A. Sunderland, "An Algorithm for Calibrating Computer Models from Simulated or Measured Data", submitted for publication in IEEE Trans. CAD
24. M.M. Pelella, et.al., "A Comparative Device and Performance Analysis Between a SiGe Epitaxial-Base HBT and a Si Double Poly I/I BJT npn Structure", IEEE BCTM Digest of Technical Papers, pp. 46-49, October 1992.
25. L. Wagner and M. Pelella, unpublished results.
26. J. Comfort, et.al., 1990 IEDM Technical Digest, p. 21; G. Patton, et.al. 1990 IEDM Technical Digest, p. 13.
27. J. Warnock, et.al., IEEE Electron Device Letters, Vol. 12, p. 315, 1991.
28. G.A. Sai-Halasz, M.R. Wordeman, and R.H. Dennard, "Alpha Particle-Induced Soft Error Rate in VLSI Circuits", IEEE Trans ED, vol. 29, p. 725, 1982.
29. C.M. Hsieh, P.C. Murley, and R.R. O'Brien, "A Field-funneling Effect on the Collection of Alpha-particle-Generated Carriers in Silicon Devices", IEEE Electron Device Letters, Vol. 2, No. 4, pp.103-105, April 1981; also, "Collection of Charge from Alpha-Particle Tracks in Silicon Devices", IEEE Trans. ED, Vol. 3, No. 6, pp. 686-693, June 1983.

30. G. R. Srinivasan, "Modeling Methodology for Both Cosmic Rays and Alpha Particle Induced Soft Fails in Integrated Circuit Chips", to be published.

31. P. C. Murley, R. R. O'Brien, H. H. K. Tang, and G. R. Srinivasan, "Soft error Monte-Carlo Model", to be published.

32. L. B. Freeman, "Soft Error Rate Investigation - Critical Charge Calculations for a Bipolar Array Cell ", to be published.

33. W. A. Klaasen, "Modeling Sea Level Soft Error Rate in Dynamic RAMS", to be published.

34. H. H. K. Tang, G. R. Srinivasan, and N. Azziz, Phys. Rev. C, vol. 42, p. 1598, 1990.

35. J.K. Ousterhout, "TCL: An Embeddable Command Language", Proc. 1990 Winter USENIX Conference, pp. 133-146, 1990.

36. J. K. Ousterhout, "An X11 Toolkit Based on the TCL Language", Proc. 1991 Winter USENIX Conference, pp. 105-115, 1991.

Technology CAD at Intel

J. Mar

Intel Corporation, 3065 Bowers Avenue, Santa Clara, CA 95051, USA

Abstract

This paper describes Technology CAD (TCAD) at Intel. The relative roles of TCAD and ECAD are clarified, followed by descriptions of where TCAD is currently applied at Intel. Following that, Intel's TCAD system to support those applications is described, including its underlying design, architecture, and components. The paper ends with some remarks on future trends.

1. Definition of TCAD

At Intel, Technology CAD (TCAD) encompasses the CAD tools used for developing and controlling IC manufacturing technologies, as well as tools for optimizing chip designs to specific IC processes. Included here are process and device modeling and simulation tools, process data analysis tools, and circuit verification and optimization tools with process-dependent modeling.

These are to be contrasted with higher level layout and functional design tools in the Electronic Design CAD (ECAD) area, which are relatively process dependent. At Intel, the boundary between TCAD and ECAD is necessarily inexact, since some process dependence exists in most ECAD tools. In addition, technology tradeoffs alter the degree of process dependence from technology generation to generation. How we handle the interface between TCAD and ECAD will become clearer in Section-3, when details of our TCAD system are described.

2. Role of TCAD at Intel

Intel's TCAD system is best understood in terms of the roles it performs. As mentioned earlier, TCAD tools are applied in three general areas: *process development, manufacturing control, and circuit optimization.* These application areas are depicted as circles and ovals in the technology development cycle diagram shown in Fig. 1. The horizontal position of the shaded rectangles represent periods of major activity in various development areas of a new technology.

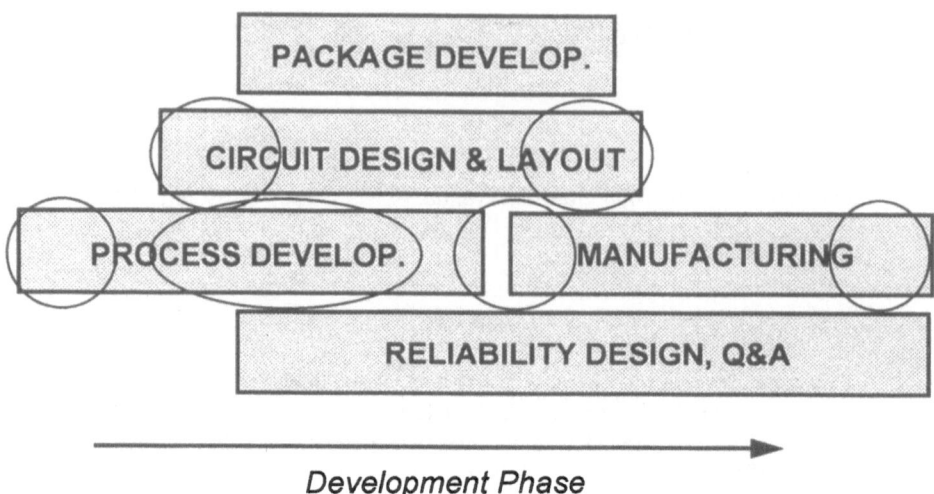

Development Phase

Figure 1. Present TCAD application areas versus technology development cycle.

2.1   Process Development Applications

In the process development area, TCAD tools are used extensively during the *conceptual design phase* of a new technology development (left-most circle in Fig. 1). Here, process and device simulators allow developers to analyze various process and technology scaling options, before committing to specific development directions and purchasing expensive wafer fabrication equipment.   They help technologists determine how far various dimensional scaling approaches can be pushed.   They also aid the evaluation of new research ideas by enabling developers to first "try them out" in computer simulations. Coupled process and device simulation tools also allow technology developers to examine the sensitivity of device behavior to process variable control (i.e., examine potential manufacturability).

Since minimal experimental data is available during this stage, applications in the conceptual phase rely heavily on the predictability of underlying physical models in the simulation tools. Predictable models allow technologists to use TCAD tools to investigate a much larger range of options than would be possible experimentally, and enable them to rule out dead-end approaches and speed the determination of the best technology approach for full-scale development.

Given all models have limitations, a major requirement is knowing where models are valid. Models used outside their region of validity generate incorrect answers and lead to nonoptimal design decisions. At Intel, major effort is expended to characterize models and determine their regions of applicability. This knowledge is integrated into simulation methodology packages to ensure tools are not used improperly.

TCAD tools are also used extensively in the *process optimization phase* of process development. This is the oval region immediately to the right of the conceptual design circle in Fig. 1. Compared to the previous stage, this stage differs by having wafers and experimental data. While basic process/device simulation tools may be adequate for determining general technology directions (for the conceptual phase), they are often inadequate to model detailed effects, second-order phenomena, and equipment related behavior. Also the predictability of models for the back-end of the process (deposition, metalization, etc.) is typically much poorer than those for the front-end (implantation, diffusion, oxidation, etc.). These model deficiencies become especially apparent when the first experimental wafers are measured and results compared with simulations. General, uncalibrated process and device simulation tools while able to predict correct trends, often fall short in accurate quantitative data predictions.

At Intel, process optimization is carried out using a hybrid modeling methodology, one involving alternating combinations of experiments and modeling. The unknown effects are lumped into adjustable model parameters in numerical process/device simulation tools, which are characterized using experimental data from the new preliminary process. The result are "locally-calibrated" models that are quantitatively accurate for moderate technology excursions, models that can be used to further optimize the process and design subsequent experiments. As additional experiments are completed and new data becomes available, these models are further refined and the procedure is repeated. This methodology is continued until the process design is fully optimized.

The use of TCAD tools in this manner significantly reduces the number of experiments required for process optimization, and both speeds and lowers the cost of the total development process. The use of locally-calibrated simulation models for process optimization also allows a more thorough optimization of the process than would be possible with experiments alone. An additional bonus of this methodology is the availability of a set of quantitatively accurate simulation models for the process at the end of the optimization. The value of those models will become apparent in the next section.

2.2 Manufacturing Control Applications

The simulation modeling package generated at the end of the process optimization phase serves as the basic models for TCAD manufacturing control applications. Since those models are quantitatively accurate over the normal range of process variations in manufacturing, they are an accurate representation of the optimized manufacturing process.

Since the simulation models are both a behavioral and wafer-level specification of the process, they contain more information than conventional process flow specification documents. They are especially useful for keeping the process at its optimum design point after the process is transferred into high volume manufacturing. The models enable process engineers to compensate for small differences in high-volume manufacturing fabrication equipment (vs. that used in development) to ensure wafers generated in the high volume process are identical to those in the optimized development process. This is especially important for design optimization, since product designs are started early in the technology development cycle (see Fig.1), and any changes to the processed wafers in manufacturing will invalidate earlier process-specific design optimizations. The circle spanning the "Process Develop." and "Manufacturing" rectangles in Fig. 1 depicts this application area.

The second area of application is manufacturing control, where TCAD models serve as extrapolatable reference models of the optimized manufacturing process. Here, calibrated process simulation models are used to simulate the target values of various process control monitors to generate references for process control measurements. Example process monitors include electrical measurements on test devices ("E-tests") and in-line measurements taken at intermediate stages of the process. The simulated process monitor values allow more accurate specification of the control limits and enable those limits to be correlated with the simulated performance of products. These models are especially useful when processes drift from their optimum design points, since the drifts will first be manifested as drifts in various process monitor values. The availability of accurate simulation models allow process engineers to rapidly diagnose and correct the cause of the drifts before product yields are impacted. In Fig. 1 this application area is depicted by the right-most circle in the "Manufacturing" rectangle.

2.3   Circuit Optimization Applications

As mentioned earlier, circuit simulation is a TCAD capability at Intel, even though such tools are applied in an ECAD environment. Compact, analytical circuit simulation models are currently the primary vehicle for coupling process-specific device behavior to the design process. They provide the base layer in a hierarchy of performance verification/optimization tools, and the layer through which most process effects are coupled to higher level design tools. The availability of accurate circuit simulation models determines when the optimization of circuits to fabrication process can begin.

The complexity of lead Intel microprocessor products require their design to begin very early in the technology development cycle, often when the process is still in the conceptual design phase. While elements of the chip design do not depend on the process, optimization of functional level layouts (chip plan) requires optimization of routing and interconnects to maximize performance and reliability. Design and optimization of standard cell building blocks also require detailed knowledge of the electrical behavior of transistors and parasitic devices. These require accurate circuit simulation models be available early in the development cycle (left-most circle in the "Circuit Design & Layout" block in Fig. 1). Accurate statistical circuit models for simulating the effect of manufacturing variations may also be required for the optimization of critical path circuit designs.

Since insufficient experimental material is available during the early development phases for experimental characterization of the circuit models, process and device simulation tools play an important role here. Process and device simulation models (both general and locally-calibrated) are used to characterize the circuit models. The simulation models supplement available data to generate the most accurate circuit simulation models for any particular stage of the design process. As the fabrication process becomes optimized and more experimental data becomes available, these models are refined. The final set of models (the right-most circle in the "Circuit Design" block in Fig. 1) are released after the process optimization is completed and the process design frozen.

3. Intel's TCAD System

Compared to ECAD, TCAD is still an immature field with relatively few generally predictive models available, and with capabilities scattered across many different tools. Most TCAD tools are difficult to use, require detailed knowledge of model limitations, and consume large amounts of CPU time to execute. This situation is in sharp conflict with the needs of users. Unlike ECAD users, TCAD users are often not computer literate and require only sporadic use of these tools. What they require are simple and robust tools.

The immaturity of the TCAD field means there are far more problems to solve than resources to solve them. The challenge is to pick the right problems to tackle - that is, to pick the problems with the greatest potential returns. How we have worked around these limitations to respond to the application needs described previously will described in the following.

3.1   Applications Focus and Technology Synchronization

TCAD development programs at Intel have a strong applications focus with explicitly defined deliverables and customers. The focus on application-based deliverables ensures that all the pieces needed for delivering a useful simulation capability are available when needed. This approach screens out development programs that only solve a small portion of a much larger and difficult problem. The addition of customers to the program definition process aids the making of engineering tradeoffs and definition of meaningful delivery schedules. Because of the cyclical nature of technology development processes, the time windows during which TCAD tools can impact a technology design are relatively narrow (inside the circled regions in Fig. 1). Technology development schedules are driven by external market requirements and technology decisions will be made with or without TCAD tools. Once specific technology decision windows are passed the opportunities for applying TCAD to that phase of the design are typically gone.

Application windows define what and when capabilities must be delivered in order to be useful. Although this focus can cause development to be concentrated on near term solutions, this has not been the case at Intel where multiple generations of IC technologies are developed concurrently. At any given time, application windows and customers exist for several generations of technologies. This has led to a situation where near-term TCAD capabilities are targeted for the nearer technologies and long-range capabilities are driven by the further-out technologies. The further-out technologies promote aggressive long range TCAD development programs, since the device goals and therefore the TCAD modeling requirements for those technologies are far more aggressive (i.e., much further beyond current capabilities) than that for nearer-term technologies.

3.2 Hierarchical Models

The range of TCAD applications described in Sec. 2 cannot currently be supported by a single class of models. Very general models, such as those in general-purpose two and three-dimensional process and device simulators, are too slow and difficult to use for many applications. Tools using locally-calibrated, analytical models are fast and robust, but too specialized for most purposes. Instead, the hierarchy of model types shown in Fig. 2 is utilized. In that figure, model classes are plotted as a function of generality and CPU-time requirements, with primary application areas listed to the left of the graph.

Numerical simulation models built around physically accurate models are able to handle problems with the widest range of boundary conditions. These tools are calibrated by optimizing imbedded models to a database containing data from a wide variety of processes. The ability of a single calibration of those models to fit all the data simultaneously tests their physical correctness. The resulting "global characterization" maximizes the predictability of those models for new, not-yet-developed technologies.

Numerical simulation modeling tools are among the most CPU-intensive and difficult-to-use TCAD tools. For example, typical three dimensional simulation tools require comprehensive specification of problem boundary conditions (information that is often unavailable), are prone to nonconvergence or incorrect answers if initial numerical grids are not chosen carefully, and may require supercomputing computing capacity for practical simulations. At Intel, these tools have been useful for initial technology design - where boundary conditions are relatively simple (since they are based on "paper" technologies) and where lack of experimental data make locally-calibrated models unusable. These simulation models have also been very useful for locally-calibrating more specialized models in the hierarchy.

The next class in the hierarchy are the *numerical/analytical models*. These are the workhorse models - offering reasonable generality and requiring moderate CPU-time.

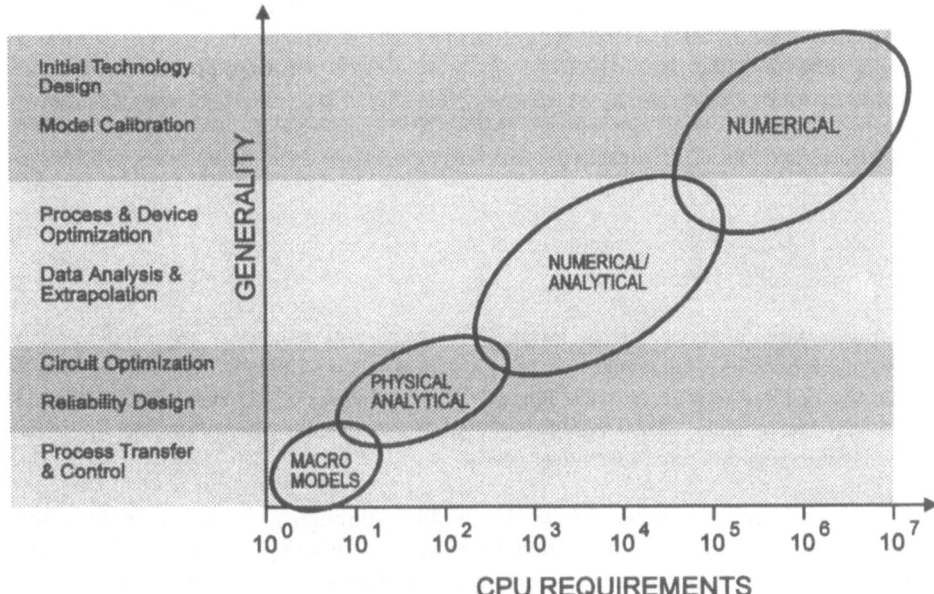

Figure 2. TCAD Model hierarchy.

They lump effects that cannot be explicitly modeled (because of lack of physical models or boundary condition data) into analytical expressions with adjustable model parameters. These models are calibrated near their region of application and used with globally-characterized models in numerical simulation tools (the "locally-calibrated" models described in Sec. 2.1 are in this class). They are widely used for process and device optimization as well as data analysis and data extrapolation. These models must be recalibrated for each region of application - either using experimental data or more general numerical simulation models.

The next level in the hierarchy are the *physically-based analytical models*. These are typically 5-6 orders of magnitude faster to compute than purely numerical models, and can be calibrated with either experimental data or simulations of models further up the axis in Fig. 2. They are used wherever model speed is extremely important. Applications include those requiring many iterative model computations, such as circuit simulation and optimization, and statistical process and device optimization. These models are also useful for interactive tool applications. A special class of these models is used for data filtering. Experimental or simulation data are fitted to analytical *intermediate characterization models*, where the parameters of those models represent filtered data for subsequent processing.

The last level in the hierarchy are the *macro models*. These are nonphysical, empirical models calibrated by data. They range from response-surface models to neural-network learning models. The lack of physics in these models means they require extensive calibration to be useful, as a consequence they are often calibrated only for narrow operating ranges. Such models are useful for manufacturing transfer and control applications (see Sec. 2.2), where process variations around model calibration points are small.

3.3   Open Tool System

The third design characteristic of Intel's TCAD system is flexibility. The rapid pace of change in technology, manufacturing, product design, computing, CAD, and the market make any rigid system design very limited. The optimum TCAD system will change as these variables change. The unpredictability of these changes means the ideal system is one that can adapt quickly to changing needs.

At Intel we have chosen to address this problem through a family of modular tools coupled through a flexible automation framework. Referring to Fig. 3, the program modules fall into five categories: simulation platforms, simulation preprocessors, simulation posprocessors, data access modules, and model characterization utilities.

The heart of the system is the box labeled *Simulation Platforms*. This box includes the basic simulation tools and models on which most simulation predictions are based (the upper two ovals in the model hierarchy in Fig. 2). They consist of a collection of internally and externally developed simulation programs, whose I/O structures have been modified to integrate into Intel's TCAD system, and whose models have been upgraded to be compatible with the best models at Intel. Tools in this box serve as the basic software platforms for building proprietary modeling capabilities. For that reason, direct control and modifiability of program source code is an important requirement for these programs.

Preprocessing Modules are simulation input processing modules. They include input parsers to convert process flow information into program readable inputs and specialized graphical user interfaces tailored for special customers. They also include script-based automation programs to execute supporting simulation runs prior to executing primary simulations. *Postprocessing Modules* are simulation output processing programs. They range from visualization tools for graphing and analyzing simulation outputs, to tools that utilize outputs from multiple simulation runs to fit analytical models. The *Data Access Modules* are utilities for accessing network databases - especially databases outside the TCAD area, where have been used to retrieve such data as manufacturing E-test data and chip layout information. Finally the *Model Characterization Utilities* are specialized routines for automating and processing data for rapid calibration of TCAD models.

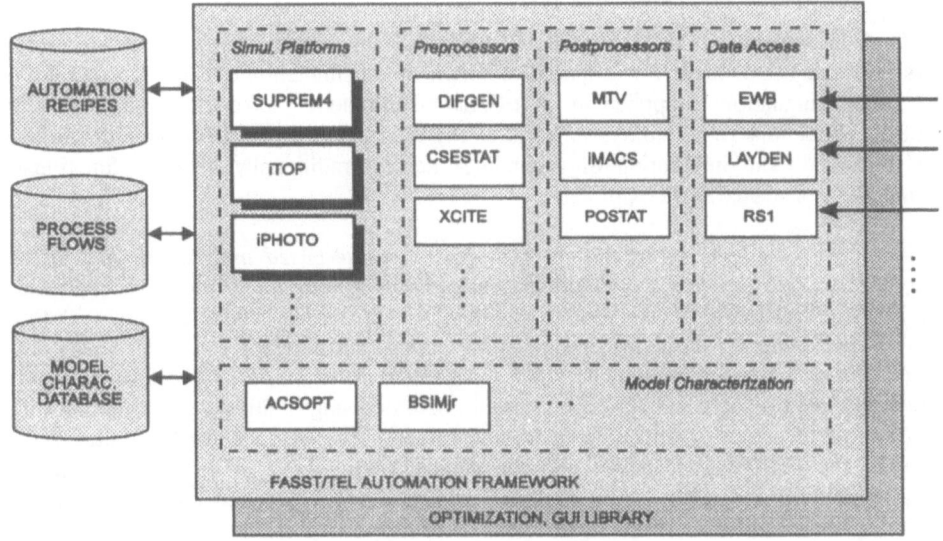

Figure 3. Open TCAD Tool System.

The glue for the system is the FASST/TEL automation framework. The automation framework is implemented in the higher-level SCHEME language and supported by a still higher-level interpretive language called TEL. The automation framework is supported by a library of C-language routines for such functions as optimization, graphical user interfaces, and data extraction. This system is routinely used to automate simulation methodologies for user applications, especially those that are executed repetitively. The automation scripts represent an additional level of modeling (application methodology model), and are often the products delivered to end-users once appropriate simulation methodologies are developed for their application.

The TCAD system is also supported by three local databases: one for automation scripts, process flows, and model characterization data. A key design goal of this system is open expandability, to enable new module capabilities to be incrementally added as needs arise. Code enhancements to common modules are always implemented in an upward-compatible manner, to ensure older capabilities can be continued to be supported.

One last feature in support of flexibility is software portability. Core TCAD code are written in standard C, C++, or FORTRAN77, to support operation in any generic UNIX environment supporting X11 Windows and MOTIF. This allows the system to be easily ported to wide variety of computer systems - ranging from 486 PC systems, Pentium[tm] CPU-based systems, RISC-based workstations, and high performance compute servers.

3.4   Cross-Functional Integration

At Intel, increasingly TCAD applications require the interaction of programs and data outside of the TCAD area. For example, although circuit simulation is a TCAD capability, the circuit simulation is applied in an ECAD environment. Likewise, manufacturing control applications require the installation of TCAD capabilities inside the manufacturing environment. The integration of capabilities across traditionally separate functional boundaries is a special system problem.

At Intel we have approached this in four ways: *engine-based integration, file-based integration, cross-functional automation,* and *computing-node integration.* Engine and file-based integration have been used to integrate TCAD's circuit simulation tools with the ECAD environment. Basic circuit simulation models and network equation solvers are developed as independent modules and maintained as a TCAD capability. Circuit simulation input parsers and waveform output postprocessing utilities are tool modules in ECAD's performance verification environment (see Fig. 4). Standardized interfaces between the ECAD environment and the engines allow these modules to be linked transparently to the user. This approach allows TCAD to independently focus on providing the circuit simulation engines needed for supporting changing technology needs (e.g., digital simulation, analog AC, interconnect, reliability, etc.), provided they all support the standardized ECAD-TCAD interface. At the same time development of the ECAD environment to support integration with higher level design tools and layout can also proceed independently of TCAD.

In the case of circuit simulation, file-based integration is also utilized for communicating model parameter information for process-dependent models. In Fig. 4, the cylinder labeled "Process Files" is a standardized database that can be accessed by both TCAD and ECAD environments. File-based communication is also used to couple manufacturing process information to TCAD process control/monitoring tools.

A third method of integration is cross-functional tool automation. In this case, special purpose remote program initiation utilities are coupled to our tool automation system to enable external programs (in ECAD, manufacturing, etc.) to be executed automatically from within the TCAD environment. These allow external program capabilities to be treated as subroutine capabilities in our automation scripts. This approach has been utilized for automated circuit optimization and applications involving mask layout information.

The final and most radical method of integration is computing node integration. Here, a complete UNIX computing system, preloaded with the required TCAD software, is installed in the customer's environment. Delivered systems are networked to central TCAD servers for maintenance and tool tracking. This approach has been used to deliver TCAD capabilities to manufacturing plants. The declining cost of computing hardware has made this approach practical.

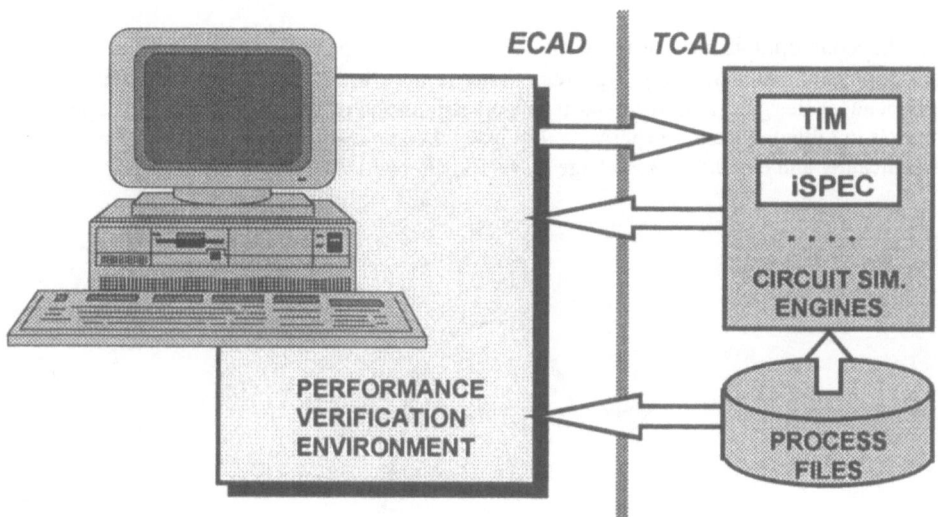

Figure 4. Example of ECAD-TCAD Integration.

4. Future Directions

Demands on TCAD are expected to grow significantly in the future. Two forces are driving this process: escalating technology costs and shrinking design windows.

The extreme cost of developing and bringing up new generations of manufacturing technology (currently well over 10^9 U.S. dollars) means fewer mistakes can be afforded. Referring to Fig. 5, process design iterations are easiest to absorb when they are made in the research and planning stage (before specialized fabrication equipment is purchased). They are 10X more expensive when they occur in the experimental optimization stage, and 100X more expensive when they occur in volume production. Currently most design decisions are made in the experimental optimization stage (the solid curve in Fig. 5). As the predictability of TCAD tools improves, it should be possible to move design iterations to the left (towards the dashed curve in Fig. 5). The resulting cost savings will probably be required to make continued technology development affordable.

The continued shrinking of transistor dimensions and the steady expansion of chip sizes has led to a major narrowing of design margins. Emerging limits in materials reliability, second-order device effects, power dissipation, testing costs, interconnects, packaging, have all imposed new limits on the technology design process and shrunk design windows. They have made it increasingly difficult to achieve the product improvements required for justifying the enormous cost of bringing up new technology generations. This situation has led to an increase in the need for global optimization of the total technology - namely simultaneous optimization of tradeoffs in process development, product design, package development, and manufacturing area. This is a reversal of a trend towards process independent design. This will lead to major expansions of the use of process dependent models in affiliated design processes. In Fig. 1, the application areas for TCAD will grow significantly and expand into Package Development and Reliability Design areas.

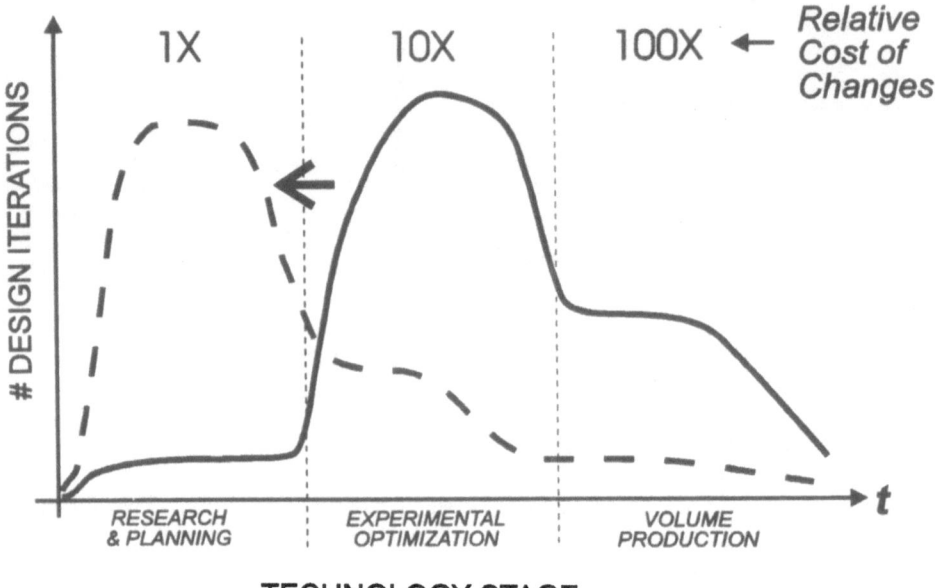

Figure 5. Long term TCAD goal.

The trends above are driving the need for better TCAD tools. The challenge is responding to those needs in a timely fashion. The development of more physically complete and accurate models are key to making tools more predictable. The establishment of industry standard TCAD integration frameworks for integration and tool development can aid both their development and use. Standardized tool development frameworks can lead to a more "plug and play" tool development approach - one that will reduce the need for making special source-level code modifications and accelerate the development of needed solutions. Establishment of industry-wide, standardized protocols for intertool communication will also aid cross-functional tool integration.

Perspective on TCAD Integration at Berkeley

A. Neureuther, R. Wang, and J. Helmsen

Dept. of Electrical Engineering and Computer Sciences, University of California, Berkeley, 231 Cory Hall, Berkeley, CA 94720, USA

Abstract

A perspective on TCAD integration is given along with highlights from work at Berkeley on applications and future systems which contain exploratory aspects of interest in TCAD development. The aspects of interest include process flow simulation, linking to the IC designer, design task interfaces, utilities for mapping data between simulators, computational geometry algorithms, and interfacing and testing centralized computational services from solid modelers.

1. Introduction

TCAD integration is a major opportunity for the process and device community especially now that simulation is viewed as very cost effective. Mention TCAD frameworks and the potential user quickly envisions a collection of interoperable functional components which go well beyond the variety of tools found in a PC software catalog. A vision of an integrated TCAD framework is a major motivator for developers and each organization needs to have its own system vision. Yet the magnitude of the integration problem and the feedback from industry that they want industrial grade code and only a particular subset of any university TCAD framework brings us back to reality. This puts us in a perennial state of dancing around with other TCAD developers trying to find where to participate be it doing exciting exploratory experiments to define future directions or be it the painfully slow definitions of standards. At the same time we are trying to sustain the integration of our own tool strengths to continue to interact with technologists on process development issues.

This workshop is an opportunity to take stock of our approaches and this presentation will overview our activities at Berkeley. The talk begins with a discussion of the overall role being played by TCAD at Berkeley and then moves to specific work being carried out on applications and future systems which contain exploratory aspects of interest in TCAD development.

2. Berkeley Perspective on TCAD

Developing common TCAD systems will be a very hot issue in the 90's as we are just starting up the learning curve. Since there are significant tradeoffs in CPU, memory, and communication bandwidth with how the shared functionality is built, experiments must be carried out to find bottlenecks and explore feasible new directions. Thus common TCAD systems will probably evolve in distinct 3–5 year phases of implementation to reach maturity. Reaching industry wide agreements on standards and implementation time-lines requires cooperation and hence TCAD is somewhat like politics and might be thought of as sharing the definition of politics as being "the art of the possible".

FACTORS SHAPING TCAD INTEGRATION

INTEGRATION IS REQUIRED FOR SIMULATING THE INTEGRATION IN IC DESIGN

TIMELY RESPONSE REQUIRES A TEAM EFFORT

SUPPORTING THE WIDE TECHNOLOGY SCOPE REQUIRES TEAM APPROACHES

INTEGRATION WORK MUST RESPECT EXISTING TCAD INVESTMENTS

COMPUTER SOFTWARE/HARDWARE PORTABILITY AND NETWORKING ARE ESSENTIAL

LINKAGE TO CIRCUIT CAD FOR LAYOUT/DEVICE/PARASITICS IS IMPORTANT

COUPLING TO DESIGN METHODOLOGY IS IMPORTANT

USERS' EXPECTATIONS ARE BEING RAISED BY GRAPHICAL INTERFACES ON PCS

MORE COMPLEX TECHNOLOGY REQUIRES MORE AND MORE FLEXIBLE INPUT

SUPPORTING DEVELOPERS OF INDIVIDUAL PROCESS MODELS IS ADVANTAGEOUS

ATTRACTING ADDITIONAL PHYSICAL MODELERS IS ADVANTAGEOUS

INVOLVING NUMERICAL ANALYSTS AND COMPUTATIONAL GEOMETRISTS IS ADVANTAGEOUS

STRENGTHENING TIES WITH HUMAN-FACTORS ENGINEERING IS ADVANTAGEOUS

Figure 1: Factors Shaping TCAD Integration

The forces shaping TCAD systems relate to the immense diversity of the task of IC design and the computer techniques used to carry it out. A number of specific examples of these forces are listed in Fig. 1. The factors might be grouped into six major categories as relating to the nature, size and scope of the task, the computer environment, linkage to overall design methodology including IC design CAD and CIM, interacting with the novice and experienced user, supporting physical modelers and simulator developers, and involving specialized experts in difficult common problems.

The functionality necessary in a modern TCAD environments is very diverse. A CAD system might be thought of as having many arms like an octopus to interconnect this functionality as is reflected by the choice of the name OCT in the Berkeley IC Design CAD. For purposes of discussion the func tionality to be integrated in TCAD might be some what arbitrarily divided into 7 major groups in Fig. 2. A modular approach is envisioned in which individual services available from each of these components is invoked through a direct and standardized function call. The component developers are thus free to use different implementation approaches to provide the best service or call on other services.

The question in developing common TCAD systems is how to balance the extremes of bogging down on standards and of running off and creating new unrestrained systems. Admittedly working on standards in the SWR experience has helped structure the problem and gelled the personal dynamics of the international community. In my view, however, hacking the code is about five times more effective than long discussions of concepts in standards. Since I do not think we can correctly anticipate structure defining issues in advance, TCAD systems will need to evolve through major stages of redefinition about every 5 years. One of the most important roles for universities is to do exploratory work on concepts and their inherent consequences on performance. A good example is the proposal for a client-server model for centralizing computational functions to support rapid development of new physical models. Coding an exploratory experiment is highly motivating to students and in making the

---

COMMON TCAD ENVIRONMENT COMPONENTS

DATA BASE AND ACCESS METHODS

USER INTERFACE WIDGETS

GRAPHICAL VISUALIZATION/CAPTURE

TASK MANAGEMENT

SYSTEM ENVIRONMENT

DATA MAPPING AND NUMERICAL UTILITIES

LAYOUT AND CIRCUIT CAD INTERFACE

PROCESS FLOW AND CIM INTERFACE

Figure 2: Common TCAD Environment Components

system work they will discover unanticipated new issues and invent new approaches.

Universities, however, cannot put all their eggs in the future system basket as we need to continue to provide technologists with application oriented simulators to bring financial resources in the door. As a result at Berkeley in integrating our applications we have adopted a strategy of using buffered interfaces to our own data representations which can easily be replaced by SWR calls when SWR compatible frameworks become available. Thus we try to play both the near term and far term by creating SWR compatible modules to explore applications and concepts which can also run as small stand alone systems.

3. Exploratory Aspects of TCAD in Berkeley Simulators

In developing simulators such as SIMPL, MASC, and PROSE to assist process and circuit designers we have gained a great deal of experience in exploring many of the common TCAD environment components listed in Fig. 2. SIMPL[1], as shown in Fig. 3 from Cole et. al[2], is a process flow manager with its own form of process flow language. It also has the ability to make bidirectional links to SAMPLE[3] for topography simulation and SUPREM[4] for impurity simulation. To facilitate this linkage a framework and set of data mapping and numerical utilities were developed known as the Berkeley Topography Utilities[5]. The graphical interface of SIMPL also has design features for identifying likely problems (HUNCH), misaligning layout (WORST) and critiquing the device cross section afterward (CRITIC).

For interactive design with the user we have developed a set of windows based on the Tcl/Tk widget set suitable for use with many of our simulators on hopefully PC's as well as workstations. These windows were developed in conjunction with the MASC[6] program which manages SPLAT[7] aerial image analysis to perform design tasks involving layout and optical exposure tool parameter interactions. An example of some of the available design widgets is shown in Fig. 4.

The PROcess Simulation Environment PROSE[8] was developed to make process simulation tools available at the level of the IC circuit designer. As shown in Fig. 5, PROSE brings the wafer cross section simulation view to the designer in an RPC-ized and SWR compatible version of SIMPL. PROSE also offers a set of design tools for addressing technology issues in phase-shift masks. New tools at the layout level allow designs to automatically be phase-shifted and then design rule checked for intra and

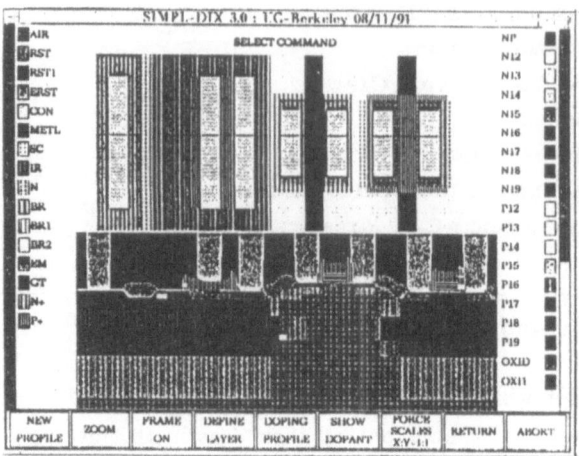

Figure 3: SIMPL System-5 Simulation of a BiCMOS process[2].

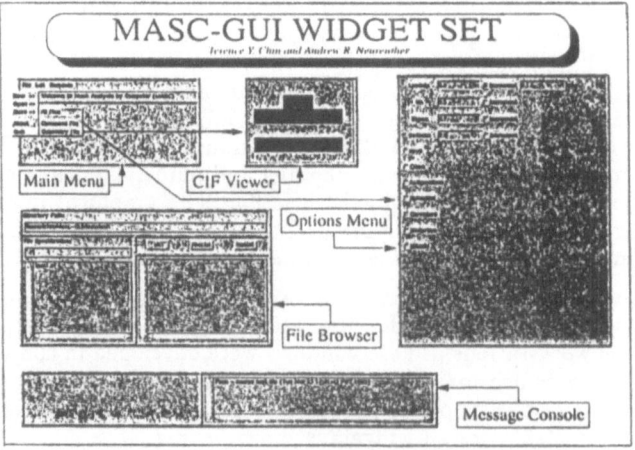

Figure 4: Example of widgets for design tasks.

inter phase level errors. For questionable areas the designer can capture the layout and send it to the MASC and PSIDOL[9] programs for design and optimization. The modified layout can then be reinstantiated in the chip layout. It is my personal belief that allowing the IC designers to be able to explore consequences of layout is one of the most important aspects of TCAD frameworks.

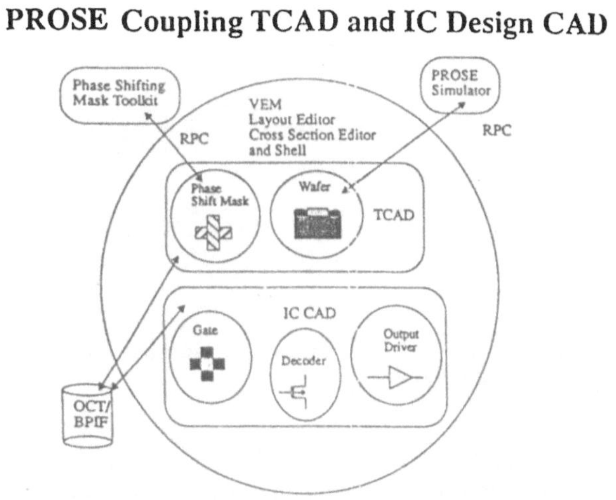

PROSE Coupling TCAD and IC Design CAD

Figure 5: The PROSE system as an application under OCT/VEM/RPC.

4. Future Aspects of TCAD Systems

Moving to 3-D represents a significant challenge and this is where much of the university effort is centered. SAMPLE-3D has been completed as a stand alone lithography and topography simulator[10,11]. Even here, seeing the breadth of the physical expertise needed to span the scope of etching, CVD, metalization, etc. which the code can support we were forced to move to a framework mode. SAMPLE-3D provides an organizational structure and computational modules for various actions which the expert in a particular technology can call on along with an occasional application specific subroutine to develop a new simulation capability.

One of the things we will see with frameworks is that experts with specialized skills will be able to make significant contributions to the field. Our own example of this is the work that John Helmsen is doing with respect to computational geometry aspects in SAMPLE-3D. John has revisited the deloop and surface advance code and is looking at visibility in etching/deposition[12,13]. He has come up with powerful geometrical constructs not only applicable to SAMPLE-3D but to geometry modelers for TCAD and likely other fields of engineering.

One of the key debates in defining future TCAD environments is the extent to which computation services might be centralized and how that might be accomplished. The IBM Yorktown simulation group view is that there should be a master geometry representation and the best way to preserve validity is to have process simulators provide action through well characterized commands to a central geometry engine.

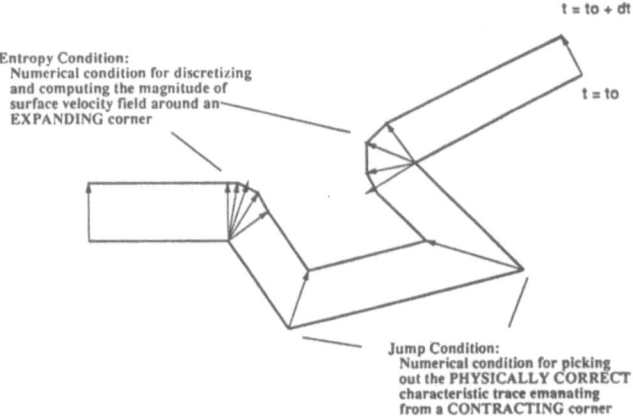

Figure 6: Examples of the way in which surface motion can be specified to preserve the the solid model representation consistency.

An example of specifying surface motion in a manner which preserves the consistency of the solid model representation is shown in Fig. 6. To accommodate physically dissimilar processes the geometry engine would be able to reorganize the view of the geometry it uses to the nature of the process. At the other extreme is my personal view that purpose built data structures and algorithms are most efficient. The best approach of course depends on how computationally intensive the task is and who is willing to write the code.

Robert Wang has been carrying out exploratory work with Michael Karasick of IBM on the suitability of the IBM Geometry Engine for topography simulation applications[14]. This activity involves developing a program interface to the Geometry Engine to implement the kinds of actions required in dissolution, etching and deposition. In addition Robert has been testing the interface and underlying solid-modeler to determine memory, CPU and communication bandwidth requirements. One of the key issues is the granularity of the queries and surface movement requests and the concept of surface monotonicity has been introduced to overcome this problem.

5. Conclusion

Although industry is consolidating process development, there will be a resurgence in TCAD frameworks due to the cost effectiveness of simulation in dealing with the increasing complexity of processing. The massive scope of a TCAD frameworks will require industry and commercial vendors to take the lead. However, universities will play a very important role in exploring new functionality frontiers and centralized computational services as the frameworks evolve through major changes about every 5 years.

Acknowledgement

Research highlighted here was supported in part by IBM through the California State MICRO Program grant 92-112 and the Semiconductor Research Corporation grants 90-MC-500 and 92-CD-008.

References

[1] E.W.Scheckler, A.S.Wong, R.H.Wang, J.R.Camagna, A.R.Neureuther, G.Chin and R.W.Dutton, IEEE Trans. on CAD, 11, (1992) 911.

[2] L.Harris and D.Cole, NASECODE, 1992.

[3] W.G.Oldham, S.N.Nandgaonkar, A.R.Neureuther, and M.M.O'Toole, IEEE Trans. on Electron Devices, Vol. ED-26, No. 4, pp. 717-722 April 1979.

[4] C.P.Ho, J.D.Plummer, S.E.Hansen, and R.W.Dutton, IEEE Trans. Electron Devices, vol. ED-30, no. 11, pp. 1438–1453, Nov., 1983.

[5] R.H.Wang, A.Gabara and A.R.Neureuther NUPAD-IV Seattle, June 1–2, 1992.

[6] D.M.Newmark, and A.R.Neureuther, SPIE 1674, (1992) 2.

[7] K.H.Toh and A.R.Neureuther, SPIE 772, (1987) 202.

[8] A.S.Wong, Ph.D.Thesis, University of California, March 1992.

[9] Y.Liu, A.K.Pfau, and A.Zakhor, SPIE 1674 (1992) 14.

[10] K.K.H.Toh, A.R.Neureuther and E.W.Scheckler, SPIE 1463, (1991) 356.

[11] E.W.Scheckler, K.K.H.Toh, D.M.Hofstetter, and A.R.Neureuther, 1991 Symposium on VLSI Technology, 97.

[12] J.J.Helmsen, E.W.Scheckler, A.R.Neureuther and C.H.Sequin NUPAD-IV Technical Digest, (1992) 3.

[13] J.J.Helmsen and A.R. Neureuther, SIPE 1927, (1993) to appear.

[14] R.H.Wang, M.S.Karasick, and A.R.Neureuther, VPAD Proceedings, 1993.

A TCAD Framework for Development and Manufacturing

D.M.H. Walker, J.K. Kibarian[†], Ch.S. Kellen, and A.J. Strojwas

Dept. of Electrical and Computer Engineering, Carnegie Mellon University,
Pittsburgh, PA 15213, USA
[†]PDF Solutions, Pittsburgh, PA, USA

Abstract

The semiconductor manufacturing engineer is often faced with the problems of
process integration, equipment control, parametric yield diagnosis, process
centering, worst-case design, and device parameter extraction. These problems
require the use of technology CAD (TCAD) tools across a spectrum of speed
and accuracy. We have developed a specialized TCAD framework to integrate
the required tools and auxiliary software into a system that is convenient for
everyday use. The initial work was embodied in the PREDITOR system, and
has now evolved into *pdFab*, a commercial product. In this paper, we will
describe the PREDITOR and *pdFab* systems, and show experimental results of
using *pdFab* on industrial problems.

1. Introduction

Traditionally, technology CAD (TCAD) software was developed as point tools with an
emphasis on process and device simulation. These programs typically required a significant level of user expertise, which limited the application of these tools to process/device
development performed by a handful of people in a given semiconductor company. There
was practically no link between these tools and the CAD systems used by the circuit
designers. The same could be said about the CIM systems used in the fabrication lines. As
a result, the impact of these tools has been rather limited and the TCAD community has
been restricted to a relatively small number of researchers and/or developers.

Recently, the cost and time requirements for development of new generations of semiconductor technology placed very aggressive demands on the TCAD tools and simultaneously
created a real opportunity for TCAD to play a more significant role. Moreover, device simulation tools have reached a high level of maturity and advances in process simulation
(especially ion implantation, diffusion and oxidation) have greatly enhanced their predictive capabilities.

However, improvements in physical accuracy and even computational efficiency are not
sufficient to increase the impact of TCAD. New solutions are necessary to enable TCAD
tools to be useful not only for new technology development but also for product develop-

ment and even manufacturing. To achieve this goal, TCAD tools must be able to perform many new tasks. In the technology development domain, the simulation capabilities must span many levels ranging from equipment simulations to process/device simulations to extraction of compact device model parameters used in circuit simulation. These tools must take as inputs the true process recipe and device layout, and propagate the data through the simulation levels in an automatic manner without painful and error prone user intervention. This requires integration of the simulation modules into a system with a common database which can be linked to the CIM system containing the process recipe and the CAD system which contains layout information. Such a system should also allow the user to specify a number of tasks such as, for example, process recipe optimization or layout design rule development.

For product development purposes other tasks might be required such as specification of realistic worst case parameters necessary for prediction of IC performance or investigation of IC layout sensitivity to process fluctuations. For manufacturing-oriented purposes the virtual fabrication line simulation capabilities which include equipment simulation are crucial.

To support the range of tasks mentioned above, a broad spectrum of simulation modules is required with various computational efficiency specifications. Hence such an integrated system must be capable of supporting simulation models ranging from very detailed physical simulations on the atomistic level to numerical (e.g. finite element or finite difference) models to analytical models. The distance scale of the models is similarly broad, from the submicron range within a single device up to many centimeters within a reaction chamber. Moreover, the complexity of the modeling problems is such that the solutions for different process steps or device types are coming from the world-wide community. This new generation TCAD system must be developed is such a way that continuous upgrades and/or replacements of simulation modules are possible in a painless manner.

Finally, to significantly expand the utilization of TCAD, a user-friendly interface is required to allow both the novice user (or the non-TCAD person such as a circuit designer or manufacturing engineer) to efficiently use the system to its full potential without a huge training effort.

In recent years systems with similar goals have been under development in many universities, industrial labs and TCAD vendors. This paper describes the status of such a system, PREDITOR, which has been under development at Carnegie Mellon University and has been recently commercialized as *pdFab* by PDF Solutions, Inc. The paper is organized as follows: Section 2 describes the semiconductor wafer representation that forms the heart of the framework. Section 3 describes the organization of framework services, and Section 4 describes the framework modeling and data analysis clients. Section 5 describes the results of using *pdFab* in the modeling and analysis of an industrial process, and Section 6 concludes with lessons to date, and directions for future work.

2.   Semiconductor Wafer Representation

As described in the previous section, a spectrum of tools is required for the range of application tasks. In order to achieve the desired range of speed versus accuracy, 1-D, 2-D, and 3-D analytical and numerical models are required. In addition, the results of these different models must be combined together, since there can be a range of accuracy needs within a single task or within a single device structure. In addition, multiple data entry and data analysis tools may be used. The need to integrate a spectrum of models and tools demands a common *semiconductor wafer representation* (SWR) to store the wafer state and related simulation and analysis results. The wafer state includes the geometry, fields, and attributes of the wafer during simulation. Fields are a range on a domain, such as the doping concentration, electric field, electric current, or stress within a geometrical region. Attributes are name-value pairs attached to geometry or field objects. Such a central representation eliminates the need for many different data translation programs to communicate among the large number of tools. In addition, a central representation can provide services to greatly simplify the development of new tools. Acceleration of tool development is essential if TCAD is to closely track the rapidly changing manufacturing technology. We have developed the Chip Database (CDB) as the wafer representation that forms the heart of an integrated TCAD system, providing services for both tool integration and development [36]. PREDITOR and *pdFab* are both examples of such systems.

The need for a common SWR has been recognized for some time. The FABRICS statistical process/device simulator [19][20][25][24] was one of the first TCAD systems to integrate multiple simulation models, and hence one of the first to have a common representation. The Process Engineer's Workbench (PEW) [34][33] provided an interactive environment for quasi-three-dimensional process/device simulation using the FABRICS models, and was one of the first systems to use a centralized wafer state with a strong link to the mask artwork. Other simulation frameworks also included some form of wafer representation [2][21][37][38][32][17][16][31]. These systems had drawbacks, primarily in being specialized to one type of simulation model. Except for frameworks targeted at lithography modeling, most of these systems did not have any link to the mask artwork. Recently the need to link the wafer representation to the mask artwork has become more widely recognized [35][12].

In order to make it easier to combine the results of independently-developed simulation models, and to avoid reinvention, attempts have been made to develop an SWR standard. Initial efforts focused on an interchange language [6][10] and later on database servers [40][39][4][41][8]. An effort is currently underway to develop a standard SWR architecture and programming interface [1].

The relationship between a wafer state and its representation as CDB objects is shown in Figure 1. The surface of the wafer is partitioned into rectangular or trapezoidal *regions*, based on the needs of the simulation models. In most of the work described here, the partitioning is done using the mask artwork. Associated with each region is a *layer stack*, which describes the sequence of layers in the region. Associated with each layer within the stack is a list of *attributes*, or name-value pairs. In CDB fields are represented as attributes. Attributes can also be associated with regions or with the entire database.

The organization of CDB is shown in Figure 2. CDB is broken into the CDB database, CDB extensions, STEPS modules, and user-defined extensions. The programming interfaces of all of these modules are visible to client models and tasks. The philosophy of the CDB architecture is to provide a set of tools for building general-purpose or specialized wafer representations, rather than a single general-purpose representation. This permits the construction of fast but non-general implementations for specific needs, as well as general implementations.

The CDB module provides the basic data structures and methods for representing and accessing simple geometry and attributes, and performing simple data extraction functions. The CDB extensions include new abstract data types useful for process and device simulations, including fields. The STEPS module is a CDB extension that provides abstract data types and methods for creating, editing, and executing sequences of *simulation steps*, which include process or device simulation steps, or data extraction steps. In effect, simulation steps encapsulate all simulation and analysis tools. The result is that CDB also serves the functions of *semiconductor process representation* (SPR) [22] and *task manager* [3]. Rather than being a passive database accessed by the client tools, CDB performs the tool invocation.

FIGURE 1. CDB representation of wafer state

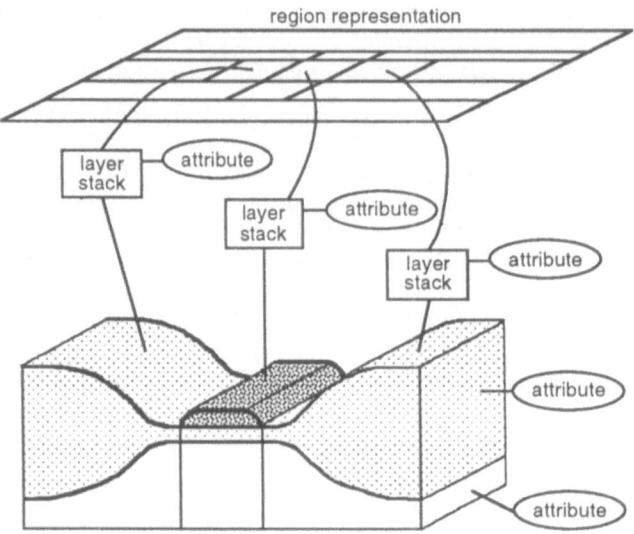

FIGURE 2. CDB organization

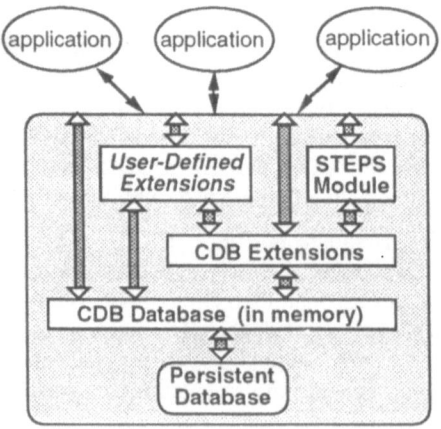

2.1 Geometry

CDB geometry is defined by the composition of the layer stacks, and the layers within the stacks. The surface representation of each layer is defined using a *layer model* attribute. Examples of layer models include flat, piecewise linear, and triangular facets. Since the layer model is attached to layers within stacks, it is possible to use different surface representations in different sections of the wafer. This is particularly convenient if different process simulation models are used in different regions. Layers can consist of any material, including air, which permits the representation of nonconvex surfaces.

2.2 Fields

CDB provides native field representations, such as analytical functions, and a variety of meshes. In addition, CDB provides a powerful mechanism, the *virtual grid*, which provides the ability to mix representations and dimensions. The virtual grid is an abstract data type for two-dimensional fields. These fields can be either the results of a full 2-D solution, or a set of 1-D fields interpolated into 2-D. The 1-D fields can be either analytical functions or meshes. A typical use of the virtual grid is to generate a 2-D doping field from a series of 1-D results, and then use the 2-D result as input to a 2-D device simulator. The advantages of this interpolation approach are that it is much faster than a full 2-D field computation, and it is possible to consistently mix analytical and numerical results, depending on the solution accuracy needed in different locations of the field.

2.3 Attributes

Each CDB database, region (layer stack) and layer can have a list of attributes associated with it. Attributes attached to the entire database are termed *global attributes*, while attributes attached to regions and layers are termed *local attributes*. CDB does not provide any predefined attributes. For each attribute type, the CDB user provides an *attribute type model* with functions to interpret the value field. Each type must also include a set of standard functions that are used by CDB to create, and copy attribute information as well as to save and restore it. Layer models and fields are examples of attributes. Fields are normally associated with an entire CDB (e.g. a mesh with the substrate doping profiles), or with a region (e.g. the doping profile in the region). However fields can also be associated with a layer, such as the current flow or doping profile in a conducting layer. Other examples of local attributes are material type or device electrode. Examples of global attributes include the process flow and mask artwork used to create the database. A given attribute can be shared among several objects. For example, several layer stacks may share the same mesh that describes the doping profile across multiple regions. The general-purpose attribute mechanism is the key to CDB extensibility.

2.4 STEPS Module

The STEPS module provides an abstract data type for simulation steps. A simulation step is a parameterized simulation routine such as a process simulation, device simulation, or data extraction step. Device simulation and data extraction steps are treated as equivalent to process simulation steps so that they can be inserted into the process recipe to perform analysis in a manner equivalent to taking in-line measurements in the real process. The STEPS module provides facilities for creating and editing simulation steps, assembling them into process recipes, and editing the recipes. These recipes can then be executed to modify the database or extract information from it.

New simulation steps are made known to CDB by writing *step models*. The step model is a data type which contains information about step parameters such as parameter names, types, units, and default parameter values, as well as a *simulation routine* to call with the parameter values when the step is executed. Functions are also provided to execute an entire sequence of steps.

Step models make it easy to provide a uniform interface for different simulation models operating at the same level of abstraction. For example, at the wafer environment level of abstraction, the essential parameters of an ion implantation are dose, energy, and species. A nearly-identical step model could be used to encapsulate a one-dimensional ion implantation step implemented using several different solution methods. The step model could add an additional layer of abstraction by implementing a generic ion implant step, and adding the implant simulation model as a parameter.

Simulation steps are the means by which physical models or external simulation tools transform the CDB state. Models encapsulated in a step can be viewed as CDB extensions, rather than CDB clients. The STEPS module also provide a flexible means for using CDB in applications outside the traditional TCAD domain. For example, in support of a functional yield simulator, a 3-D circuit extraction step and a Monte Carlo defect generation and fault analysis step have been developed.

2.5   Representation Type Conversions

CDB does not provide an underlying universal representation for geometry and fields. Instead, a lazy evaluation approach is used to perform type conversion as needed by the simulation models. Each simulation step must check the representation type, and call the appropriate conversion routine if necessary. It is our experience that clusters of simulation models use the same representation, so the cost of conversion is relatively low. Type conversions are also used to interface to external simulator file formats. For example, in order to plot a 1-D doping profile, it would be converted to the "data type" used by the graphing program.

2.6   Global Attributes

Attributes associated with the entire wafer are termed global attributes. In addition to storing the process recipe, they act as a database for storing simulation results closely associated with the wafer state. Global attributes are used to store the results of task-level simulations. Examples of task-level simulation include Monte Carlo process and device simulations, factorial designs, sensitivity analysis and latin hypercubes. In task simulation, the process and device simulation is repeated many times with the inputs varying slightly from one run to the next. For a simple layout containing two to four transistors, the three dimensional CDB representation of the devices can consume 0.5-2MB of data. Storing all the data generated during a 100-sample Monte Carlo run would require 50-200MB. As with manufacturing data, it is necessary to discard some information. Unlike manufacturing data, where often only summary statistics are available, CDB uses the global attributes to store all process control settings, as well as all CIM database level attributes (i.e., in-line measurements, I-Vs., SPICE models, and simulated electrical tests) for each run. Since the process control settings for each run are stored, it is possible to regenerate the detailed three-dimensional representation for any of the individual runs. Just as with all other CDB data types, each global attribute has translators associated with it. Hence, Monte Carlo simulation results "know" how to be translated into correlation plots, while sensitivity runs know how to be translated into bar charts. By using the global attributes, CDB can summarize data while still be able to regenerate the details. The type conversion paradigm used throughout CDB makes it possible to associate analysis functions with only certain classes of data types, preventing incorrect analysis of results.

2.7   Relationship Between CDB and the SWR and SPR Standards

CDB and the SWR standard are both designed to represent the wafer state during process and device simulation, and to provide functions to make it easier to develop new simulation models. There are several significant differences between CDB and SWR:

1. Scope of area represented. CDB can represent a few to hundreds of transistors. This is necessary if one what to predict spatial variations across a chip, or yield/defect simulation. To accomplish this, CDB uses the layout to split the chip into regions which can be (but are not necessarily) represented with 1-D models. The SWR is intended primarily for detailed process and device simulation of a few transistors.

2. Geometry representation. CDB uses the layer model to represent material boundaries while SWR uses a cell complex representation. While both are equivalent, the former is more appropriate for the algorithms used in lithography simulation (since a pointer to a string function is easily accommodated), and the latter lends it self to 2-D thermal process simulation. Since CDB uses pointers to functions for boundaries, it can easily be modified to point to an SWR geometry server.

3. Field representation. SWR fields are primarily designed to represent meshes. CDB is designed to handle a spectrum of representations from 1-D, mix analytical/numerical, and 2-D meshes. This is appropriate given the need to incorporate a spectrum of simulators.

In summary, from a representational standpoint, CDB and SWR should be viewed as complementary. The largest difference between the two is not the representation but how they are used. The SWR standard is envisioned as a slave interface, with the client tools driving the SWR. In contrast, CDB performs the tool invocation, sending information to and from the client tools. The advantage of the CDB approach is incorporation of off-the-shelf tools is much easier since the tools do not even need to understand the details of CDB. The disadvantage of CDB is that since its built-in layer models and field representations are limited, they must be extended to incorporate each new class of tools. For example, once an interface to a particular 1-D process simulator was built, all 1-D process tools could be accommodated. The same is true for 2-D process or 2-D device simulators. If a new 2-D lithography tool were added, CDB would need to have new layer models since the basic layer models would not be sufficient. However, since results are always stored in CDB once a tool is hooked into CDB, all other client tools can make use of the results.

We would like to eventually provide a CDB programming interface for the SWR standard. This would permit tools that comply with the standard to be used without modification. We can build this interface with suitable CDB extensions.

As is the case with the SWR standard, the representation of process recipes in CDB is very similar to the objects and programming interface proposed for the SPR standard. The primary differences are: the SPR standard includes more levels of model and recipe abstraction than are currently available in CDB; the standard explicitly supports dynamic type creation while CDB does not; and the standard provides a passive simulator-driven interface while STEPS extracts parameters from the process recipe to the simulation model. We would like to eventually support the SPR standard by providing an interface to exchange recipes between an SPR server and STEPS.

3. A Technology CAD Framework for Development and Manufacturing

Because of the wide range of process engineering tasks, a flexible TCAD integration and development framework is required to adequately support all the different simulation and analysis clients of the Chip Database. We have developed two such frameworks - PREDITOR and its commercial successor *pdFab*. Block diagrams of these systems are shown in Figures 3 and 4. As can be seen, CDB is at the center of these frameworks, completely encapsulating the simulation and analysis tools. In terms of standard CAD framework terminology, individual simulation models are *components*, which are encapsulated by CDB to form *resources*, such as the simulation steps, which are then used in modeling and analysis *tasks*, such as worst case device parameter extraction. Access to the CDB resources and tasks is provided by framework services which include process recipe and mask artwork manipulation, simulation control, and data visualization. The primary difference between the PREDITOR and *pdFab* frameworks is that PREDITOR has been organized as a flexible development framework for experimentation, with a CDB server process communicating with user interface processes, while *pdFab* has focused primarily on enhancement of modeling capabilities and analysis tasks, while incorporating new PREDITOR developments.

FIGURE 3. PREDITOR block diagram

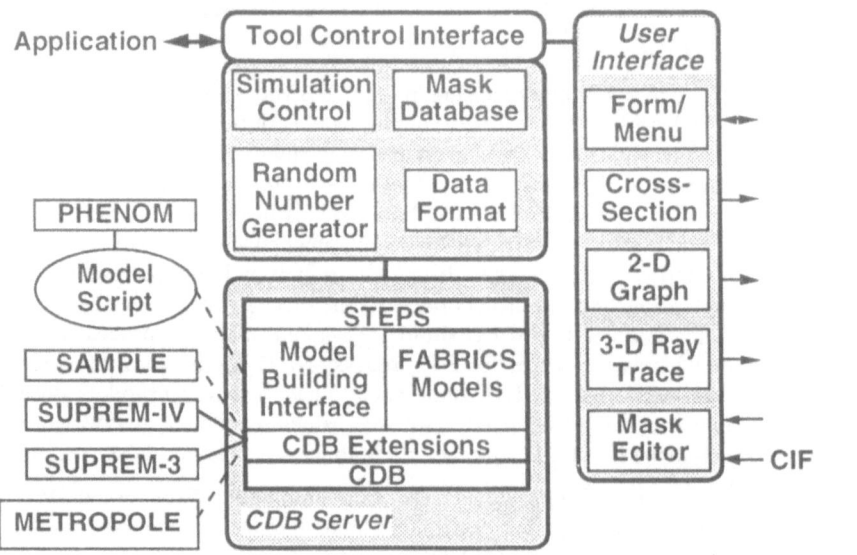

PREDITOR is organized with a CDB server communicating with separate process editor, mask editor, graphing, and cross-section programs. Each program includes a Tk/Tcl interpreter [28][29] communicating with the TCP/IP protocol, so the system is easy to modify, and programs can be located on different computers. For example, CDB and the simulation models can be located on a compute server, while the editors and plotting programs reside on the user's desktop system. Tcl was designed as an extension language, so it can easily be used to add additional interface programs, or to write scripts for specific tasks. For example, a task to compute the threshold voltage of a MOSFET can be specified in Tcl as the sequence of process and device simulations, parameter extractions, and plots to execute. This same command interface permits other application tools to use the PREDITOR system as a virtual fabrication line, such as for process and device optimization.

FIGURE 4. *pdFab* **block diagram**

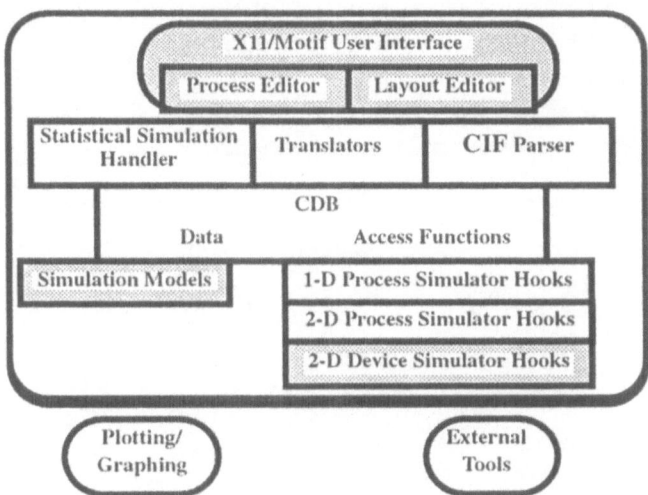

PdFab is organized as a CDB library linked to an X11/Motif graphical user interface, with separate graphing and cross-section program. In a network environment, external simulation models can be located on separate compute servers. In addition, the *pdFab* program itself can run on a compute server, communicating with the user system via the remote X protocol. The device parameter extractor task uses a Tcl interpreter, so it can easily be modified for different device models. *PdFab* provides a command line interface so that it can be easily controlled as a virtual fabrication line. *PdFab* is currently used in this mode within the Odyssey CAD framework [5].

3.1 Process Recipe

As described in Section 2, the process recipe is stored in CDB in the STEPS module. In order to permit interactive modification of the recipe, the framework must provide a process editor. This editor is used to specify all process and device simulation, parameter extraction, and measurement steps. PREDITOR uses a Tk/Tcl-based process editor. When PREDITOR is initialized, the editor queries the CDB server for the list of supported simulation steps, the required parameter fields, and default values. The user can then assemble a recipe by filling out forms for each step and adding them to the recipe. The root process recipe window and a form for an ion implantation step are shown in Figure 5. Note that the form includes entries for parameter variations (i.e. standard deviations).

When a simulation request is made to the CDB server, it queries FLO in order to obtain an updated copy of the process recipe and load it into the STEPS module prior to simulation. In *pdFab*, the Motif-based process editor has editing and display capabilities similar to those of PREDITOR. It uses the STEPS module directly to access and store the process flow, so no synchronization is needed prior to simulation.

FIGURE 5. Process recipe window and ion implantation form

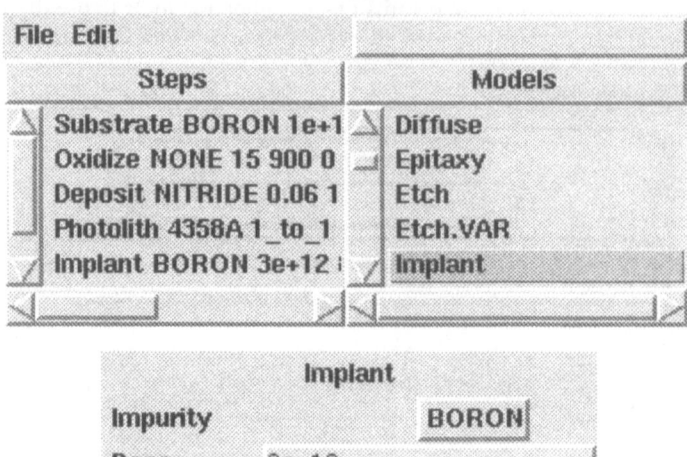

3.2   Mask Artwork

The mask artwork of the structures to be simulated are specified in two ways: by a layout editor or by a CIF layout file. The layout editor is designed for interactive specification of small pieces of artwork, such as for a single transistor or gate, and so it is sometimes referred to as a "little layout editor". The PREDITOR layout editor operates as a separate program communicating with the CDB server. The *pdFab* layout editor is part of the Motif interface communicating with the CDB library. The layout editor is also used to specify measurement points, cutlines, electrodes, etc. Figures 6 and 7 show the PREDITOR and *pdFab* layout editor windows. Prior to simulation, the mask data is transferred to the CDB as a global attribute, so that it can be accessed by the lithography simulation steps. The data is organized as a list of mask layers, each of which consists of a list of mask polygons on that layer. This organization was chosen under the assumption that the simulation models would not need to directly search the mask attribute. When selecting a point or cutline on the artwork, the search is done within the layout editor, which can use a more sophisticated data structure. As noted in Section 2, within CDB, regions that result from lithography steps are stored as corner-stitched rectangles, so they can be searched quickly.

Some tasks require analysis of larger sections of artwork, such as an entire test structure or macrocell. These are more conveniently specified with a full-featured layout editor in a commercial ECAD system. These layouts are transferred into PREDITOR and *pdFab* using the CIF layout language. The CIF parser fully-instantiates and loads the mask art-

work directly into CDB, where it can later be viewed by the layout editor. It is not practical
to make sole use of a full-featured layout editor, since this would require the TCAD system
to be part of a full ECAD framework. For small layout problems, these editors are more
difficult to use than a little layout editor. It is also difficult to integrate them into the framework since all communication must be done by reading and writing CIF files, and the full-
featured editors are not designed to accept commands from other programs. In addition,
full-featured editors do not in general have the ability to interactively specify points and
cutlines, except as special labels.

FIGURE 6. PREDITOR layout editor

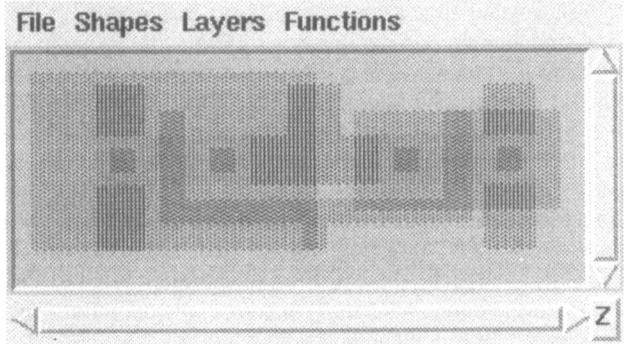

FIGURE 7. *pdFab* layout editor

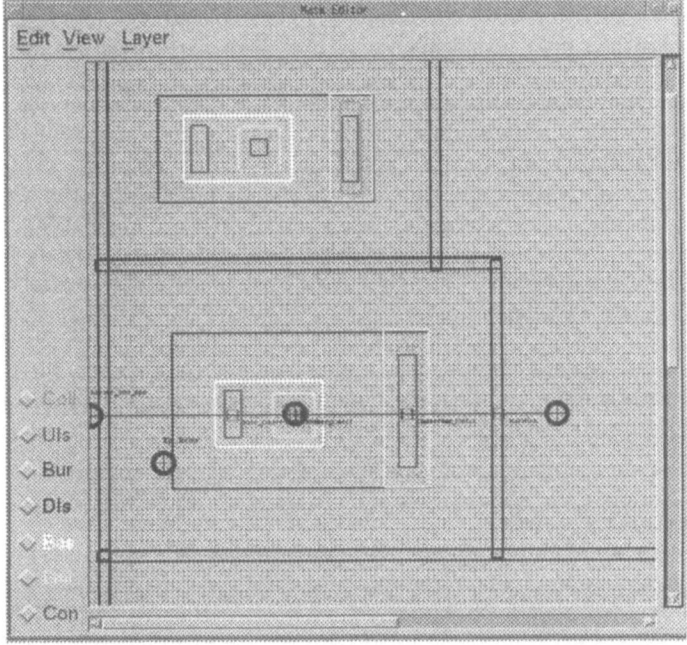

The CAD Framework Initiative will eventually specify a programming interface standard for the exchange of mask artwork. This will permit all compliant layout editors, both full-featured and little layout editors, to be used interactively for artwork manipulation and location specification.

3.3   Simulation Control

Both PREDITOR and *pdFab* provide user and programming interfaces for simulation control. The interactive user interface permits the user to select the type of simulation experiment that will be carried out, e.g., whether a process simulation will be deterministic or statistical, the sample size for statistical runs, etc. The programming interface provides a means for adding task-level clients. An example of a task would be to run the appropriate process and device simulations, and parameter extractions to compute worst case MOSFET threshold voltages. The programming interface is implemented with the Tcl extension language. This choice is especially appropriate for PREDITOR since Tcl is already used for communication between the user interface programs and the CDB server, so no additional effort is required to provide it. Since Tcl is an extension language, additional task scripts can be readily added to the systems. In the case of PREDITOR, new Tk graphical interfaces to these scripts can be added without modification to the rest of the system. Part of a script for a *pdFab* MOSFET threshold voltage extraction is shown in Figure 8. Even when using fast analytical models, the overhead of the interpreted extension language is negligible, since task-level commands specify macroscopic operations such as to simulate a particular process recipe, or execute a device simulation.

3.4   Visualization

There are four basic types of visualization in PREDITOR and *pdFab*: chip top views, chip cross-sections, chip perspective views, and data plots. The chip top view is provided by the layout editor, which is used both for the artwork specification, and for simulation feedback, e.g. drawn versus actual linewidths. Cross-sections are displayed using the *xmdraw* program. The draw cross-section command dumps out an ASCII file of the geometry and fields along the cutline, which is then read by xmdraw. An example xmdraw cross-section of a MOSFET is shown in Figure 9. A similar capability for perspective views of the database geometry is provided by passing a description of the CDB geometry to the RAYSHADE ray-tracing program, and then to its output browser. An example perspective view of a CMOS AND gate is shown in Figure 10. Two-dimensional data plots are provided by the *xgraph* graphing program. A file is prepared by the appropriate measurement or analysis tool, and passed to xgraph. Figure 11 shows a plot of the doping profiles in the middle of a vertical NPN BJT. Xmdraw, RAYSHADE, and xgraph are all encapsulated by CDB, being executed in response to a visualization command, rather than acting as servers. The reason for this is that like external simulators, these external visualization tools execute as a program that reads a file, displays a graphics window, and then exits when the "quit" button is pressed.

4.   Tasks and Clients

To solve common process engineering problems, task-specific CDB extensions have been added to PREDITOR and *pdFab*. To support these tasks, a set of equipment, process and device simulators have been added as CDB clients.

FIGURE 8. Simplified Tcl script for threshold voltage extraction

```
#= Calculates the threshold voltage for MOS transistors by finding the x-intercept
#= of the regression line through a user-specified region of the Vgs x Id curve.
#= Parameters:
#= Vds, Vbs, Vgs_low, Vgs_high, num_points
#= Outputs:
#= /iv/$dev_name/Is_NMOS -- 0 if NMOS, -1 if PMOS, /iv/$dev_name/Vth -- threshold voltage

set dev_name $PDFAB_Device
# Is this an NMOS? Set to 0 if NMOS -1 if PMOS
set Is_NMOS [string first "NMOS" $dev_name]

set_default Vds 0.5
set_default Vbs 0.0
set_default Vgs_low 0.0
set_default Vgs_high 2.5
set_default num_points 2

#Change signs if it's a PMOS
if {$Is_NMOS == -1} {
set Vds [expr -$Vds]
set Vbs [expr -$Vbs]
set Vgs_low [expr -$Vgs_low]
set Vgs_high [expr -$Vgs_high]
}

source [info library]/Vth.tcl
source [info library]/least-squares.tcl

GAcreate /iv List 1
GAcreate /iv/$dev_name List 1

...
GAset /iv/$dev_name/num_points/ $num_points

# begin algorithm
set Vs [expr 0.0-$Vbs]
set Vgs_step [expr ($Vgs_high-$Vgs_low)/($num_points-1.0)]
for {set i 0} {$i < $num_points} {incr i} {
set Vgs [expr $Vgs_low+$i*$Vgs_step]
lappend Vgs_list $Vgs
set currents [2D_solve $dev_name \
[expr $Vgs+$Vs] $Vs [expr $Vds+$Vs] 0.0]
set Id [lindex $currents 0]
lappend Id_list $Id
}
set regression [ls_regress $Vgs_list $Id_list $num_points]
set m [lindex $regression 0]
set b [lindex $regression 1]
GAcreate /iv/$dev_name/Vth Double
GAset /iv/$dev_name/Vth/ [expr -$b/$m]
```

4.1 Statistical Simulation: Monte Carlo

PdFab supports a number of task-level simulation modes, including Monte Carlo (both standard MC and latin hypercubes), sensitivity simulation, fractional factorials, and lot splits. We explain in detail the justification for Monte Carlo simulations since they are critical for variation prediction.

Our goal in this work is to use process and device simulation to determine the distribution of device characteristics (e.g., Idmax, and Vt) based on the distributions of in-line measurements and process controls. Since the functional relationship between process controls and devices characteristics is not well understood, it is not possible to simply transform the in-line and process control distributions to determine the device distributions. Monte Carlo simulation provides the most accurate method to mapping process controls variations to device variations. In the Monte Carlo simulation approach, the user specifies the number of runs to simulate. For each run, every process control is randomly sampled from the distribution specified by the user. The new process flow is then fed to the process simulation tools to simulate the manufacturing process. In this way, the small variations in controls which typically occur in manufacturing are mapped into variations in the transistor's structure. Each of these new structures is then passed to the device simulators to compute device characteristics. This process is repeated for every run as shown in Figure 12.

FIGURE 9. xmdraw cross-section of a MOSFET

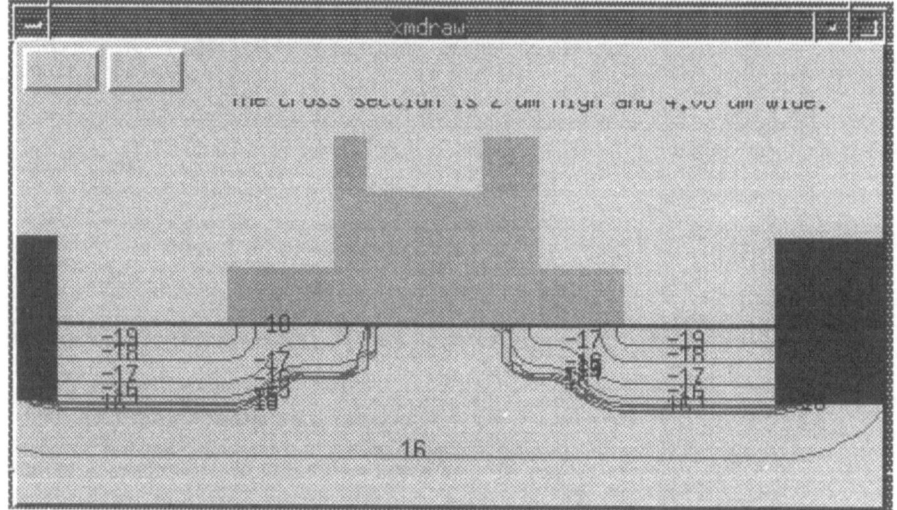

FIGURE 10. RAYSHADE perspective view of a CMOS AND gate

FIGURE 11. Doping concentrations in a vertical NPN BJT

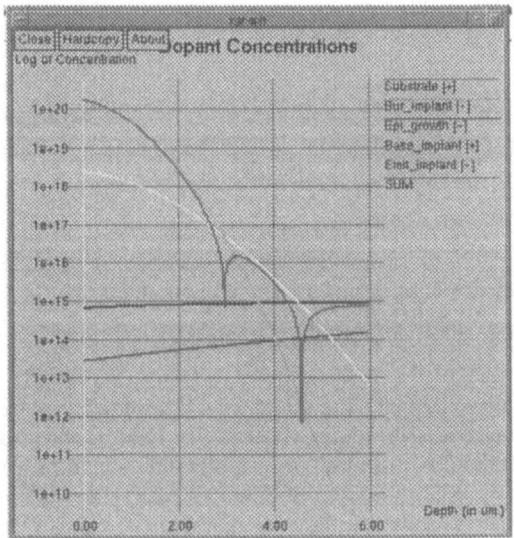

FIGURE 12. Flow chart for Monte Carlo simulation

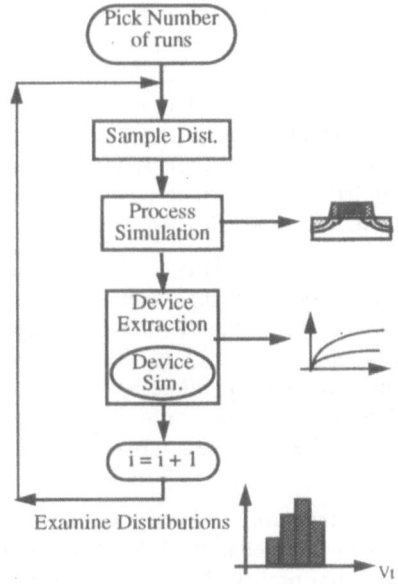

4.2 Data Translation

At the completion of the Monte Carlo simulation, distributions of device characteristics, their inter-correlations, and their correlations with in-lines and process controls can be plotted. To achieve reasonable confidence in the predicted correlations, the number of runs must be fairly large (typically 50-200).

The advantages of using Monte Carlo simulation to compute correlations and standard deviations are:

  ■ A large sample can be generated in a short amount of time

A Monte Carlo simulation with a sample size of 100 can be generated in 4 to 30 hours of simulation time, yet it represents the production of 100 lots.

  ■ A single process flow is used

In a pilot line, process control skews corrupt the electrical test database since they represent changes which do not happen during normal manufacturing.

  ■ In-line, electrical tests, and actual process control settings are all collected in a single database

Many times, in-line measurements are stored in a CIM database while electrical tests are stored in a second database. Combining and cross-correlating this information is difficult. Moreover, during Monte Carlo simulation we know the actual process control setting for each run. Hence it is possible to determine the correlation between an individual control and device characteristics. However, the very nature of process control variation is that it cannot be measured. With measured data it is not possible to be able to determine the correlation between the variations of an implant and device characteristics.

4.3   Wafer-Level Simulation

New processes rely on single wafer processing and large diameter wafers to achieve run-to-run uniformity at affordable costs. As a result, increased emphasis has been placed on equipment modeling for predicting within-wafer uniformity. Presently there is no commercial system available that can simulate equipment models in the process flow, or predict wafer maps of transistor characteristics given the spatial dependencies of process controls. By making use of CDB's global attributes, both the wafer map (centimeter-scale) as well as the detailed device structure (micron-scale).

In-line information from equipment models is used in wafer-level TCAD simulations to predict the eventual distribution of device characteristics, as shown in Figure 13. Thus, process control can be used to determine whether the equipment is performing to the desired level of quality (equipment control) and whether the product in production will likely meet specification (product control). *pdFab* has already been used to determine process control limits for in-line measurements to assure that device specifications are met [11]. By extending *pdFab* to handle equipment models, it will be possible to use it for the control of unit process steps.

With the addition of equipment modeling and wafer simulation, users will be able to simulate process integration. As demonstrated in the MMST project [23], changes in equipment settings will result in different uniformity characteristics. By including equipment models in the simulation of the process flow, the effect of equipment changes on the uniformity of transistor characteristics can be quantified.

As an example, consider the following experiment performed with *pdFab*. The experiment assumes that gate oxidation temperature, photoresist thickness, and implantation dose are spatially dependent. In most process simulators, process controls are considered to be constants. In *pdFab* they are assumed to be independent random variables, thus we were able to give each process control a spatial dependence and a random component. Figure 13 contains an example of how *pdFab* is used for this task. Often equipment models do not exist. In these cases, engineers can run experiments on the actual equipment and then build models using PHENOM to capture the physical basis of equipment's behavior.

FIGURE 13. Mixed equipment, process, and device simulation to predict wafer maps and uniformity

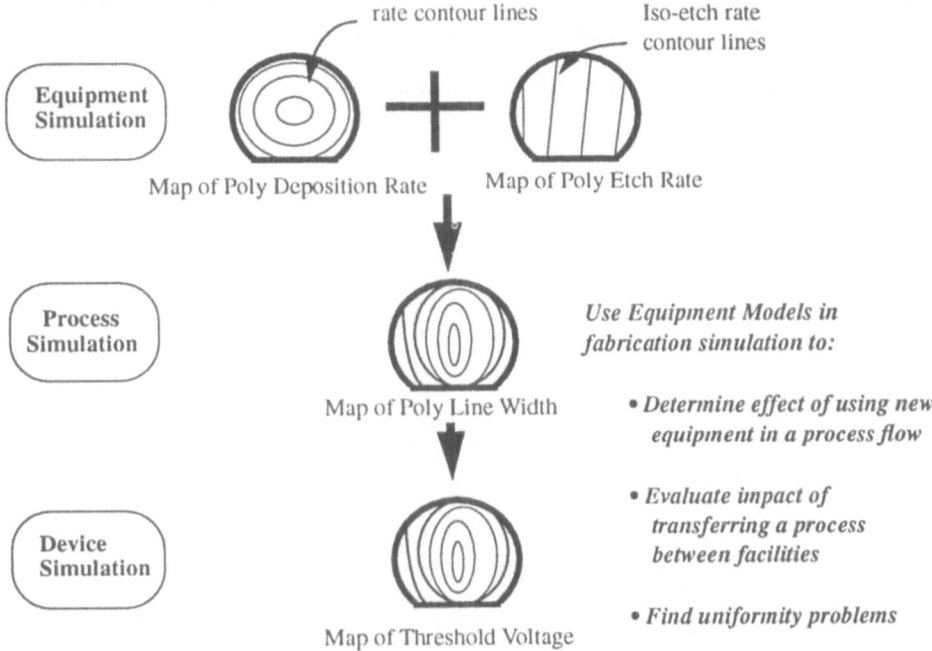

4.4 Task Level to Tool Control

During a simulation, a process recipe must be mapped into simulator commands. A distinction is made between simulators that model multiple steps (typically process simulators) and simulators which model a single step (device simulators). An entire process recipe may be mapped into multiple simulation tools. For every wafer or device attribute there are at least two models which can compute the value. This is demonstrated in Figure 14. An individual step can be mapped to multiple tools. For example, a process recipe can be mapped to FABRICS-II compact analytical models, or combinations of SUPREM-III and FABRICS-II, or SUPREM-IV without changing the recipe. The simulator specific commands are separated from the recipe.

For device simulation, a similar situation exists. Any device characteristics of interest can be computed by compact models or user-defined scripts. Those scripts are written as Tcl programs and as such, are not specific to a particular tool. During the process definition, the user can select the tool from the menu. In this manner, the script can be representative of the true measurement technique, and the simulator and mesh-specific commands can be separated. Hence, a device characteristic can be computed by a compact model, or scripts which can control a number of numerical simulators.

The advantage of this approach is two fold. First, values which seem suspect to the user can be checked by selecting a different tool option from a menu, and second, compact models can be used for large statistical simulations. It is the second point that is important for manufacturing-based simulation. The compact models have three advantages over numerical models: they are faster, rarely have convergence problems, and preserve gradient information. Hence, they can be used to determine the relationships between device parameters (e.g. TF and Beta). Once an extreme point is determined, numerical simulators can be used to compute more accurate values. By providing a spectrum of tools, the user can make the appropriate compromises to accomplish the task at hand.

FIGURE 14. Mapping a process flow to simulator commands

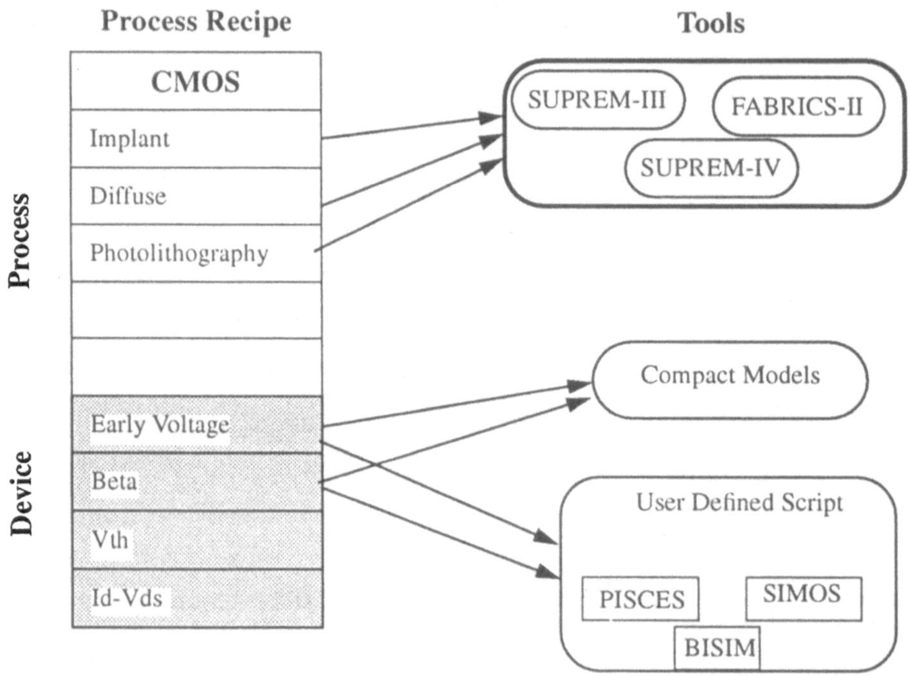

4.5 Equipment and Process Simulation

Both numerical and analytical equipment and process simulation models are necessary to support the tasks. Equipment models provide the relationship between equipment settings and the device structures, which is necessary for failure diagnosis and control. Process models provide a mapping between wafer environment and structure, or specify the structure directly as a wafer effect model. In the former, times and temperatures are the model parameters. In the latter, the specific wafer change is the parameter. Both analytical and numerical models are necessary to provide the combination of speed and accuracy. As discussed in Section 2, the CDB virtual grid provides a means of interpolating from a series of 1-D fields to a single 2-D field. The analytical models supported are those from the FABRICS-II and PEW simulators. The numerical process simulator is SUPREM-III [13]. In the CDB extensions, mask exposures for each layout region are automatically used to generate the individual SUPREM-III decks for each unique region. Common initial exposure sequences are reused to minimize total simulation time. Two-dimensional process simulation is supported using SUPREM-IV [15]. *pdFab* supports both the TMA and Silvaco versions of SUPREM-III and SUPREM-IV.

Topography simulation is an essential part of the integrated process simulation system. A number of tasks ranging from process integration to layout design role development require accurate predictive simulation of topography changes caused by steps such as deposition, lithography and etching. A complete lithography simulation program called METROPOLE [18] has been developed at CMU and will be integrated into the next generation of the framework. METROPOLE implements a true vector 2-D model, based on efficient solution of Maxwell's equations by the waveguide method. METROPOLE is capable of simulating photoresist deposition, bleaching, post-baking and development for dimensions below 0.5um. It can model the stepper optics, non-planarity of lithographic mask (especially important for phase-shifting masks) and arbitrary nonplanarities of the substrate wafer. This program has been verified extensively for a number of tasks such as alignment modeling, defect simulation and technology feasibility studies. Since it takes as inputs arbitrary layer stacks with the detailed description of shapes and produces as output the realistic window in the photoresist layer, it is ideally suited for use in PREDITOR and *pdFab*.

Although, as we described earlier, the focus of PREDITOR and *pdFab* is the simulation of devices within one chip, due to the CDB global attributes it is possible to enter, for example, the spatial distribution of the simulation parameters within a wafer or even from wafer to wafer. This permits extending process simulation to include equipment model which operate on a centimeter scale. For example, it is possible to predict nonuniforimty in the thickness of an LPCVD polysilicon layer by modeling the deposition rate distribution as a function of process recipe parameters and spatial coordinates on a given wafer. A comprehensive equipment simulation system which utilizes the FLUENT computational fluid dynamics program [9] has been developed which permits accurate simulation of the physics and chemistry of industrial deposition systems. To date it has been applied to problems such as PECVD of silicon nitride and CVD of tungsten. Such accurate physical models are then used to generate simplified, but still physically-based, analytical models using PHENOM [7][30]. These PHENOM-generated equipment models are then automatically included in the library of process models. Models described in C require relinking of the libraries. Models described in Tcl can be used immediately. An equipment model for LPCVD of polysilicon has been included in the library.

4.6   Device Simulation

In order to map from device structures to device parameters, transistor I-V characterisitics must be computed via device simulation. The simulation clients supported by *pdFab* include SIMOS, a 2-D numerical solver for MOS [26]; BISIM, a 2-D solver for BJT's [27], and a compact analytical solver for BJT SPICE model cards. TMA Medici and PISCES are also supported. All of these simulators take the doping field from the virtual grid as input data.

4.7   Parameter Extraction

Device I-V characteristics are translated into compact circuit simulation models via parameter extraction. In PREDITOR, a simple SPICE level 2 model extractor was provided. In *pdFab*, extraction scripts are specified via the Tcl-based extension language . An example of such a script is shown in Figure 8. The extraction programs are separate from the device simulation controls and meshing. Hence, they can be re-used for many devices and simulators without changing them. They can be stored in a *pdFab* library so others can use them. Standard scripts are provided, such as Vt, VAF, and BSIM model extraction. *PdFab* has physically-based SPICE model extraction for statistical analysis. These routines support industry-standard SPICE models such as BSIM, but make use of deterministic physical extraction. This method is described in more detail in the next section.

4.7.1 SPICE Model Extraction

In order to provide meaningful statistical information, the BSIM model parameters must be extracted in a physically-based manner. Parameter optimization, as shown in Figure 15, can result in a non-unique parameter set. This adds variance to the parameter distributions. Hence optimization-based methods cannot be used for statistical analysis. The *pdFab* parameter extraction methodology incorporates both two-dimensional dopant layout information and efficient two-dimensional numerical device simulation performed using SIMOS or Medici.

FIGURE 15. Flow chart of conventional optimization based device characterization procedure

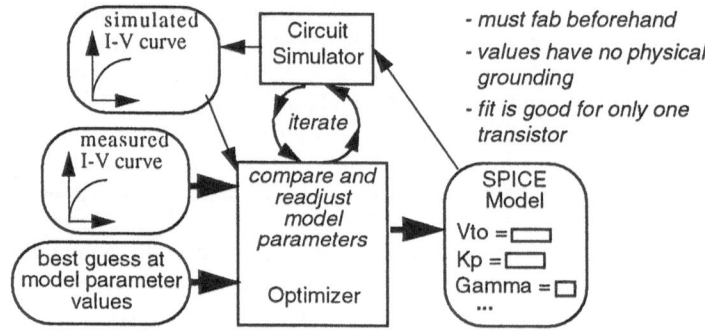

Several BSIM model parameters have physical definitions and can be determined directly from the dopant profile. Among these model parameters are the channel length and width, gate oxide thickness, drain and source resistance, and the capacitance parameters For short channel devices, the threshold voltage is strongly effected by the drain and source depletion regions. Thus, two-dimensional numerical device simulation is required to accurately extract the threshold voltage model parameters. Although the difference in the Fermi potential between the monosilicon channel and polysilicon gate can be physically extracted from the dopant profile, the other threshold model parameters are determined via 2-D device simulations. Similarly, parameters for the average channel mobility, gate field mobility reduction factor, the I-V relationship in the linear and saturation operating regions, and subthreshold drain current are each extracted in turn using a total of 20 2-D device simulations. This model parameter extraction approach does not make use of any special 2-D device simulator features, and so it can be implemented with any of the device simulators within *pdFab*.

5.   Experimental Results

PdFab has been applied to the analysis of BJT, CMOS and BiCMOS processes. Instead of supporting a single approach to manufacturability analysis, *pdFab* is used as a tool to support many different strategies. In this section we show the results of using *pdFab* to determine the manufacturability of a National Semiconductor CMOS process, using a Monte Carlo simulation approach to predict correlations and variances of device characteristics. The process was a 0.8um CMOS process with LDD NMOS and LDD/buried channel PMOS devices. The goals of the analysis were to determine whether the process could be controlled, and to compute realistic worst case limits for device SPICE models.

The process was entered into *pdFab* and simulated with SUPREM-III as the process simulator. The analytical doping profiles were calibrated to the SUPREM profiles by adjusting the diffusivity of boron, arsenic and phosphorus, and the segregation coefficient of boron and phosphorus. By doing so, the analytical profile models could be used during the Monte Carlo simulation. The lateral diffusivity of boron and phosphorus were adjusted to achieve the correct effective channel length. *PdFab*'s ability to incorporate a spectrum of models was critical for this activity. Given the number of Monte Carlo runs made (and the sample size for each), it would have been prohibitively time consuming to perform all simulations using SUPREM-III.

Standard deviations were then assigned to key process controls based on specification limits. For implantations, the standard deviation of the dose and energy was set to 5% of the process control setting. Initially, the standard deviations of furnace temperatures and poly linewidth were set to match specification limits of the oxide thickness and critical dimension. On inspection of CIM data from a pilot line, it was found that actual deviations differed from the specification limits. In subsequent experiments the standard deviations of furnace temperature and poly linewidth were set to match measured data. The standard deviation of the furnace temperature was set to two degrees for high temperature steps. Using a two-degree standard deviation, the simulated standard deviation of gate oxide thickness was close to measurements. The poly linewidth standard deviation was set to 0.039um.

5.1   I-V Prediction and SPICE Models

Once entered, the process was simulated with all control settings at their designed values. *PdFab* generated BSIM models (HSPICE level 13 model) for both the NMOS and PMOS transistors. The I-Vs generated with the BSIM model were compared to curves measured from "typical" transistors in Figures 16 and 17. Since we do not know the exact conditions that were used to fabricate the typical transistor, the purpose of comparing these curves is not to determine a goodness of fit measure. Rather, we demonstrate that the simulated I-Vs are similar to the typical measured transistor. The BSIM model was extracted by using results from the SIMOS device simulator. The extraction computation (including the process and device simulation) takes 15 CPU-minutes on an IBM RS6000/550.

5.2   Statistical Data

With satisfactory results from the simulation of the designed process, a Monte Carlo simulation of 100 samples was run. Rather than doing a full BSIM extraction for every point, the threshold voltage and Id Max (5X5 current) were simulated at every point for both NMOS and PMOS transistors. The Monte Carlo simulation with 100 process steps required 8 hours of CPU time on the IBM RS6000/550. The simulated distributions of device parameters were compared with the specification limits and distributions of measured data. In this process, the specification for each device parameter represent the mean value, +3 sigma and -3 sigma points on the distribution. In Table 1 the percentage difference between the predicted 3-sigma limits, the specification limits, and the measured 3-sigma limits are presented. The measured data consisted of approximately 1000 sample points per parameter collected from during fiscal year 1993. From the table, it is clear that there is good agreement between the predicted distributions and both the measured distribution and specification limits. There is slightly better agreement with the measured distribution than the specification limit.

FIGURE 16. NMOS I-V curves - measured and BSIM

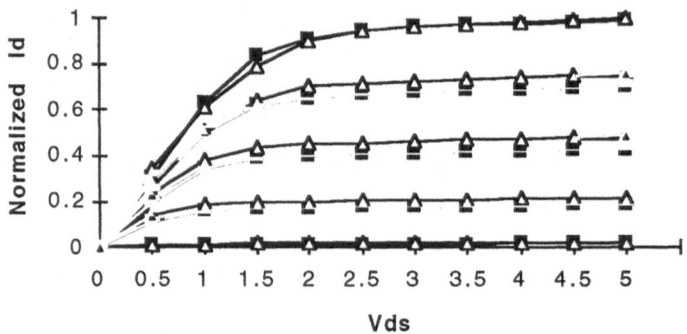

FIGURE 17. PMOS I-V curves - measured and BSIM

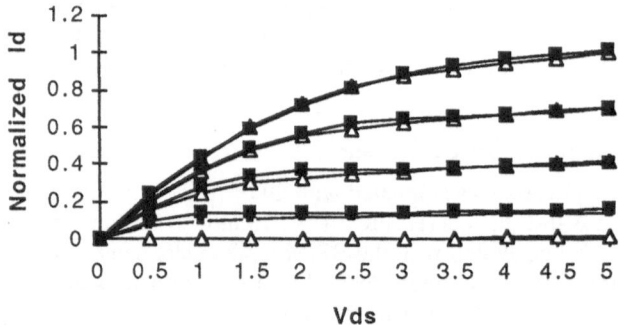

Also computed are the correlations between Vt and Id were also computed, as shown in Table 2. Given the sample size, all correlations, with the exception of the correlation between Vtn and Idp, are within the statistical confidence limit of the measured values. It is interesting to note that the correlation between devices (Idn and Idp) is greater than the correlation within the devices (Id vs. Vt). This information is used when generating worst case corners for design.

TABLE 1. Comparison of simulated and measured 3-sigma limits to specification limits

Measurement	%Error from Specification	%Error from Measurement
Id Max	-0.02%	5.7%
Id Mean	6.4%	6.3%
Idn Min	12%.	7.6%
Idp Max	-8.6%	0.0%
Idp Mean	-1.6%	5.3%
Idp Min	9.29%	13%
Vtn Max	-12.6%	-6.2%
Vtn Mean	-4.1%	-2.6%
Vtn Min	-8.0%	-0.2%
Vtp Max	-5.5%	-3.7%
Vtp Mean	1.9%	4.6%
Vtp Min	3.1%	6%

TABLE 2. Predicted and measured (in parentheses) correlations

	Vtn (meas)	Idn (meas)	Vtp (meas)	Idp (meas)
Vtn	1.0	0.61 (0.73)	0.13 (0.21)	0.12 (0.42)
Idn		1.0	0.0 (0.13)	0.71 (0.8)
Vtp			1.0	0.57 (0.54)
Idp				1.0

5.3 Device Correlations with In-lines and Process Controls

Correlations between device characteristics and in-line measurements were computed for poly linewidth critical dimension (poly CD), Tox, and sheet resistances. While poly CD and Tox are measured in-line, the sheet resistance is not. Using pilot wafers, the sheet resistance measurements could have been made in-line on a batch basis.

The predicted and measured correlations were compared. There is good agreement between measured and simulated correlations. Due to limitations in the CIM database, it was not possible to compute the Tox measured correlations. Note that the linewidth variations dominate both Idp and Idn while neither Vtp or Vtn are strongly correlated with poly linewidth variation. It is the common dependence on linewidth variation that creates the correlation between Idn and Idp.

5.4 Applications to Process Control

The results in the previous section demonstrate that *pdFab* can predict the device parameter distributions observed at wafer electrical test. This prediction is based on process control variations. Hence, *pdFab* is providing the service of virtual factory by mimicing not only the ideal behavior but also the manufacturing variation that will certainly occur during production. This service is the needed component for the design of process control strategies.

TABLE 3. Predicted and measured (in parentheses) correlations between device parameters and in-line measurements

	Poly LW (meas)	Gate Ox	Nwell Rsh (meas)	Pwell Rsh (meas)
Vtp	-0.03	-0.05	0.77	-0.08
	(-0.21)		(0.59)	(-0.10)
Vtn	0.02	0.86	0.33	-0.36
	(0.39)		(0.02)	(-0.20)
Idp	-0.76	-0.11	0.24	0.11
	(-0.76)		(0.21)	(0.31)
Idn	-0.76	-0.52	-0.31	0.21
	(-0.75)		(-0.07)	(0.45)

Process control strategies are designed to assure that devices meet specifications at wafer electrical test inspite of the process variations. Typically, in-line measurements are measured and specification lmits are established to assure that the devices will meet specification limits at electrical tests. Picking the correct in-lines to measure and design specification limits for those in-lines is difficult since it has not been possible to determine the correlations between parameters aprior to manufacturing many lots.

Since *pdFab* can be used to determine the correlations between parameters, it is now possible to determine process control strategies based on modeling. The first step in process control design is determining if the in-line measurements provide enough observability to determine device variations. With the correlation data presented in the previous section, its possible to determine if the in-line measurements are sufficient.

The CMOS process flow specifies that gate oxide thickness and poly CD are to be monitored during production. The goal is to assure that the device meets the desired specifications. Since the poly CD and Tox are uncorrelated, we can compute the percentage of the variance of the four monitored device characteristics (Vtn, Vtp, Idn, and Idp) that can be predicted via in-line measurements by summing the percentages for each individual measurement. The results are summarized in Table 4.

After the gate oxidation in-line is measured, a large percentage of the variance of Vtn has been determined. Hence, this in-line can be used to control Vtn. The poly CD measurement only accounts for 58% of Idn, but if this information were used along with the Tox in-line, 85% of the Idn variation could be predicted once the poly CD is measured.

The observability for the PMOS transistor is not as good. Only 59% of the Idp variation and 0.4% of the Vtp variation can be observed from measuring the two in-lines. This is because a large amount of the PMOS transistor's variation is controlled by variations in the buried channel dopant concentration. We determined this by examining the correlation of Vtp and Idp with the N-well sheet resistance Rsh). Since we cannot get an in-line measurement of the N-well Rsh after gate oxidation, it is not possible to monitor Vtp and Idp using in-line measurements. The only way to guarantee that these two device characteristics are within the specifications limits is to explicitly measure them at electrical test. Hence quality assurance can only be achieved via direct testing. The conclusion of this analysis is that while close-looped control can be used to improve the NMOS transistor, the PMOS transistor can only be manufactured via open-loop control.

TABLE 4. Percentage of the device variance predicted by in-line measurements

	Tox	Poly	Poly & Tox	Nwell Rsh
Vtp	0.3%	0.09%	0.4%	59%
Vtn	74%	0.04%	74%	11%
Idp	1%	58%	59%	9.6%
Idn	27%	58%	85%	5.7%

5.5   Application to Worst Case Circuit Models

Typically, worst case models are released to circuit designers which represent the corners of the process where circuits are likely to fall out of specification. For the National process, each transistor is released with two worst case models, weak and strong. The weak transistor has both a high threshold voltage and low drive current while the strong transistor has high drive current and low threshold voltage. The NMOS and PMOS models are treated as being independent (i.e, it is necessary to simulate the weak-strong condition).

By examining the predicted correlations, we see that the NMOS and PMOS transistors are significantly related. Hence the strong-weak case is unlikely, as shown in Figure 18. From the distribution we picked points to represent realistic worst case corners. For example the corner which has the strongest NMOS transistor also had the strongest PMOS transistor. The selected worst case drive current corners are marked with circles in Figure 18.

Typically, it is assumed that the maximum drive current transistor also has a low threshold voltage while the minimum drive transistor has a high threshold voltage. The implied assumption is that the correlation between Vt and Id is 1.0. As shown in Figure 19, this is not the case. On the scatterplot we have drawn a line. If the assumption were correct, all data points would have to lie on this line. In Figure 19, we have circled two cases were the transistor has typical threshold voltage, but the drive currents are close to opposite ends of the distribution. We generated BSIM models at the corners circled.

6.   Conclusions

Our experience with PREDITOR and *pdFab* has demonstrated that technology CAD tools hold wide promise in manufacturing, and providing the link between design and manufacturing. The experimental results show that the distribution of device characteristics and their correlations can be accurately predicted. The statistical data can be used to determine the adequacy of in-line measurements for process control, and to generate a realistic set of worst-case device parameters.

It has been estimated that a $3B market in technology CAD tools could be supported if their use increased manufacturing yields by 10% [14]. In this paper we have shown that when assembled into the appropriate framework, such promise has a realistic chance of being fulfilled. The essential elements of this framework are a flexible wafer representation, and a spectrum of tools organized to accomplish the desired tasks. Directions for future work include enhancement of the framework for tool and task development, expansion of the framework clients to include more sophisticated simulators, and the addition of tasks such as process diagnosis as an aid for rapid parametric yield learning.

FIGURE 18. NMOS and PMOS Idmax correlation plot

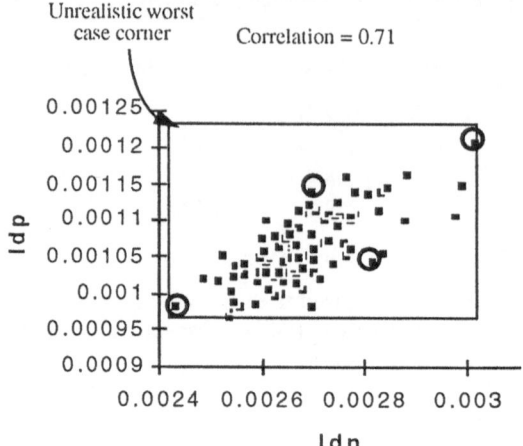

FIGURE 19. Idp - Vtp correlation plot

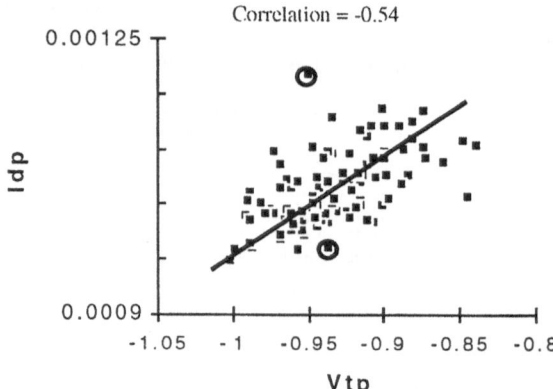

Acknowledgements

David M. Svoboda of Carnegie Mellon University is responsible for the development and maintenance of the PREDITOR graphical user interface. David A. Hanson, Hung Sheng Chen, and Dah Bin Kao of National Semiconductor are responsible for the CMOS process simulation and data analysis. The authors would like to thank them for allowing us to use their results.

References

[1]   A. Wong (ed.). *Semiconductor Wafer Representation Architecture Document V1.0*. Technical Report CFI Document TCAD-91-G-1, CAD Framework Initiative, July, 1992.

[2]   D. S. Boning and D. A. Antoniadis. MASTIF - A Workstation Approach to Fabrication Process Design. *IEEE International Conference on Computer-Aided Design (ICCAD) Digest of Technical Papers*, pages 280-282. November, 1985.

[3]   D. S. Boning, ed. *Technology CAD Framework Architecture*. Technical Report S90013, Semiconductor Research Corporation, May, 1990.

[4]   D. S. Boning, M. L. Heytens and A. S. Wong. The Intertool Profile Interchange Format: An Object-Oriented Approach. *IEEE Transactions on Computer-Aided Design of Integrated Circuits and Systems*. 10(9):1150-1156, September, 1990.

[5]   J. B. Brockman and S. W. Director. The Hercules CAD Task Management System. *IEEE International Conference on Computer-Aided Design (ICCAD) Digest of Technical Papers*, pages 254-257. IEEE, November, 1991.

[6]   S. G. Duvall. An Interchange Format for Process and Device Simulation. *IEEE Transactions on Computer-Aided Design of Integrated Circuits and Systems*. 7(7):741-754, July, 1988.

[7]   D. Collins and A. Strojwas. A Methodology for the Development of Semiconductor Equipment/ Process Models. *Symposium on Automated Integrated Circuits Manufacturing*. Proceedings of Fifth Symposium(90-3):66-77, March, 1990.

[8]   F. Fasching, C. Fischer, S. Selberherr, H. Stippel, W. Tuppa, and H. Read. A PIF Implementation for TCAD Purposes. *Simulation of Semiconductor Devices and Processes (SISDEP) Vol. 4*, pages 477-482. September, 1991.

[9]   Fluent, Inc. *FLUENT User's Manual, version 3.0* March, 1990.

[10]   M. D. Giles and S. R. Nassif. *Unified System for Process and Device Simulation: Proposed Database Interface Format*. Technical Report, AT&T Bell Laboratories, Allentown, PA, June, 1988.

[11]   D. A. Hansen, J. K. Kibarian, H. S. Chen, and D. B. Kao. The Use of Technology CAD in MMC-MOS Process Development - Manufacturability and SPICE Model Prediction. *TCAD Workshop, National Semiconductor Corporation*. May 26-27, 1993.

[12]   C. Hegarty, T. Feudei, N. Hitschfeld, R. Ryter, N. Strecker, M. Westermann, and W. Fichtner. An Approach to Three-Dimensional VLSI Process Simulation. *Proceedings of the 178nd Meeting of the Electrochemical Society*, pages 565-575. May, 1993.

[13]   C. P. Ho, J. D. Plummer, S. E. Hansen, and R. W. Dutton. VLSI Process Modeling - SUPREM-III. *IEEE Transactions on Electron Devices*. ED-30(11):1438-1453, November, 1983.

[14]   J. Hogan. DAC92 Panel - Manufacturing Interface. June, 1992.

[15]   M. E. Law and R. W. Dutton. Verification of Analytic Point Defect Models Using SUPREM-IV. *IEEE Transactions on Computer-Aided Design of Integrated Circuits and Systems*. 7(2):181-190, February, 1988.

[16]   K. Lee and A. Neureuther. SIMPL-2: (Simulated Profiles from the Layout - Version 2). *IEEE Transactions on Computer-Aided Design of Integrated Circuits and Systems*. 7(2):160-167, February, 1988.

[17]   P. Lloyd, H. K. Dirks, E. J. Prendergast, and K. Singhal. Technology CAD for Competitive Products. *IEEE Transactions on Computer-Aided Design of Integrated Circuits and Systems*. 9(1):1209-1216, November, 1990.

[18]   K. Lucas and A. J. Strojwas. A New Vector 2D Photolithography Simulation Tool. *International Electron Devices Meeting Technical Digest*, pages 7.5.1-7.5.4. IEEE, December, 1992.

[19]   W. Maly, A. J. Strojwas, and S. W. Director. Fabrication Based Statistical Design of Monolithic ICs. *Proceedings of the IEEE International Symposium on Circuits and Systems*, pages 135-138. April, 1981.

[20]   W. Maly and A. J. Strojwas. Statistical Simulation of the IC Manufacturing Process. *IEEE Transactions on Computer-Aided Design of Integrated Circuits and Systems.* 1(3), July, 1982.

[21]   J. Mar, K. Bhargavan, S. G. Duvall, R. Firestone, D. J. Lucey, S. N. Nandgaonkar, S. Wu, K. Yu, and F. Zarbakhsh. EASE - An Application-Based CAD System for Process Design. *IEEE Transactions on Computer-Aided Design of Integrated Circuits and Systems.* CAD-6(6):1032-1038, November, 1987.

[22]   M. McIlrath (ed.). *Current Concepts in Semiconductor Process Representation V0.2.* Technical Report CFI Document (TCAD-91-T-2), CAD Framework Initiative, February, 1991.

[23]   P. K. Mozumder and L. M. Loewenstein. Method for Semiconductor Process Optimization Using Functional Representations and Selectivity. *IEEE Transactions on Components, Hybrids, and Manufacturing Technology.* 15(3):311-316, June, 1992.

[24]   S. R. Nassif, A. J. Strojwas and S. W. Director. FABRICS II: A Statistically Based IC Fabrication Process Simulator. *IEEE Transactions on Computer-Aided Design of Integrated Circuits and Systems.* CAD-3(1):40-46, January, 1984.

[25]   S. R. Nassif, A. J. Strojwas, and S. W. Director. FABRICS II: A Statistical Simulator of the IC Fabrication Process. *IEEE International Conference on Circuits and Computers Proceedings*, pages 298-301. September, 1982.

[26]   M. S. Obrecht. SIMOS - A Two Dimensional Steady State Simulator for MOS Devices. *Solid State Electronics.* 34(7), 1991.

[27]   M. S. Obrecht and J. M. G. Teven. BISIM - A Program for Steady-State Two Dimensional Modeling of Various Bipolar Devices. *Solid State Electronics.* 34(7), 1991.

[28]   J. K. Ousterhout. Tcl: An Embedded Command Language. *Winter USENIX Conference Proceedings.* 1990.

[29]   J. K. Ousterhout. An X11 Toolkit Based on the Tcl Language. *Winter USENIX Conference Proceedings.* 1991.

[30]   D. Collins. *The PHENOM User's Guide, v1.34* September, 1991.

[31]   E. W. Scheckler, A. S. Wong, R. H. Wang, G. Chin, J. R. Camagna, K. K. H. Toh, K. H. Tadros, R. A. Ferguson, A. R. Neureuther, and R. W. Dutton. A Utility-Based Integrated Process Simulation System. *Symposium on VLSI Technology, Digest of Technical Papers*, pages 97-98. June, 1990.

[32]   M. R. Simpson. PRIDE: An Integrated Design Environment for Semiconductor Device Simulation. *Workshop on Numerical Modeling of Processes and Devices for Integrated Circuits (NUPAD) III*, pages 57-58. June, 1990.

[33]   A. J. Strojwas and S. W. Director. The Process Engineer's Workbench. *IEEE Journal of Solid-State Circuits.* SC-23(2):377-386, April, 1988.

[34]   A. J. Strojwas and S. W. Director. A Process Engineer's Workbench. *Proceedings of the IEEE Custom Integrated Circuits Conference*, pages 329-332. IEEE, May, 1987.

[35]   W. Fichtner. From Layout to Circuit: Multi-Dimensional Process and Device Simulation - Current Status and Open Problems. *1993 International Workshop on VLSI Process and Device Modeling.* May, 1993.

[36]   D. M. H. Walker, C. S. Kellen, D. M. Svoboda, and A. J. Strojwas. The CDB/HCDB Semiconductor Wafer Representation Server. *IEEE Transactions on Computer-Aided Design of Circcuits and Systems.* 12(2):283-295, February, 1993.

[37]   J. S. Wenstrand, H. Iwai, and R. W. Dutton. A Manufacturing-oriented Environment for Synthesis of Fabrication Processes. *IEEE International Conference on Computer-Aided Design (ICCAD) Digest of Technical Papers*, pages 376-379. November, 1989.

[38]   J. S. Wenstrand, H. Iwai, M. Norishima, H. Tanimoto, T. Wada, and R. W. Dutton. Intelligent Simulation for Optimization of Fabrication Processes. *Workshop on Numerical Modeling of Processes and Devices for Integrated Circuits (NUPAD) III*, pages 15-16. June, 1990.

[39]   A. S. Wong, D. S. Boning, M. L. Heytens, and A. R. Neureuther. The Intertool Profile Interchange
Format. *Workshop on Numerical Modeling of Processes and Devices for Integrated Circuits
(NUPAD) III*, pages 61-62. June, 1990.

[40]   A. S. Wong and A. R. Neureuther. *The Intertool Profile Interchange Format: A Technology CAD
Environment Approach.* Technical Report C90361, Semiconductor Research Corporation, July,
1990.

[41]   A. S. Wong and A. R. Neureuther. The Intertool Profile Interchange Format: A Technology CAD
Environment Approach. *IEEE Transactions on Computer-Aided Design of Integrated Circuits and
Systems.* 10(9):1157-1162, September, 1991.

Technology CAD at Stanford University: Physics, Algorithms, Software, and Applications

R.W. Dutton and R.J.G. Goossens

Center for Integrated Systems, Stanford University, Stanford, CA 94305, USA

1. Introduction

The evolution of technology computer-aided design (TCAD) over the past two decades has been remarkable. During this period we have wittnessed the ubiquitos use of circuit simulation for IC design, the broad acceptance and use of device analysis for technology design and the innovation of new classes of simlators for process and equipment modeling. Much of this expanded role for TCAD has been driven by the IC revolution itself where device dimensions have been reduced by more than 20 fold and gate densities per chip have increased by more than 10,000 times. The growing need for and role of TCAD is mandated by requirements on control over the manufacturing process, electrical reliability, as well as business aspects like minimizing the time to market. Moreover, the design issues of both high-speed and low power -- each from a very different perspective -- require greater care and precision in IC design.

The purpose of this paper is to provide an overview of the TCAD efforts at Stanford. Over the past two decades the efforts have broadened from research in individual tools (i.e. SUPREM and PISCES) and their physical models towards integrated software systems and applications. Here we will provide a discussion based on the latter rather than former point of view. In Section II we use specific problems to illustrate progress in both tool development and applicability to practical IC design problems. This includes key issues related to consortia-based collaborations and the growing need for infrastructure to support both TCAD software as well as graphics and other interfaces. Section III turns to recent advances in physical models with emphasis on calibration strategies. The discussion of algorithms and the role of parallel computation is presented in SectionIV. Finally we give a summary of highlights presented in this paper.

2. Application-Driven TCAD

2.1 Aladdin-CAD

As suggested in the Introduction, TCAD is driven by a need to accuratelyand quickly predict factors that improve performance, reliability and yield for IC technologies. Over the past two years we have been involved in a collaborative project with several industrial partners and funded by the State of California's Department of Commerce. The Aladdin-CAD Project (*Analysis of advanced devices based on industry-networked CAD*) has focused on interesting device and technology innovations as well as improvements related to tool usability and robustness. The following examples demonstrate progress in this application-driven approach.

High-density silicon devices

Random Access Memory (RAM) and especially Static or SRAM have been a key pacing factor in microprocessor system performance. The trade-offs involved in increasing density and speed while achieving high yields has been a major challenge to the industry. During the course of the Aladdin-CAD project we have worked closely with Cypress Semiconductor in the modeling and simulation of their next-generation SRAM technology. While the details of their technology and devices are proprietary, the following inverter example gives a clear indication of the issues both in modeling and user interfaces.

Figure 1   3D Solid Model of an inverter structure generated with a Stanford developed geometric-modeling tool, based on ACIS [1] for the solid model and AVS [4] for visualization.

Figure 1 shows a 3D solid model of a CMOS inverter, half of a simple SRAM cell that has been created using a Stanford-developed tool based on a commercial geometry-modeling library, ACIS from Spatial Technologies [1]. This result clearly demonstrates to us the important role played by the underlying information models. Specifically, the structure's layout was defined using the Cadence layout system and the standardized CIF format was used to provide a set of surface polygons, which the ACIS-based tool used to create the solid geometry model. From this geometric model, two consistent alternative data representations were created, namely one conforming to the HDF-Vset standard [2] to interface with visualization and the other to the SWR 3D prototype based on the present standard [3] to support gridding for device analysis. Figure 1 was created using AVS [4], an advanced graphically programmable environment for the manipulation and visualization of scientific data. It was in fact through such visualization and exploration of many of the complex features of the SRAM structure that our collaborators at Cypress were able to understand and unravel design rule trade-offs and potential problems that could have affected yield.

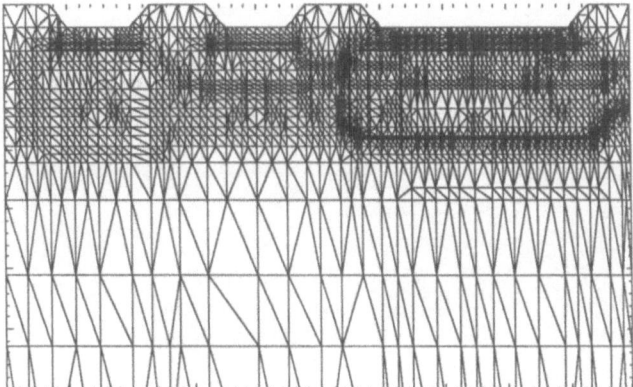

Figure 2   Inverter cross section. Grid was created using SMB.

Figure 2 shows a stylized cross-sectional view of one of the inverter pair shown above and the analysis grid created automatically by SMB. This is a gridding tool that is based on Meshbuild from ETH, Zurich [6] and has been enhanced by Stanford [5]. Using this gridded structure we could analyze several features of the cell performance using PISCES-MP [7]] or other device analysis programs. Specifically, the cell switching properties as well as the latchup immunity have been studied using different stimulus and boundary conditions, including circuit elements for mixed-mode analysis. Such large structures require very large grids and hence can only be supported by the computational resources offered by large-scale parallel systems. In Section IV we will discuss some of the issues involved in using parallel computation with multi-processors, based on our experience with a multi-processor-version of PISCES. It is evident however that the ability to use in excess of 100,000 grid points in a 2D device simulator is an essential ingredient in the accurate analysis of multi-devices cross section that are extracted from standard cells.

High-voltage silicon devices

There exists a variety of special-purpose devices that essentially require simulation and modeling as part of the design process. High voltage device design is one such class and Stanford has made a number of contributions to this field [8]. Again in the context of the Aladdin-CAD project, we have worked closely with an industrial partner, National Semiconductor, to help in the design of a next generation high-voltage bipolar process. The focus of this design effort was to assess for this new process technology the sensitivity of critical device-performance parameters with repect to critical dimensions and in particular to assymmetry due to mask misalignment. In particular, a key issue with high-voltage devices is to correctly predict breakdown characteristics. This is difficult owing to not only unique physical effects but also numerical challenges to correctly control the boundary conditions throughout the tracing of the I-V curves.

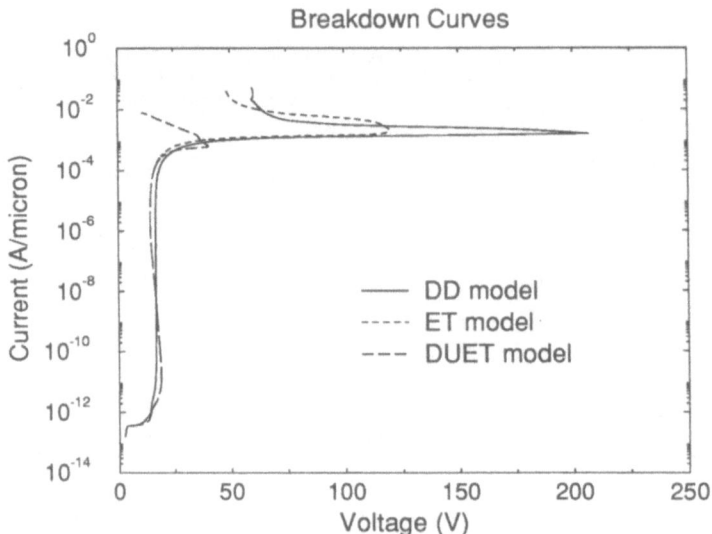

Figure 3   I-V breakdown curves for a high-voltage diode from DD, ET and DUET
                  simulations.

Figure 3 shows the breakdown characteristics of a diode structure being studied by National
Semiconductor and the comparison of results obained with different physical assumptions
used during the analysis. The tracing of the I-V curves was controlled automatically, based
on new algorithms where the device dynamic impedance and external load are matched at
each bias point to insure local stability during every simulated bias step [9]. The curves
compare the analysis results of conventional drift/diffusion (DD), energy transport
(ET) [10] and a new dual energy transport (DUET) formulation that has been implemented
in PISCES [11]. The key difference between the ET and DUET models is the consistent
treatment of the coupled carrier and lattice temperatures during the analysis. The results
show that the inclusion of lattice heating into the breakdown calculation leads to a dramatic
reduction in breakdown voltage for such devices due to thermal breakdown.

Compound material heterojunction devices

There are a growing number of applications where silicon technology cannot meet the system requirements. Both ultra-high-speed devices and opto-electronics are illustrative of
such applications. In connection with Hewlett-Packard's Optoelectronics and High-Speed
Devices Research Laboratories, we have extended the capabilities of 2D device simulation
(PISCES) and demonstrated excellent results in prediciting technology dependences of advanced devices fabricated in GaAs-based materials systems. Figure 4 shows the vertical
band diagram of a typical, albeit idealized, MODFET device, which shows both variations
in bandgap and band discontinuities for a InGaAs/InAlAs device.

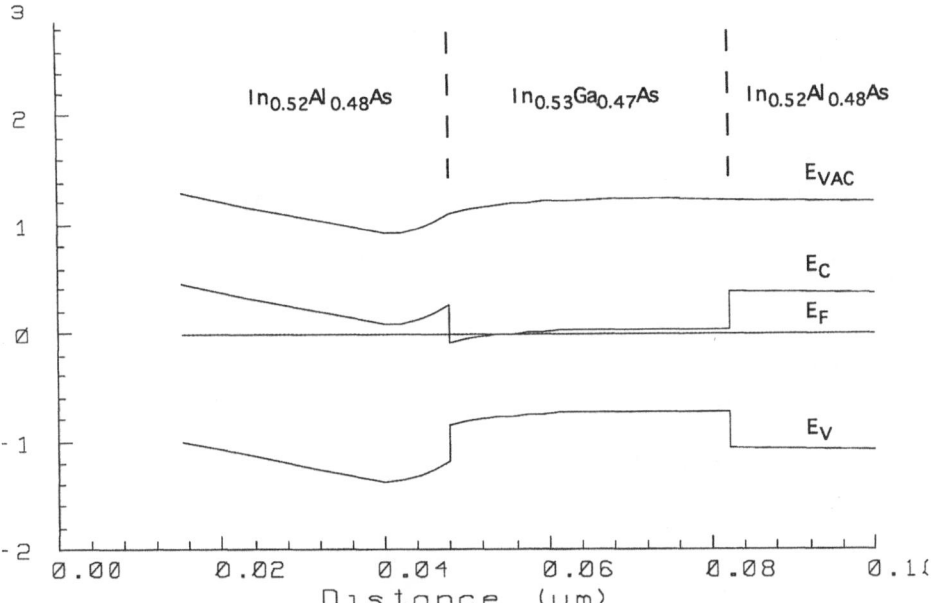

Figure 4   Band diagram along the bisector of the gate contact at thermal equilibrium. Note that the vacuum energy level is deliberately shifted down-
ward by 3.5 eV to highlight the bandgap region.

The MODFET takes advantage of the high electron mobility in InGaAs (can be as high as 18,000 cm^2/Vs in InAs) to realize high-speed devices -- in this case an n-channel MODFET uses the low bandgap (0.72 eV), high mobility, InP lattice-matched material, $In_{0.53}$-$Ga_{0.47}As$, as the channel region (as well as the source/drain regions). The channel region is lightly doped to $5x10^{15}$ cm^{-3}, and the channel conduction is formed due to the spill-over of electrons from an adjacent wide bandgap (1.45 eV) $In_{0.52}Al_{0.48}As$ region doped to $6x10^{18}$ cm^{-3}, which can clearly be seen from the band diagram (Figure 4) at the thermal equilibrium. The band diagram is taken along the vertical direction perpendicular to the the the gate contact towards the substrate. The free surface of InAlAs layer between the gate and source/drain contacts can have Fermi-level pinned by specifying the surface trapping density. The output I-V characteristics (drain current versus drain/source bias) are shown in Figure 5.

2.2 Manufacturing Science

The above examples have emphasized both silicon and compound-material technologies and the critical role of device simulation in the design of devices using these technologies. In the area of process simulation and equipment modeling, in collaboration with Texas Instruments on the *Microelectronics Manufacturing Science and Technology* (MMST) project, we have achieved very exciting results in the control and calibration of models for rapid thermal processing (RTP). Stanford has pioneered the concepts of the "virtual factory" and its connection to the "programmable factory", wherein computer simulation and automated control provide key links between the two systems. Such tight linkage between the virtual and programmable factory is needed to enable automatic generation of control

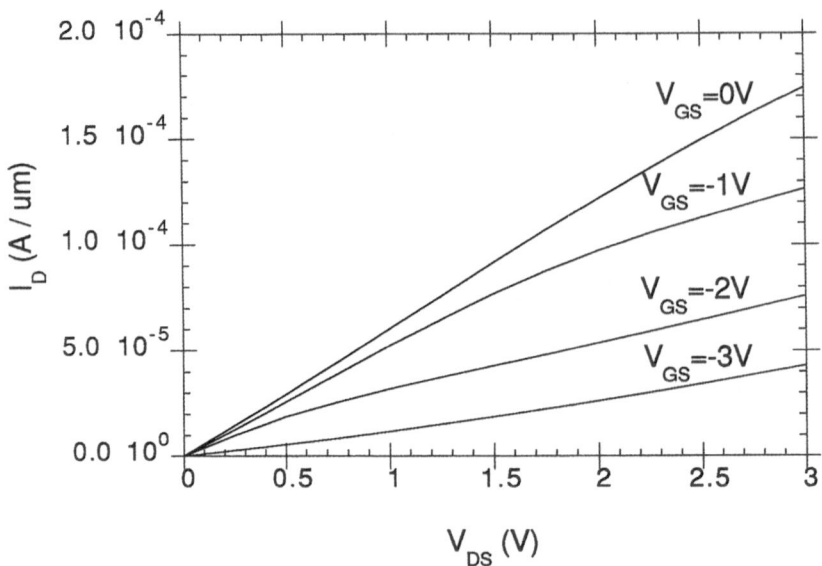

Figure 5   Output characteristics of MODFET with V_{GS} changes from 0 to -3V in increments of -1V.

methods that support high-yielding and flexible manufacturing. In August of 1992, Stanford provided a key demonstration of these concepts at our annual summer workshop by designing and manufacturing a trench-based device using a prototype integrated virtual and programmable factory.

Equipment control

The trends in single-wafer processing technology and RTP in particular are very promising. The key issues to be addressed with such new technologies are the controllability and reproducability of the resulting device characteristics. Over the course of the MMST program we have worked closely with TI in developing and characterizing the control of temperature profiles in an RTP system that TI has built. In the left side of Figure 6 the RTP system is shown schematically. It illustrates the complexity of the interactions that must be modeled, measured and controlled. In particular, the lamp configuration consists of three concentric rings, each with independent control. Using a unique acoustic surface wave temperature sensing system and combination of feedback and feedforward control, excellent control of both absolute temperature and ramp rates has been achieved [12],[13]. The right side of Figure 6 shows schematically a typical control sequence of power for the lamp rings and the temperature versus time sequence. Having such accurate controllability and observability of the processing conditions are key not only for making the results reproducible, but also in supporting the simulation models to be discussed next.

Process models

The modeling of diffusion processes in silicon is strongly influenced by both wafer and am-

RTP System with Multi-Zone Lamp & Multivariable Control

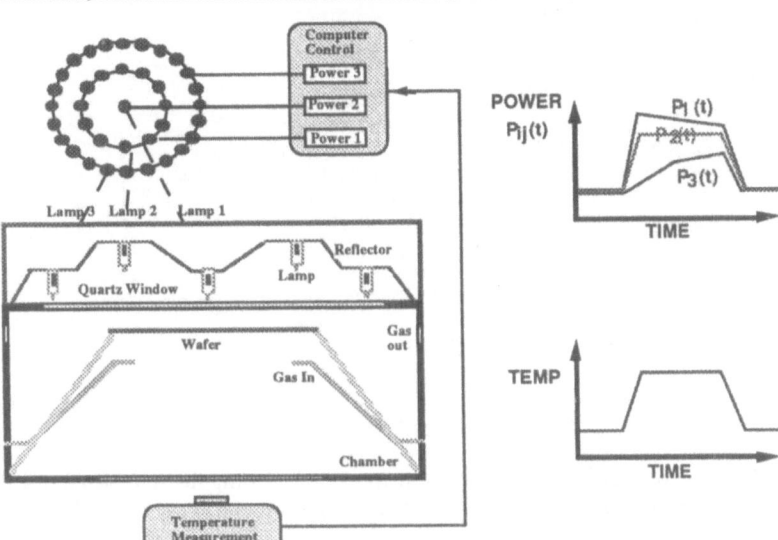

Figure 6   Schematic diagram of the TI RTP system and the interactions that must
                be modeled, measured, and controlled.

bient conditions. Especially in the RTP environment, the details of the initial profiles, damage profiles and the thermal cycles all contribute in determining the final impurity profiles. The capabilities discussed above for automated control of the RTP system are key to achieving the modeling capabilities now discussed. We have carefully investigated the role of implant damage and annealing conditions on profile characteristics for shallow junctions and have used a powerful new simulation methodology to extract the model parameters that accurately fit the experimetal data. In the context of RTP models it is expected that "self-calibrating" systems will be needed in order to track process and equipment variations. The results presented here have helped to quantify effects of the extended defect structures and extracted values of interstitial diffusivity. Figure 7 shows simulation results using SU-PREM where a new model for interstitial trapping is used to alter the interstitial profile and boron diffusivity in an epitaxially grown marker layer [14]. These results along with data from a number of other experiments were simultaneously optimized using parallel simulations on the Intel iPSC/860. Figure 8 shows schematically the flow diagram of the optimization process. A total of more that 500 simulations were performed over 48 hours in order to optimize 2 coefficients and simultaneously fit 5 experimental profiles.

2.3   TCAD Integration

The integration of TCAD software systems is an area of growing concern and necessity. While Stanford continues to prototype systems that demonstrate proof-of-concept, we have deliberately choosen to collaborate through consortia and partnerships that promote an open plug-and-play methodology. We perceive the development of the *Semiconductor*

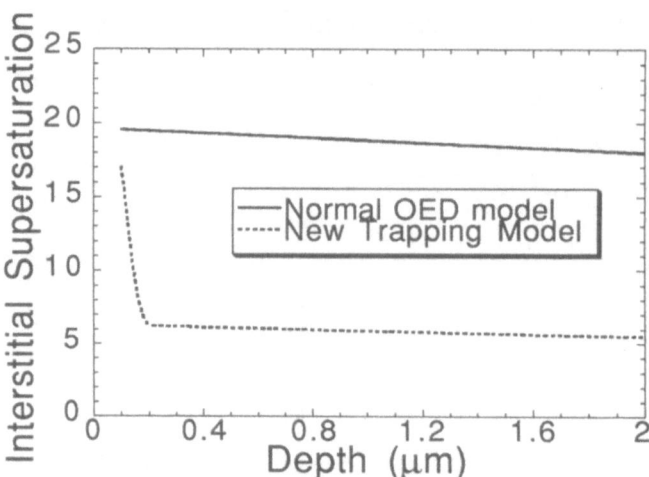

Figure 7   Simulation results for a new model describing the interstitial profile and
                  boron diffusivity of an epitaxially grown marker layer [14].

Wafer Representation (SWR) being developed through a technical subcommittee of the
CAD Framework Iniative (CFI) as a key activity in creating both a robust TCAD information model as well as the utilities that will broadly support tool development and integration. In fact, it is the objective of CFI to promote the role of commercial vendors as
providers of tools that comply with the standards. One prototype software integration
project at Stanford, SEWB (*Simulation Engineering Workbench*) [15], illlustrates how vendor tools and the standardization of information models (including SWR) provide an exciting testbed for TCAD development and innovation. Work in the area of scientific
visualization demonstrates how the same plug-and-play approach leverages exciting new
development outside the direct scope of TCAD.

SEWB

Statistics is coming to play a major role in both design and characterization of ICs. SEWB
provides a unified environment, based on a consistent use of framework tools, in which to
explore these device limitations. Within this system, interactive graphical device specification is provided by a using the Cadence layout system to sketch the structure. Meshing is
provided by SMB. Subsequently, SEWB can automatically create device simulation input
specifications for Stanford's PISCES as well as for similar vendor tools [15]. SEWB provides the posibility to parameterize the device definition and will manage the generation of
a whole statistical design experiment on these variations, the generation of the corresponding input decks as well as the subsequent simulation of these simulation decks across a heterogeneous collection of workstations including a network-connected parallel machine
such as in our case the Intel iPCS/860 discussed in connection with the work on the MMST
project.

Statistical characterization of devices and the associated process technology alone is not
enough. A variety of cases exists where testing and experimental verification of the under-

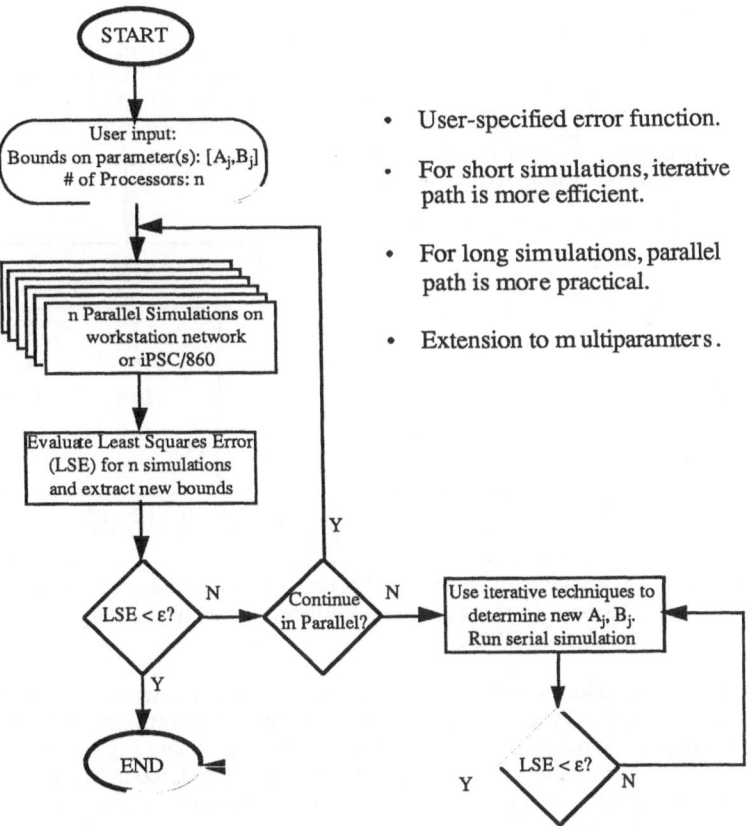

* User-specified error function.

* For short simulations, iterative path is more efficient.

* For long simulations, parallel path is more practical.

* Extension to multiparamters.

Figure 8   Flow diagram for simultaneous optimization of multiple process-model parameters using distributed computing.

lying device models are necessary. In the earlier example we considered the specification of simulation input without concern for testing these device models. We are now investigating the use of SEWB in connection with the calibration of device models for circuit simulation. Specifically, Dr. Don Scharfetter has developed the BSIMJr tool [16] as an interface and model development environment to facilitate the characterization of physical effects in MOS devices. However, already the extraction of something as "simple" as the effective mobility requires the generation of a large number of simulated I-V data, namely curves in the linear regime for various backgate biases and all relevant device dimensions. This is where SEWB comes to our help, i.e. in the automated creation and execution of the complete simulation space. This is illustrated in Figure 9.

Scientific visualization

Visualization of simulation results and other forms of scientific data has become a field of research in its own right and TCAD is an avid user of such graphical capabilities. In Section IV we will discuss some of the progress in computations needed to support large scale

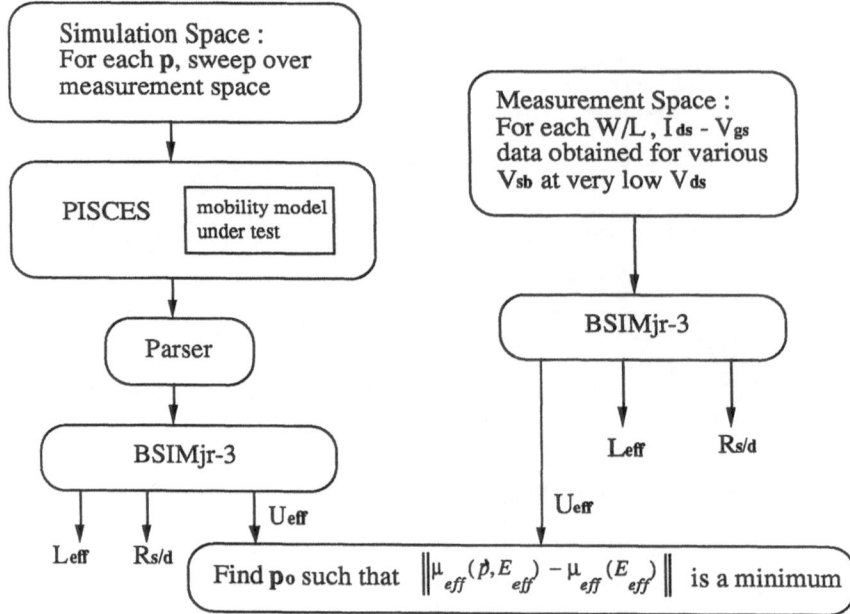

Figure 9   Schematic diagram of the connection between SEWB, PISCES, and
                BSIMjr. In this case the optimization part of the diagram is provided ex-
                ternally to BSIMjr.

TCAD modeling. Here we show the results of collaborative work with IBM Research in
the visualization of data sets containing huge volumes of information where interpretation
of the results can be quite difficult. Specifically, Figure 10 and Figure 11 show two views
of current flow patterns in a bipolar transistor that was simulated in 3D using both the Intel
iPSC/860 at Stanford as well as a much larger Delta machine sited at Caltech. Further details of the simulations are discussed below. Here we focus specifically on the information
that can be extracted from the visualization.

Figure 10 shows the vector flow of electron current in the bipolar transistor from emitter to
collector. In addition there is a cut-plane showing electrostatic potential across the middle
of the device. The results show an interesting effect in this submicron-scale emitter structure where there is an apparent "current focusing" effect. Namely, the current flow is actually concentrated in the collector region rather that spreading out. This seems to be due to
the focusing influence of the potential contours as can be easily seen from the figure. This
phenomenon is typical for submicron structures where lateral device dimensions have been
scaled more aggressively than vertical ones, leading to inherently two-dimensional device
behavior. Figure 11 shows another vector flow pattern, this time for the holes in the base
region. Here we see that the flow of holes from the base contact into the intrinsic base region follow the paths of lowest spreading resistance which in some cases require that the
flow goes around the edges of the emitter. This figure clearly demonstrates the necessity
to include both surface dimensions in the simulation, justifying the need for a 3D device
simulator. These figures were created using the Power Visualizer System (PVS) from IBM

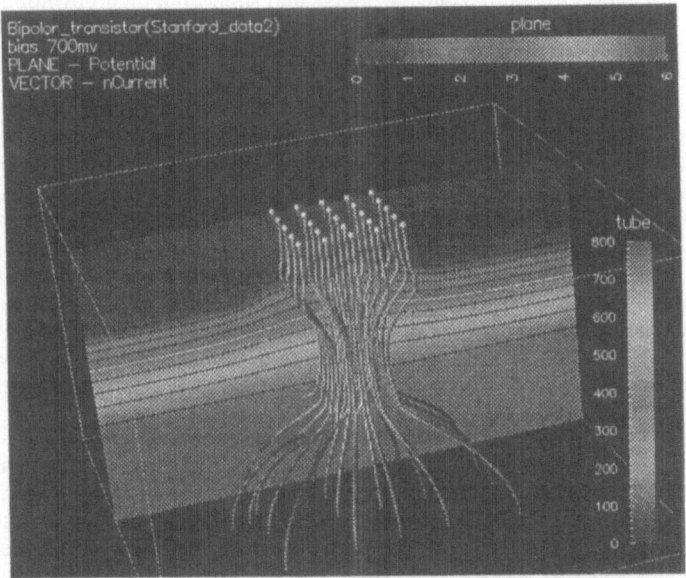

Figure 10   Electron current density in a submicron bipolar transistor. Notice current focusing in the collector region. Visualization done using PVS [17].

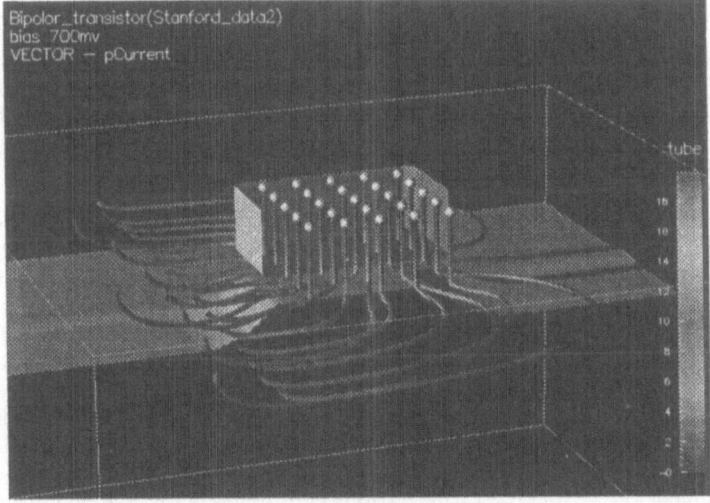

Figure 11   Hole current density in the same structure as in Figure 10. Notice how the holes tend to flow around the intrinsic base, minimizing total resistance.

by Dr. Ed Farrell, who provided extensive advise and support in their development [17].

3. Physical Models

Accurate physical models are essential for all levels of TCAD modeling. Compact models for circuit simulation have traditionally been central to circuit design and have been broadly discussed in the literature and at technical meetings. Since device characteristics can be easily measured, compact models are comparatively easy to calibrate and test. Process models are at the other end of the spectrum in that they are rather complex to calibrate and data can often be obtained only indirectly. While 1D process models for steps such as oxidation and diffusion have been studied extensively, no routine calibration of 2D data for diffusion is presently available. Device modeling lies somewhere between the above two extremes in that many effects can be characterized through either terminal behavior or secondary quantities including various optical and thermal effects. In this section we consider several examples of advances in both physical models and characterization with emphasis on device transport effects.

The reduced dimensions of both MOS and bipolar devices have given rise to a number of non-thermal equilibrium effects that influence both performance and reliability. While the so-called "Drift-Diffusion" (or DD) carrier transport formulation has been quite effective in modeling most transport effects, provided that the carrier mobility is modeled empirically, effects such as gate and substrate current in MOS devices require informtion about the carrier temperature and impact ionization. Such information can only be obtained by including higher moments of the Boltzmann Transport Equation (or BTE) in the simulation. Over the past two decades there have been a number of formulations proposed and implemented in order to solve the equations governing the higher moments of the BTE. The two best known of these are the Hydro-Dynamic (or HD) and Energy Balance/Transport (or ET) models. The HD solves for the complete velocity vector as well as carrier energies, thus requiring 11 equations whereas the ET formulation adds only the carrier tempertures and hence has 5 coupled equations. Over several years in collaboration with Professors Hess and Ravaioli at the University of Illinois, we have investigated and extended the ET formulation as well as studied the issues of calibration of impact ionization using both full band and simplified Monte Carlo (or MC) techniques.

In the discussion of high-voltage devices, the results for breakdown were presented showing a comparison simulations based on DD, ET and DUET formulations of the transport equations. Details of the equation formulations and simulator implementation are to be found elsewhere [10]-[11]. The main point to be emphasized in this section relates to the physical interpretation and calibration of the ET formulation and comparisons with HD and MC results. As stated in these papers, the key difficulty in accurately modeling the non-thermal equilibrium behavior of carriers arises because of the gradient terms involving carrier temperature. We have found that although it is widely accepted that the non-physical overshoot effects in the drain regions of MOS devices can be adjusted in the HD formulation by means of the heat coefficient, in fact the optimum value is not unique and depends on device dimensions. Analysis of the various contributions to the total current near the drain region where the non-physical overshoot can be observed was presented earlier [18].

In order to calibrate various physical models used in the continuum formulations such as DD, HD or ET, Monte Carlo analysis has proved to be a valuable tool. We previously reported collaborative work with Professors Sangiorgi and Venturi in improvements of the computational effeciency of BEBOP based on a simplified band MC analysis method [19]. In the area of improved physical modeling for simplified MC analysis, we have used the full band MC results of Bude [20] to calibrate the impact ionization model in BEBOP [21]. In order to reduce the full band model results to the isotropic formulation used in BEBOP it was necessary to integrate over iso-energy surfaces physical quantities for the full band model, such as the density of states or the average ionization rate. These integrated values can then be used to construct the corresponding values for the isotropic-band models. This is shown in Figure 12 for the impact ionization rate as a function of electron energy.

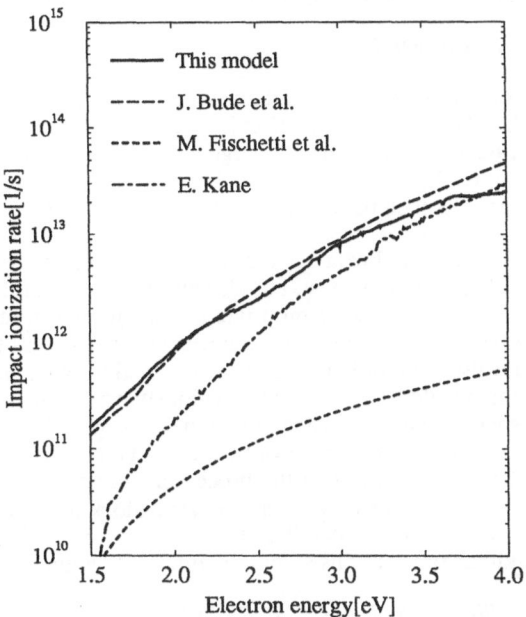

Figure 12   Impact ionization rate versus energy for various models. Full-band-structure data are translated into isotropic-band data for use in BEBOP. For details on the models, see [21] and references therein.

4.   Parallel Computation

The previous sections have motivated the need for both accurate and efficient TCAD simulations. Accuracy is required to support IC design objectives in achieving performance and reliability. The role of efficiency is both economical and in support of accuracy. The complexity of TCAD modeling and the interdependence of model parameters, often in highly nonlinear ways, require extensive simulations to both explore the design space and to optimize models. Moreover, the support of statistical analysis is critical in determining process windows. All these arguments support the need for substantial computer resources for TCAD in the context of statistical design.

Another area of growing importance and concern is 3D simulations. While individual 3D simulations have been used often for both MOS and bipolar device studies, there is now a growing list of interacting device effects that require not only 3D analysis but often coupled boundary conditions (mixed mode) with circuit elements, ambient conditions such as temperature and even inputs to support sensor applications. The results presented in Section II concerning visualization of 3D bipolar results indicate state-of-the-art results in very large-scale simulations. Over a period of seven years and three-and-a-half generations of Intel iPSC computers, we have explored algorithms that exploit coarse-grain parallelism for such computations. Here we will highlight recent trends based on both the 32-node Intel iPSC/860 at Stanford and experiences using the 520-node "Delta" machine at Caltech.

3D Device Simulation Using STRIDE

STRIDE is our 3D finite-volume device simulator that was envisioned as a parallel program from the outset [22]. The focus in this effort is on iterative linear solvers and preconditioning schemes, nonlinear strategies and convergence, and the interactions with domain decomposition. The results shown in Figure 10 and Figure 11 were computed using STRIDE based on advances in such parallel algorithms as discussed below.

Large-scale 3D simulation leads to linear systems of equation that are only tractable with iterative linear solvers. Many variants exist, but all with one common theme: they do not work unless they are combined with an appropriate preconditioner to improve the conditioning of the system. Over the past two years, significant progress has been made in this area. In particular the development of ILU-V has been a breakthrough [23]. High-level injection bias conditions that would be inaccessible previously, are manageable with this new preconditioner. In cases where previous schemes would converge, we see a reduction in the number of linear iterations by at least a factor of two. Key in the development of preconditioners for parallel systems is the way that the processor boundaries are handled. In our scheme, the preconditioning is can be done on a processor-block basis, which dramatically reduces communication and increases parallel efficiency, but only if the matrix is completely assembled including contributions across the processor boundaries.

High-Resolution Device Simulation Using PISCES-MP

PISCES-MP is a parallel version of our standard PISCES-2B code, based on version 9009. Here, the emphasis is on developing a programming paradigm for parallelization of so-called "dusty decks". The use of a specific variant of the SPMD approach (Same Program, Multiple Data) allowed us to create a parallel version of PISCES while only touching a limited fraction of the existing code while still achieving excellent speedups. This is important to preserve the knowledge base that is inherent in the standard version through more than a decade of intense industrial use.

The fact that PISCES-MP is a 2D simulator implies that the grid count will never be as massive as for the 3D code Stride. As a consequence, it is possible to continue to use direct linear solvers instead of the less robust iterative solvers. A completely new version of a distributed multifrontal solver was developed as the numerical heart of PISCES-MP, which allows us to run simulations with up to about 30,000 grid points with good numerical efficiency on the present Intel iPSC/860 machine using 32 nodes. An indication of the achieved speedups is given in Figure 13. In order to explore very-high-definition device CAD with grid counts in the 100 000+ range, one would have to use more CPU's than the 32 available

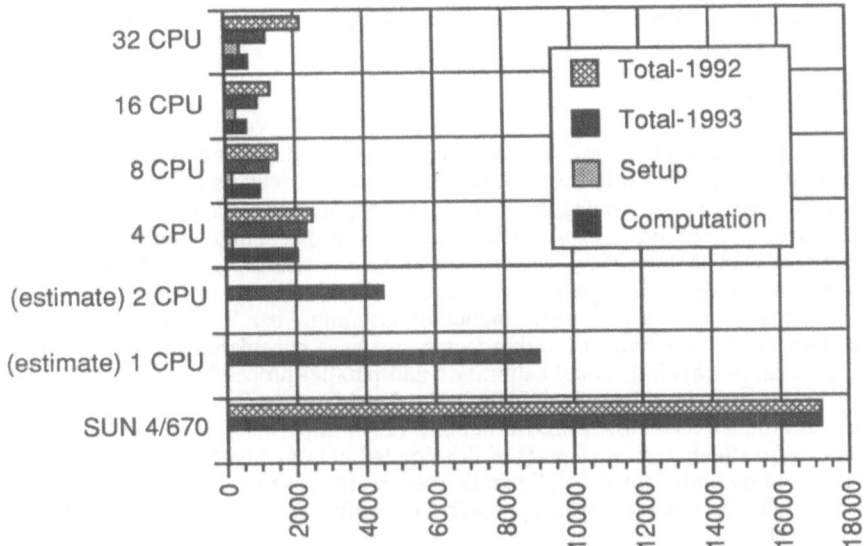

Figure 13   Total execution time for a 30,000-equation problem with Pisces-MP.
                 Times are for calculating a Gummel plot in 13 bias steps.

to us at Stanford. Benchmarks show that on such a larger machine, in the later stages of the
direct solver the communication would become prohibitive. This is illustrated in Figure 13
through the increase in total execution time for more than 16 processors. Therefore, over
the past 6 months, a hybrid solver was developed. Initially direct elimination is employed.
The code automatically detects at which stage the communication would become ineffi-
cient. At that point, the newly developed hybrid solver switches to an iterative scheme and
preserves efficiency. Preliminary results indicate that at that stage the condition number of
the remaining system has improved quite drastically compared to the original complete system, implying that a simple block-diagonal preconditioner is good enough at this stage.

Library-Based High-Performance PDE Solvers

Two years ago, an effort into adaptive parallel finite-element methods for device simulation
started in our group. The initial program aims at providing a general yet accurate and robust
formulation for tackling even the most complicated physical models such as the hydrodynamic model. Fueled by progress in the area of computational fluid dynamics, an entropy-
variable-based Galerkin-Least-Squares FEM formulation was developed, that indeed provides these features for single-carrier problems [24]-[25]. Presently, work is in progress to
add grid adaptivity as well as support for two-carrier problems.

Meantime, this effort has grown into a broad-based collaborative effort into general library-
based FEM PDE solvers for high-performance massively parallel platforms. The intent is
to provide a suite of library tools in support of general PDE solvers specifically targeted at
massively parallel architectures. Many of the code elements are generic and not application
specific: gridding, domain decomposition, loading schemes, nonlinear solvers, assembly,

linear solvers and preconditioners, and visualization are all common. Interactions are developing between collaborators that linear-algebra groups, application groups in various engineering domains, as well as hardware vendors.

5. Summary

This paper provides an overview of several aspects of TCAD work at Stanford. We emphasize both the application-oriented approach being used as well as the key factors that we see as being critical to advancing the field. On the applications side we have looked at several different areas of technology development (ie high speed, high voltage and compound materials) as well as some of the new issues related to virtual/programmable factory. In the process of discussing these applications we also addressed some of the user and information model issues such as: geometry modeling, gridding, visualization and automation of simulation control. Looking at limiting factors, we have considered two aspects of particular importance--physical model calibration and parallel processing as a means to overcome computational bottlenecks of TCAD. On the side of calibration, the effective use of statistical simulations is emphasized as being of major importance. Parallel computation requires careful attention to all aspects of the simulation tasks involved. We have presented results based on both "dusty deck" parallelization efforts as well as highly tuned parallel 3D codes. Finally, a library-based approach for building further parallel application codes is advocated.

Acknowledgement

The authors wish to thank a number of Stanford colleagues for numerous contributions including figures. Specifically, we thank: Dan Yang, Datong Chen, Zhiping Yu, Krishna Saraswat, Robert Huang, Aon Mujtaba, Chiang-Sheng Yao and Bruce Herndon. In addition, the contributions of our industrial partners in Stanford's Center for Integrated Systems and specifically those of Dr. Don Scharfetter (Intel), Dr. Ed Farrell (IBM), Dr. Reda Razouk (National Semiconductor), Dr. Waguih Ishak (Hewlett-Packard), Dr. Tony Alvarez (Cypress), and Dr. Robert Doering (TI) are acknowledged. Finally, sponsorship from ARPA/CISTO, Semiconductor Research Office, and Army Research Office are gratefully acknowledged.

References

[1]   ACIS; Spatial Technology Inc., Boulder, CO.

[2]   L. Bishop, U. Ravaioli, P. Fu, D. Yergeau, Z. Sahul, D. Yang, and R. Goossens, "HDF-Vset File Format for Visualization of Mesh-Based Simulation Data", Technical Report, Stanford University, Oct. 1992

[3]   CFI Document Number TCAD-91-G-2, "Semiconductor Wafer Representation Procedural Interface (PI), " Version 1.0, July 1992.

[4]   AVS; Advanced Visual Systems Inc., Waltham, MA.

[5]   D. Yang, K. Law, and R. W. Dutton, "Optimal Moving Meshes for Process and Device Simulation," NUPAD IV (Numerical Process and Device Modeling Workshop) Digest, pp. 181-186. Seattle, May 31--June 1, 1992.

[6]   "Meshbuild - A 2-D Mesh Generator," Integrated Systems Laboratory, ETH Zurich, Switzerland.

[7]   B. Herndon, A. Raefsky, R. J.G. Goossens, and R. W. Dutton, "PISCES-MP - Adaptation of a Dusty Deck for Multiprocessing," Presented at NASECODE-VIII, May 19-22, 1992.

[8]   R. W. Dutton, J. D. Plummer, Power Semiconductor Devices and Circuits, "Tool Integration for Power Device Modeling Including 3D Aspects," Edited by A. A. Jaecklin, Plenum Press, 1992.

[9]   R. Goossens, S. Beebe, Z. Yu, R. W. Dutton, "An Automatic Biasing Scheme for Tracing Arbitrarily Shaped I-V Curves," Submitted to IEEE Transactions on CAD.

[10]   D. Chen, E. C. Kan, U. Ravaioli, C. Shu and R. W. Dutton, "An Improved Energy Transport Model for Hot-electron Device Simulation," IEEE Elec. Dev. Lett., vol. 13, no. 1, pp. 26-28, Jan. 1992.

[11]   D. Chen, Z. Yu, R. Goossens, K.-C. Wu, and R.W. Dutton, "Dual Energy Transport Model with Coupled Lattice and Carrier Temperatures," SISDEP'93 Digest, Vienna 1993.

[12]   P. Dankoski, P. Gyugyi, and G. Franklin, "An Extensible Software Design Applied to Rapid Thermal Processing," MRS Spring Meeting, Symposium on Rapid Thermal and Integrated Processing, San Francisco, April 12-15, 1993.

[13]   F. L. Degertekin, J. Pei, Y. J. Lee, B. T. Khuri-Yakub, and K. C. Saraswat, "In-situ Temperature Monitoring in RTP by Acoustical Techniques," MRS Spring Meeting, Symposium on Rapid Thermal and Integrated Processing, San Francisco, April 12-15, 1993.

[14]   R. Huang, and R. W. Dutton, "Experimental investigation and modeling of the role of extended defects during thermal oxidation", Submitted to Journal of Applied Physics.

[15]   W. T. Wong, J. Y-C. Pan, and J. D. Plummer, "A Virtual Factory-Based Environment for Semiconductor Device Development," Proceedings of the IEEE International Manufacturing Science Symposium (ISMSS), San Francisco, CA, June 15-17, 1992.

[16]   Private communication, Dr. Don Scharfetter, Intel Corporation, concerning BSIMJr. tool.

[17]   Collaboration with Dr. Ed Farrell, IBM Research, visualization data to be published.

[18]   D. Chen, E. Sangiorgi, M. R. Pinto, E. C. Kan, U. Ravaioli and R. Dutton, "Analysis of Spurious Velocity Overshoot in Hydrodynamic Simulations of Ballistic Diodes," NUPAD IV (Numerical Process and Device Modeling Workshop) Digest, pp. 109-114, Seattle, May 31--June 1, 1992.

[19]   S. Sugino C.S. Yao and R.W. Dutton, "Parallelization of Monte Carlo Analysis on Hypercube Multiprocessors and on a Networked EWS System," SISDEP '91 Digest, Vol. 4, pp. 275-284, Zurich, Sept. 1991.

[20]   J. Bude et al., Phys. Rev. B, **45**, p.10958, 1992.

[21]   C-S. Yao, D. Chen, R. W. Dutton, F. Venturi, E. Sangiorgi, and A. Abramo, "An Efficient Impact Ionization Model for Silicon Monte Carlo Simulation," VPAD Conference Technical Digest, p. 42, May 1993.

[22]   K. C. Wu, G. R. Chin, R. W. Dutton, "A STRIDE Towards Practical 3D Device Simulation -- Numerical and Visualization Considerations," IEEE Trans. CAD, Vol. 10, No. 9, pp. 1132-1140, Sept. 1991.

[23]   Z.-Y. Zhao, Q.-M. Zhang, G.-L. Tan, and J.M. Xu, "A New Preconditioner for CGS Iteration in Solving Large Sparse Nonsymmetric Linear Equations in Semiconductor Device Simulation", IEEE Trans. on CAD, vol. 10, p.1432, 1991.

[24]   N. R. Aluru, K. H. Law, P. M. Pinksy, A. Raefsky, R. J.G. Goossens, and R. W. Dutton,

"A Finite-Element Formulation for the Hydrodynamic Device Equations," To be published Computer Methods in Applied Mechanics and Engineering.

[25]   N. R. Aluru, , "Space-time Galerkin/Least Squares Finite-Element Formulation for the Hydrodynamic Device Equations," 1993 International Workshop on VLSI Process and Device Modeling (1993 VPAD) Digest, pp. 16-17, Nara Japan, May 14-15, 1993

An Integrated Design Environment for Semiconductors

P.A. Gough

Philips Research Laboratories,
Cross Oak Lane, Redhill, Surrey RH1 RHA, UNITED KINGDOM

Abstract

In January 1989 the Integrated Device Design Environment (IDDE) software was released to Semiconductor Development sites within Philips. This paper reviews the original goals that underpinned the design of the TCAD framework, how well in the intervening 4 years it has met these objectives, what lessons have been learnt and what issues, from an industrial perspective, are still outstanding. Further, this work attempts to show that *usability* is an important component of TCAD systems and deserves far more attention than it currently gets.

1. Introduction

The development of the Integrated Device Design Environment [1] (IDDE) was largely borne out of the frustration experienced by device and process engineers attempting to use, in combination, multiple simulation tools e.g., SUPREM III [2], SUPREM IV [3], COMPOSITE [4], CURRY [5], TRIPOS [6]. Such a range of software, originating from many sources has a number of problems. Specifically these are:

- Each program has its own form of input. Often this is text based via keywords and parameters. An engineer is therefore required to convert a conceptual image of a device or process into a rather unforgiving syntax. Unsurprisingly such a conversion is a source of error[7] and imposes a high access threshold.

- To use several simulation programs requires fluency in a number of input languages. This imposes a substantial learn/relearn load on users, particularly those requiring casual use of such tools.

- Transfer of data between programs is often difficult. Users are constantly confronted with output from one program that is incompatible with another. Even when conversions do exist they tend to be rather ad-hoc being developed on an as needed basis.

The simulation tools had other difficulties, but those highlighted were identified as the main causes for impeding the integration of simulators into the device design methodology. This was particularly true for development areas as, in contrast to research, there is less opportunity to experiment with new approaches to design and a degree of impatience for a return on time invested in new methods. Therefore, it was clear that there were potentially large benefits to be gained through the development of a TCAD framework that incorporated both an enhanced interface and improved integration of the component simulators.

The Integrated Device Design Environment (IDDE) represents an attempt to produce a system for semiconductor design that addresses the problems stated above. That is, its prime purpose is to make simulators accessible to device designers and increase the efficiency of engineer/simulator interaction. IDDE is a workstation based system (originally Apollo based but recently migrated to X-Windows) that utilises the graphical capabilities of these machines to provide a highly interactive and, in comparison to text based alternatives, an intuitive interface to the user. The system additionally incorporates a number of standard interface files (c.f. the profile interchange format of Duvall [10]) to improve interprogram communication and produce an open system that can assimilate future modelling programs. The systems is sufficiently general to be used in both research and development areas.

The following sections describe in some detail the components of the IDDE system, the underlying data structures, how well during the four years of it's use it has fulfilled its original purpose and what lessons for future systems can be drawn from this work.

2. IDDE

2.1 Features

The major features of the IDDE system are:

2.1.1 Generality

A wide variety of devices are studied within Philips. These range from sub-micron bipolar transistors, to high voltage power devices to new and novel structures the subject of research. A prime goal of the IDDE project was to ensure that such diversity could be handled within the environment and so the software should not be restricted to a class of devices.

2.1.2 Integrated Framework

A number of standard files have been defined within IDDE and it is through these that the tools and simulators communicate. These features lead to an open system with the following benefits:

- By conforming to an agreed standard new tools and simulators can be efficiently integrated into the system.

- It provides a basis for design automation.

- Remote sites can easily exchange data even if they are using different simulators.

2.1.3 Graphical interfaces

An area in which IDDE differed from the contemporary environments was in emphasising an object-based graphical interface supporting direct manipulation (other systems have since appeared [11]). The main features of the interface are:

- The interface is highly graphical. For example a device is represented in schematic form familiar to designers. It is our experience that such graphics are mandatory if the system is to be "engineer friendly".

- The use of direct manipulation techniques. Although a costly paradigm to implement [8] the direct interaction with objects is considered superior (see empirical studies [12] [13]) to a menu based systems.

- Icons and Menus support other aspects of the user dialogue.

2.2   Architecture

The basic architecture for IDDE is shown in Figure 1. The arrows show the principle direction of data flow and the bold boxes are the main software components developed for IDDE.

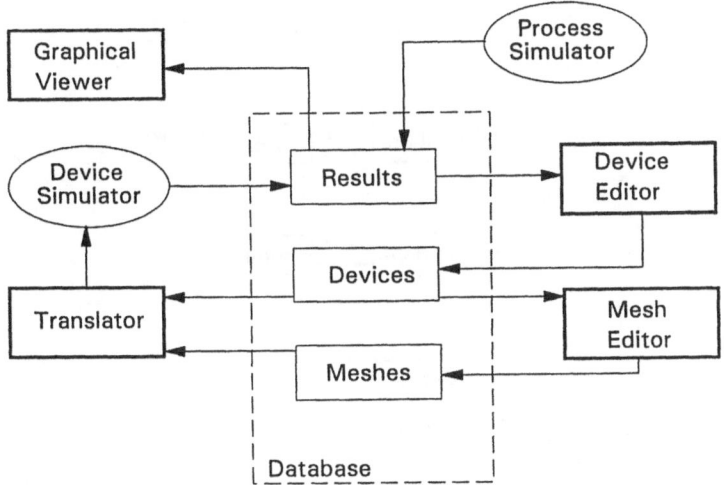

Figure 1: IDDE Architecture

A typical scenario for the system's use would be the generating of a number of 1D, 2D doping profiles and perhaps oxide regions which are stored in the database in a standard results format. A device can then be defined interactively using the regions and doping profiles previously created. A mesh can then be defined over the device, again interactively. The device and mesh data can then be translated into a variety of inputs for device simulation. The output from all of these simulators can then be viewed using the graphical viewer.

Integrated within the current version of IDDE are the process simulators: SUPREM-III, SUPREM-IV, COMPOSITE and the device simulators: HECTOR [14], CURRY, PISCES [15], TRIPOS, PADDY [16] and TESSA [17]. The components of the system are described in the following sections.

2.2.1   Device Editor

The device editor allows the user to define the geometry and physical properties of a device in 2D and 3D. This can be viewed as a tool for manipulating a hierarchical data structure that represents the device. If our original aim of generality is to be met then this representation must be sufficiently flexible to accommodate arbitrary structures.

In IDDE a device is represented as a series of overlapping regions. These regions are composed of one or more shapes (termed segments) which have a particular material type (contact, dielectric, semiconductor). Each region or segment can have associated with it an arbitrary list of attributes that characterise it's physical properties (e.g. bandgap, mobility model, permittivity). A schematic representation of this decomposition is shown in Figure 2. In 2D the shape of a segment is given as a polygon. In 3D a combination of extruded polygons is used.

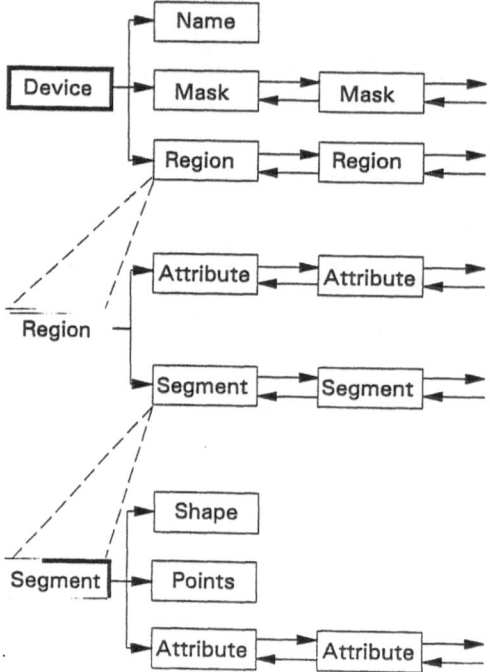

Figure 2: Device Representation

As overlapping regions are permitted a method of resolving conflicts is required i.e. when items overlap which object dominates. An item's position in the definition list is used to determine this, the region defined later is selected (the so called "painters algorithm").

Figure 3 shows a typical screen from the device editor. This consists of a central schematic image of the device with different regions (e.g. silicon, aluminium) denoted by colour (for purposes of this paper these have been converted to grey shades) with the junctions also delineated. Via the surrounding menus, various passive (e.g. zooming, panning) and active (e.g defining a new region) functions can be activated. More device specific aids are a probe that reports doping at a point in the device and the facility to view the doping profile along an arbitrary line defined by the user; in addition, a user can extract quantities such as sheet resistivities and gummel numbers from this line.

Direct manipulation is the main instrument of interaction. A user clicks on a region and it is highlighted and a panel activates giving information e.g. region type, attributes. The user can then, using the mouse, pick the object up and move it, stretch it and rotate it, essentially adopting a paradigm close to that used in object based drawing packages. Through the region panel attributes (e.g. doping profile) can be assigned to a region. The assignable attributes are stored in a ASCII template file allowing them to be updated without recompiling the editor. Given the rapid developments in physical models for semiconductors this is an important aspect that has been well used.

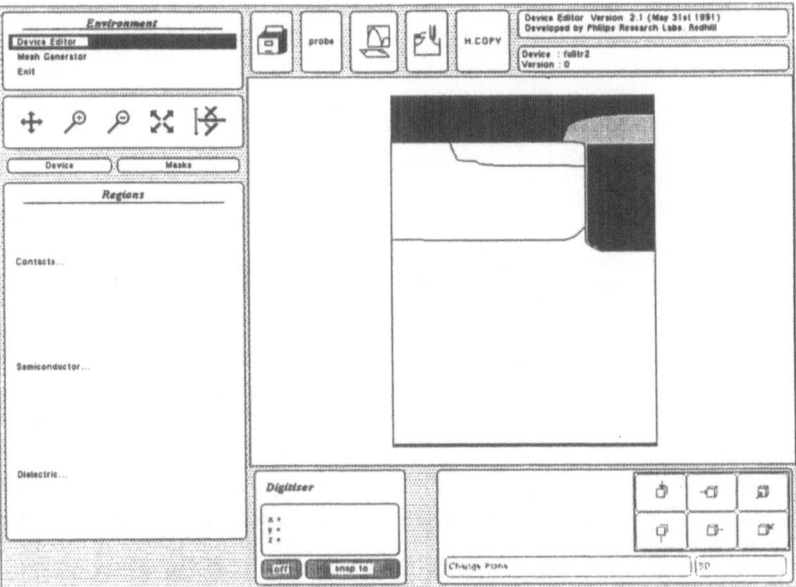

Figure 3: Device Editor

Probably the most important feature for the device designer is the facilities for defining and manipulating doping profiles. The simplest option is the definition of analytic forms (constant, Gaussian, complementary error function) these 1-D profiles can be combined with a mask to produce 2-D and 3-D distributions. A more powerful option, and of prime importance for integration, is the ability to import data from 1-D and 2-D process simulators. The 1D profiles can be rotated (to represent back diffusion), translated and combined with a mask and analytic function (to represent the sideways diffusion) to produce 2-D and 3-D profiles. 2-D profiles can also be imported. The high computational cost of such simulations often restrict the simulation region to a small critical area. Consequently the editor provides options for translation, rotation, reflection and extrapolation to place and orient a 2-D profile. A feature that has proved popular with users is the ability to combine a 2D profile with a mask. When activated the editor performs the copying, reflection and extrapolation automatically to form a complete profile. Figure 4 illustrates this facility.

The device shown in figure 3 is a TRENCHFET. This is a majority carrier power device with essentially a vertical gate formed by a trench. The gate oxide lining the trench and the polysilicon refill were imported from SUPREM-IV as was the arsenic and boron doping profiles. The slight upturn observed in the boron profile at the gate is due to Boron segregation during gate oxide growth an important effect that is strongly 2-D. The source and drain contacts were drawn using the editor.

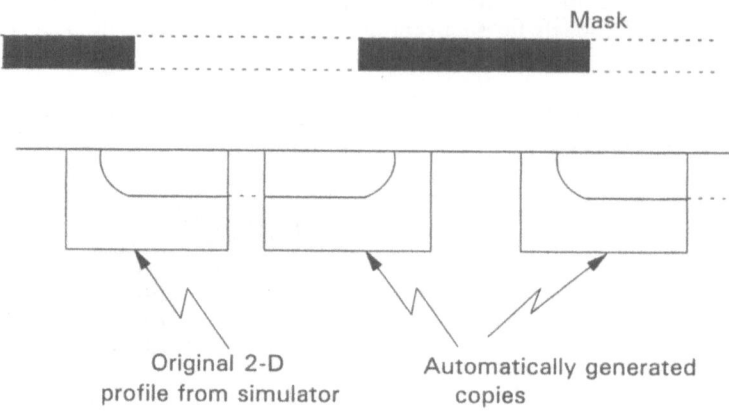

Figure 4: Coupling 2-D Profiles with a Mask

3-D devices can also be defined (for use with TRIPOS and TESSA) with the device editor. The 3-D shapes permitted are a rectangle or polygon extruded into an orthogonal plane. This is not as general as the 2-D case but complex shapes can be constructed by combining the elemental volumes. The alternative of developing a full solid modelling capability could not be justified. The definition and editing of the 3-D structure is done through orthogonal planes. Through these cross-sections the standard editing functions are available, though they now deal with volumes rather than areas. Figure 5 shows a box view of a 3-D structure (the corner of an IGBT cell) in the device editor.

2.2.2 Mesh Generator

The fully automatic generation of a simulation grids for general devices is not yet reality. However, certain guidelines or concepts of 'good practise' have emerged. Specifically [18] [19] [20]:

- Clustering meshlines in the regions where key quantities vary rapidly.

- Ensure that adjacent mesh spacings do not vary significantly in magnitude. That is, the mesh should be relatively smooth.

As there is no strong consensus on how to construct a good mesh for all simulations, the mesh generator provides the user with a range of options for constructing a rectangular grid. These vary from automatic to manual facilities. In addition, on entering the generator, meshlines that define the device geometry e.g. those that go along an interface or contact edge, are calculated and displayed thus eliminating the need for the user to complete this task. The automatic facilities allow the users to define a mesh according to the distribution of a variable (e.g. refine in regions where the doping profile or some other variable, such as electric field, vary rapidly) the system is described more fully in [1]. Semi-automatic features include adding a

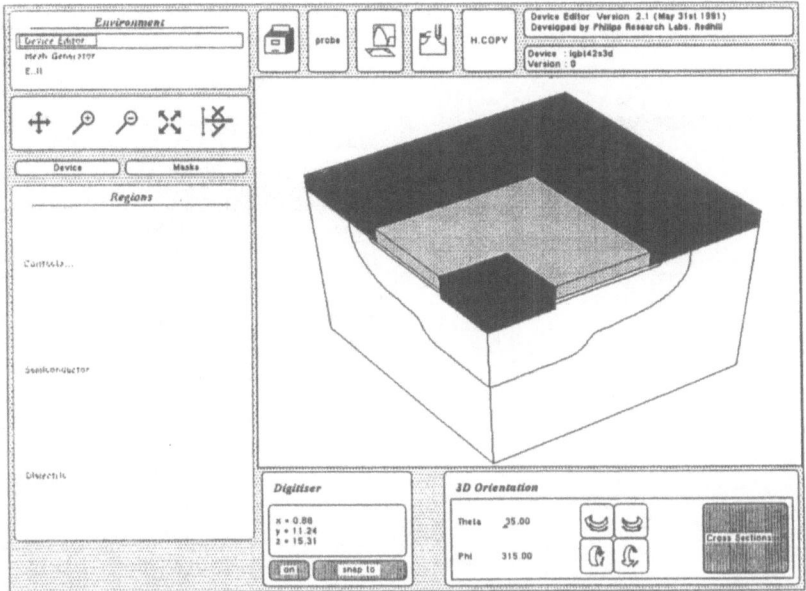

Figure 5: Box View of 3-D Device

uniform mesh, regionally biasing it, deleting multiple meshlines etc. The manual features are simply adding, deleting and moving individual meshline. All these features are executed using direct manipulation. Generally a user tends to use the automatic features to obtain a first level grid which is then refined using the semi-automatic and manual features. This is how the mesh shown in Figure 6 for the TRENCHFET was completed. In the same manner, 3-D rectangular grids can also be generated.

2.2.3 Graphical Viewer

The output from the various device simulators, the results of calculations, are entered into the database in the form of a standard results file. This file can represent n-dimensional scalar and vector data on a general grid and can be used to store data such as electron concentration, potential and doping profiles. The graphical viewer is a visualisation tool that can read the results file format and display the data in various forms. These forms are:

- Line and Colour Contour Plots.

- 3-D relief plots.

- Vector Plots.

- Box Plots for 3D device structures.

- 1-D Cross-Sectional Plots.

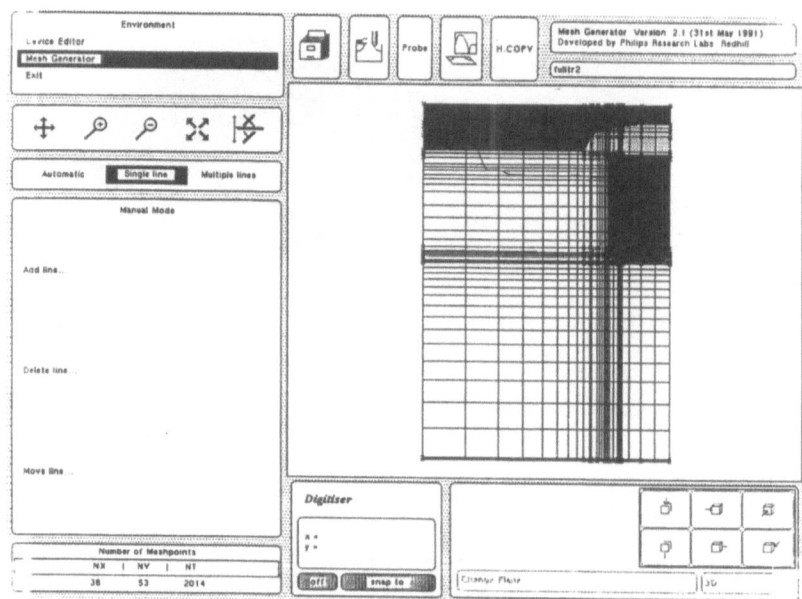

Figure 6: Mesh Generator

Such features are now standard in the TCAD field. The real benefit of the viewer is that output from all six integrated device simulators can be viewed in a single system.

A bitmap animator is also available for viewing transient results.

2.2.4 Translator

The translator is a rather prosaic but important piece of software. It converts the device and mesh files generated by IDDE into the input for the supported device simulators and is therefore a key component in providing an integrated system. As the underlying device representation is quite general the major effort when writing a translator, to a specific input format, is to account for any restrictions of the device simulator (e.g. HECTOR can only support rectangular contacts) and warn users if the translation is compromising to accommodate a simulator. Currently the translator supports the device simulators HECTOR, TRIPOS, PISCES, PADDY and CURRY. The TESSA simulator which was developed later reads the IDDE device and mesh formats directly and therefore requires no translation.

The effort required to integrate a new device simulator into the system depends on the programs capabilities. However, our experience indicates that 1 week's effort is typical.

2.2.5 Data Format

The device, mesh and data files that effectively form the database used within IDDE are all ASCII files. The format adopted for these files [9] is a keyword based LISP-like style similar to that originally proposed by Duvall [10]. The format has since been used in other semiconductor support tools [11]. A short extract from a device file is shown below:

```
( region
     ( type semiconductor )
          ( name "bulk silicon" )
               ( attributes
                    ( doping
                         ( profile 1d
                              ( dopant "phosphorus" )
                              ( filename "emitter.sup3" )...
```

Even though large files are created e.g. the data file containing doping and region information from a SUPREM-IV profile translated into the IDDE format, the time required to load the files has not been a problem and so binary files are presently not used.

3. Evaluation

3.1 Evaluation for Release

During the development of the IDDE system which began in August 1987 until beta release in August 1988 local evaluation was performed by members of the discrete device research group. Although the code was quite immature at this stage such was the disdain for text based input that there was no shortage of volunteers to try the evolving system. This effectively enabled us to perform rapid prototyping of the interface which, as a consequence of this feedback, changed considerably during the development period.

In August 1988 the beta release of the IDDE software was installed in a development site for evaluation. This evaluation took a number of forms:

- Observation of users completing a given task.

- Bi-weekly meetings for users to present feedback.

- The completion of a questionnaire by users at the end of a three month trial period.

It is important to realise that the goal of this evaluation was to assess the "usability" of the software which is distinct from software testing. In the latter the software is compared against the requirements specification. A piece of software may well meet it's requirements but this does not mean that it will be easy to use by an engineer, hence the need for a measure of usability [21]. As far as we are aware this was the first attempt to look at usability in the TCAD field.

In the initial study 4 novice users from the development site were observed, by an IDDE designer, drawing a two dimensional bipolar transistor, meshing this and translating this into input for a device simulator. The observers recorded at what points users had problems or required assistance. These points were raised together with other comments from the users, who in addition to running test problems were also running practical devices. Alterations and extensions to the interface were made as a result of these discussions. On average a new user required 40 minutes to complete the task of generating error free input for a device simulator. This compares with a time of 2 hours for an expert user to construct text based input. At the end of the trial period users, now much more familiar with the system, repeated the exercise and completed the process in 15 minutes. At the time we were concerned that such a small sample of users would not be adequate to assess the system, however, recent work in the field of usability engineering [22] [23] has indicated that such a sample size is reasonable.

The main findings of the 3 month study were:

- A significant increase in the efficiency of data preparation for device simulation. A novice user with IDDE could out perform an expert using a text based system.

- Fewer errors were made in defining a device when using the device editor in IDDE. This is attributed to the graphical feedback.

- The rate at which new users learned how to use the device simulators increased noticeably compared to those using text based formats.

- The ease of translation between packages encouraged users to simulate with packages they were previously not familiar with.

- The tighter integration of process and device simulators lead to a noticeable increase in the use of simulated process data in device simulation.

- A perceptible change in the attitude of development engineers towards device CAD. It is now perceived as something positive and a tool which they themselves can use.

3.2   What Have We Learnt in the Meantime?

In January 1989 the IDDE system was generally released within Philips. Since this inception its use has grown and it is now actively used in 3 research sites, 6 development sites and 1 university covering the full range semiconductor devices from power to IC's.

During the 4 years of use feedback has been gathered on the system. Encouragingly this information strongly supports the initial findings described in the previous section. Some other features that have emerged from extended use, these are:

- The use of multiple device simulators. Having access to a number of device simulators some users became aware that certain simulators were better at certain types of simulation. For example avalanche simulations with CURRY and electrothermal calculations with TESSA. Therefore users would switch simulators depending on the type of phenomenon they were investigating.

- The generation of device libraries. Most sites quickly realised that many simulations required a simple modification of an existing structure and therefore once drawn up these devices should be placed in a library that other users could access. This lead to a further decrease in preparation time for a device simulator.

- The device file became a means of remotely exchanging information on a structure even to engineers that were not performing simulations. That is, a device file could be sent to a remote party who could then examine profiles and dimensions in the device editor.

- As far as we are aware no user having used IDDE has returned to a conventional text method of input.

The complexity of simulations has, along with computing power, generally increased, certainly when compared to the type of simulation that were performed with IDDE when it was first introduced. We were therefore interested in how well moderate users of IDDE (those that use it in short bursts with extended breaks which is typical of development work usage) coped with these more complex types of structure, as the initial study did not address these. To examine this a set of users were were asked to generate a number of devices. The most complex structure was an IGBT which contained:

- Boron and Arsenic profiles from SUPREM-IV

- Polysilicon and oxide regions also simulated by SUPREM-IV

- 1-D Boron and Phosphorus profiles from SUPREM-III that model the out diffusion from the epitaxial layers.

The user had to translate and manipulate these items to generate the final IGBT cell structure. Devices of such complexity would not be tackled by development engineers without the use of a graphical device editor as the chance of introducing an error via a text based input is extremely high and potentially very costly if a number of simulations are performed before it is discovered. For the IGBT task the moderate users required between 50 minutes to 75 minutes to complete the task. The time for an expert user was 25 minutes. This result is encouraging as it indicates that even casual users (one of the groups particularly targeted by IDDE) can generate quite involved devices without a significant penalty in efficiency.

Along with positive feedback on the system there has of course been negative comments. In many respects these are the more interesting as they give direction to future work. The main criticisms of the IDDE system are:

- The system is incomplete. Users requiring extended features of a device simulator (e.g. a transient simulation) have to edit the the translated device file. This "stepping-out" of the environment is not liked by the user. What is required is a tool such as an analysis editor which would allow the user to graphically specify the type of analysis, voltage or current steps, external circuits etc. Output from this could then be translated, with the device information, into the input for a device simulator. Although the engineers also use the process simulators directly there was less strong feeling on providing a graphical front-end to these. This appears to stem from the fact that the input to a process simulator is similar to a standard process sheet.

- The system does not provide for multiple simulation runs.

- A more sophisticated database management system is required to help an engineer manage the many results that are generated from using multiple simulators.

- Recent users are more familiar with graphical software having been exposed to it on PC's and workstations. While this makes some aspects of IDDE easier, problems occur as it was not develop in accordance, as there was none, with a style guide. Therefore certain users would like to see aspects of the interactions more closely follow the current conventions e.g. a drop down menu bar along the top of an application.

- Engineers want support, preferable on-line, while they are defining a device as to the appropriate choice of physical models. For example users are often confused about when to apply various mobility models.

4.   Future Directions

From our experience with IDDE a number of options have been identified that attempt to address the perceived short-comings in the existing TCAD systems.

4.1   Open Framework

From an industrial perspective an open framework is highly desirable. The freedom to mix software from multiple vendors and to integrate internally generated software provides a means of defending investment in this area. Although it is currently possible to use vendor software in IDDE the responsibility for integration currently rest with ourselves. With the advent of an open system vendors will, provided there is sufficient market pull, be disposed to ensuring conformance of their software. The efforts of CFI are therefore to be supported even though an effective standard may only be available in the medium term.

4.2   Incorporating Domain Knowledge

To effectively use simulators in the TCAD field a user must have a good understanding of the physical models used and, importantly, their limitations. For example if you wish to simulate the gain of a bipolar transistor you have to include SRH and Auger recombination and bandgap narrowing. The parameters that are used for the bandgap narrowing depend on the type of mobility model that are used. Similarly in process simulation a transient enhanced diffusion model may need to be included in a simulation for a short low temperature diffusion. It is our experience that many new users have great difficulties with this and as a result execute many unphysical simulations. While they generally learn from this process it is costly in terms of computing and engineer time.

A possible solution to this problem is the incorporation in a TCAD system of some form of domain knowledge. A possible form for this could be an expert system that could comment on a the consistency of a device representation. Many of the rules are relatively simple e.g. when simulating a bipolar device included SRH and auger recombination, and so a practical knowledge base could be constructed.

4.3   Visual Programming

If a user wishes to depart from the standard path through a TCAD system e.g. to perform multiple coupled process and device simulations with a specific post-processing filter then the user generally has to describe the problem in a script language [25].

While such languages are suitable for experts they will generally not be used by engineers who are casual users of TCAD. A simpler programming paradigm is therefore required. Visual programming [26] offers such benefits. Research [30] has shown that visual languages can be highly effective in supporting non- or novice programmers in narrow, specialised domains. This matches closely the situation of engineers in the TCAD domain.

Practical examples of such systems in other fields are AVS [27] for visualisation and [28] for GIS. As visual programming follows a dataflow paradigm its application would be greatly facilitated by a standard representation for the components (e.g device, mesh, data) that will be manipulated.

4.4 Usability

Great strides have made in the field of software usability in the past decade (see [24] for a snap shot of this active field). However, much of this work has currently gone unexploited by the TCAD community. The two topics mentioned previously, domain knowledge and visual programming, are examples of aiding usability but I mention it separately as there are many other aspects that can be applied to this field. For example recent research [29] has indicated that it is possible to ascertain a users level of expertise by automatically monitoring how a user interacts with a system. Consequently, appropriate help, arrangement of menus can be adapted to better support the interaction.

It is important to realise that software usability stands in the critical path for efficient use of a TCAD system and therefore should be considered an important component of a TCAD framework. With few exceptions the developers of todays tools have largely based their user models on themselves assuming a great deal of domain knowledge. This poorly represents the practising device designer with the consequence that TCAD tools are still not fully utilised.

5.   Conclusions

Reviewing the 4 years of IDDE use it is clear that the project achieved its original goal of providing more efficient assess, and use, of process and device simulators. Further, it has played a key role in making TCAD accessible to engineers and for facilitating the assimilation of such tools into their device design methodology. In a wider context, despite it's shortcomings, it has shown that a TCAD systems is much more than an abstract representation of components and a database. Support tools and in particular *usability* stand in the critical path to an effective TCAD environment. The time has come to seriously consider the engineer not just the physics and the algorithms.

6.   Acknowledgements

The author would like to thank the members of the Hazel Grove development team and colleagues at PRL for evaluating and providing valuable feedback on IDDE. In addition I would like to acknowledge and thank M.Driessen and A.Heringa who ported IDDE to X-Windows.

References

[1] P. A. Gough, M. K. Johnson, P. Walker, H. Hermans, *An Integrated Device Design Environment for Semiconductors*, IEEE Trans. Computer-Aided Design, Vol.10, pp.808-821, June, 1991.

[2] S. E. Hansen, *SUPREM III User's Manual*, Stanford Univ., 1985.

[3] M. E. Law, C. S. Rafferty, R. W. Dutton, *New N-Well fabrication techniques based on 2-D process simulation*, IEEE Trans. Electron Devices, Vol.27, pp.518-521, Dec. 1986.

[4] J. Lorenz, el al., *COMPOSITE - A complete modelling program of silicon technology*, IEEE Trans. Computer-Aided Design, Vol 4, Oct. 1985.

[5] S. Polak, et al., *The CURRY algorithm*, Simulation of Semiconductor Devices and Processes, Pineridge Press, Vol.2, p.131, 1986.

[6] K. R. Whight, et al. *Computer aided design of multiple guard ring systems for the passivation of high voltage planar semiconductor junctions*, Proc. NASECODE III, Boole Press, 1983.

[7] N. T. Gladd, *Object-oriented interface system for particle-in-cell simulations*, IEEE Trans.Electron Devices, Vol.35, Nov. 1988.

[8] B. A. Myers, *User-Interface tools: Introduction and survey*, IEEE Software, Jan., 1989.

[9] P. A. Gough, *Device, Mesh, and Results file formats within IDDE - A proposed standard*, PRL Tech.Note. No.2801, April 1989.

[10] S. G. Duvall, *An interchange format for process and device simulators*, IEEE Trans. Computer-Aided Design, Vol.7, July, 1988.

[11] M. R. Simpson, *PRIDE: An Integrated design environment for semiconductor device simulation*, IEEE Tran. Computer-Aided Design, pp.1163-1174, Sept. 1991.

[12] B. Shneiderman, *Direct manipulation: a step beyond programming languages*, IEEE Computers, pp.57-69, August, 1983.

[13] I. Benbasat, P. Todd, *An experimental investigation of interface design alternatives: icon vs. text and direct manipulation vs. menus* Int. J. Man-Machine Studies, Vol.38, pp.369-402, 1993.

[14] P. A. Gough et al., *Fast switching lateral insulated gate transistor*, IEDM Tech. Dig., pp..218-221, 1986.

[15] M. R. Pinto, C. S. Rafferty, R. W. Dutton, "PISCES II: Poisson and continuity equation solver", Stanford Electron.Lab., Tech.Rep., Sept. 1984.

[16] E. Roks, et.al. *A bipolar floating base detector (FBD) for CCD image sensors*, Proc. IEDM, pp.109-112, 1992.

[17] P. A. Gough, P. Walker, K. R. Whight, *Electrothermal Simulation of Power Semiconductor Devices*, Proceedings of ISPSD'91, pp.89-94, 1991.

[18] P. A. Markowich, *The Stationary Semiconductor Device Equations*, Springer-Verlag, 1986.

[19] P. Pichler et.al., *Simulation of Critical IC Fabrication Steps*, Vol.32, Oct. 1985.

[20] M. Smooke et.al., *Two-Dimensional fully adaptive solutions of Solid-State Alloying Reactions*, J. Comp. Physics, Vol.62, pp.1-25, 1986.

[21] J. Nielsen, *The usability engineering lifecycle*, IEEE Computer, Vol.25, pp.12-22, Mar. 1992.

[22] J. Nielsen, *Usability engineering at a discount*, In *Designing and Using Human-Computer Interfaces and Knowledge Based Systems*, G. Salvendy, M. J. Smith (Eds.), Elsevier Science Publishers, Amsterdam, pp.394-401,1989.

[23] J. Nielsen, *Big paybacks from discount usability engineering*, IEEE Software, Vol.7, pp.107-108, May 1990.

[24] *Human Factors in Computing Systems*, Proceedings of INTERCHI'93, Addison Wesley, 1993.

[25] F.Fasching, et.al. *A New Open Technology CAD System*, ESSDERC'91, Elsevier, pp.217-220, 1991.

[26] S. K. Chang *Visual Languages: a Tutorial and Survey*, IEEE Software, pp.29-39, Jan. 1987.

[27] C. T.Upson, et.al. *The Application Visualization System: A Computational Environment for Scientific Visualisation*, IEEE Computer Graphics and Applications, pp.30-42, July 1989.

[28] F. Paterno, et.al. *The Design and Specification of a Visual Language: an Example for Geographic Information Systems Applications*, Proceedings of Eurographics UK 1993, March 1993.

[29] K. P. Vaubel, C. F. Gettys, *Infering User Expertise for Adaptive Interfaces*, Human Computer Interaction, Vol.5, pp.95-117, 1990.

[30] D. D. Hils, *Visual Languages and Computing Survey: Data Flow Visual Programming Languages*, Jour. Visual Lang. and Computing, Vol.3, pp.69-101, 1992.

The SATURN Technology CAD System

W. Jacobs

SIEMENS AG, Corporate Research and Development,
Otto-Hahn-Ring 6, D–81739 München, GERMANY

Abstract

This paper describes the SATURN TCAD system developed at SIEMENS. The architecture of the TCAD system and the philosophy behind it is explained and demonstrated by examples. An overview of the available simulation tools is given.

1. Introduction

In microelectronics, a TCAD system is designed to support the design and manufacturing of integrated circuits from a technological and physical point of view. The physical processes of the different manufacturing steps are simulated with the help of physical models. The geometrical structure and the distribution of dopants within the devices are obtained as result. To investigate the static and dynamic behavior of a device towards different circuit environments and bias conditions, the semiconductor device equations are solved. Thus, the device characteristics are obtained which allow parameters for compact circuit models to be extracted. These parameters serve as technology-specific input for circuit simulation and allow circuits to be designed at a very early stage without requiring measured data to exist.

The SATURN TCAD environment is used in process and device development. The design process is usually performed in a closed loop in which the device simulation serves to characterize the process flow and the layout geometry. The process flow and the geometry are repeatedly altered till the device reaches its optimum with respect to specified objectives. In addition, this optimum should be as insensitive as possible to process and layout variations.

In connection with the MEGA project the SATURN project was started in 1986. It aimed at developing a system that fulfilled the requirements mentioned above within a foreseeable period of time. No such system was available either commercially or from universities at that time. It was to be used for technological applications as well as in research and development. New programs were to be integrated in the whole system quickly in order to investigate new model approaches.

At the time, the hardware platform was determined by mainframes running the proprietary SIEMENS operating system BS2000. To automate simulation tasks, all programs were to be controlled by application-specific procedures. In order to facilitate migration of the SATURN system to UNIX workstations, which were steadily increasing in performance, a command-line interpreter [19] was developed in BS2000 the command structure and run control of which were modeled on the UNIX C shell. A menu-driven user interface was neither envisaged nor wanted. After migration to UNIX workstations (WS30, HP9000), all SATURN functions were transformed into stand-alone programs that run under direct control of the Berkley C shell.

In fact, a number of SATURN-integrated simulators had already existed as stand-alone programs, which had been run in individual environments. These were to remain stand-alone program modules and to be developed as such for obvious reasons. The SATURN system was to adapt to the simulators, not vice versa. For the purpose of integrating the simulators in the whole system, clearly defined interfaces for attachment to the run control (shell) and the SATURN file system were introduced. On the one hand, these interfaces helped to ensure data consistency, on the other, they facilitated migration between different hardware and software environments. Apart from that, the individual simulation programs are included in the traditional way: they obtain information from input or control files and write the results into corresponding output files.

The SATURN TCAD system was developed to perform so-called simulation experiments rather easily. An experiment is controlled by design parameters. The result of an experiment is described by response parameters, which are automatically related to the corresponding design parameters. All parameters are available in a database as well as on the shell level.

The choice of design parameters is free and depends on the application. If, for example, the process simulation is to be calibrated with measured profiles, the parameters chosen correspond to the model parameters of the underlying physical process. To optimize the process flow and the layout of the device, process and layout-specific quantities, such as furnace temperature and spacer width, are chosen as parameters.

Section 2 summarizes the simulation programs which are the building blocks of the TCAD system. Special emphasis is laid on those programs developed or further developed at SIEMENS.

Section 3 explains the system architecture of SATURN. The application- and simulator-independent approach to keep track of simulation variants across an arbitrary program flow as well as the method of tool integration are described more detailed.

Section 4 gives two application examples to illustrate the main features of SATURN.

2.   Building Blocks

Fig. 1 gives a schematic overview of the SATURN system and its installed programs, which make up the building blocks. Process and device simulation levels are completely integrated. The circuit simulator TITAN [9] is connected via the MOS parameter extractor JANUS. The postprocessor SATGRAF is used to create various graphical representations of virtually all interchange data like distribution of dopants, potential, densities or currents available throughout the system.

Depending on the problems being investigated, three different design threads are available.

The first one handles all cases, where planar structures are sufficient. It is used for characterizing and optimizing active MOS transistors. It consists of the process simulators MIMAS I and POSEIDON, the "profile manager" PHOEBE, which composes a complete MOSFET doping structure from source/drain and channel profiles, and the device simulator MINIMOS 4.

The second design thread is used for general devices and non-planar MOSFET structures. It comprises the general purpose process simulator MIMAS II [6], developed at SIEMENS, and the device simulation programs MINIMOS 5 and GALENE. SATURN also provides interfaces to accept data from other process simulators (e.g. SUPREM IV).

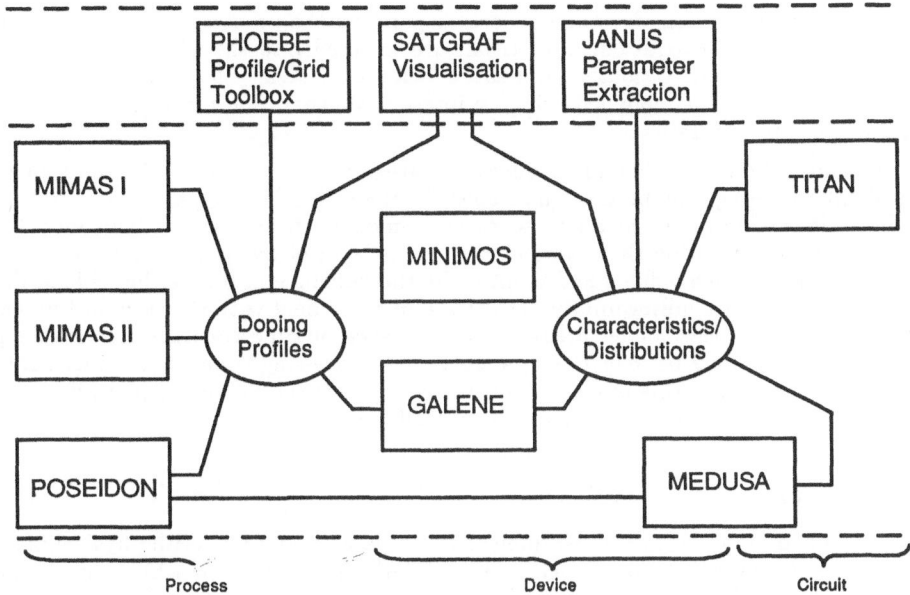

Figure 1: Building Blocks of the SATURN TCAD system

The third design thread is mainly used for bipolar process and device development and contains all programs for characterizing and optimizing one-dimensional doping profiles. The characterization is performed by the mixed-mode simulator MEDUSA [17]. Several one-dimensional profiles can be connected by suitable transportances to take into account the lateral voltage drop due to majority carrier flow.

2.1 POSEIDON

POSEIDON is a process simulator for one-dimensional problems. It is compatible to SUPREM III on the input deck level, but comprises advanced models for oxidation, diffusion als well as improved algorithms (eg adaptive mesh). At the same time it is capable of simulating complete process flows. An overview of these models is given in this section.

Three hierarchical levels of equations for describing dopant diffusion can be selected from the input deck level. The lowest level uses effective dopant diffusivities in equivalence to the models used in the well-known SUPREM III process simulator [1]. The second level includes the simulation of point defect (silicon interstitial and vacancies) distributions, where dopant diffusivity is calculated from the local point defect supersaturation (Hu's model [2]). On the most complex level, the coupling between point defect and dopant diffusion is included, taking into account the pairing between interstitials or vacancies and the dopant species (pair diffusion model [3]).

The initial distribution for point defects can be chosen from being the distribution of the previous processing step, the equilibrium concentration or an externally generated profile. The point defect distribution resulting from damage created during dopant implantation can be calculated from analytical expressions which have been deduced from Monte Carlo calculations [4].

The boundary conditions for point defects depend on the process conditions. When the silicon surface is oxidized, interstitials are injected into the silicon which gives rise to the well-known Oxidation Enhanced Diffusion (OED) effect for B, P and As dopants. This effect is modelled using an interstitial generation term which depends on the oxide growth rate.

For dopants, the crucial boundary effect is segregation at interfaces. In addition to the standard segregation model, which specifies the ratio of the dopant concentration on both sides of an interface, a new segregation model is implemented. This model is based on the assumption that there exists a monolayer of trapping sites for dopants at the interface which is filled and emptied by the dopant flux from either side of the interface. While at equilibrium the results are in agreement with the standard model mentioned above, this model predicts the existence of time constants for reaching equilibrium that may be longer than some of the processing times in a modern sub-μm technology. Also, this model satisfactorily explains the pile-up of phosphorus at the interface, as has been verified experimentally [5].

2.2 MIMAS

MIMAS I and MIMAS II are two-dimensional process simulators. With physical models at a level comparable to SUPREM III, MIMAS I is designed to achieve high computational efficiency for planar structures, featuring a fully adaptive finite-difference grid and fast numerical solving schemes, suited also for vector processors. This program is used for the calculation of MOSFET structures in cases where non-planarity can be neglected (e.g. for the calculation of basic device parameters like V_{th} or I_D^{sat}).

MIMAS II is a general purpose non-planar process simulator. The finite element method is used for describing the mechanical properties and oxidant diffusion during oxidation processes as well as the point defect and dopant diffusion during furnace processes. New approaches are used to overcome the problems which come along with the moving boundary between silicon and silicondioxide [7, 8].

To describe dopant diffusion, most of the models described for POSEIDON are available. The fully coupled pair diffusion model is being implemented. Deposition and etch processes are simulated by a string algorithm similar to the SAMPLE program [16].

2.3 MINIMOS

For MOSFET structures, the MINIMOS program [13] is the most efficient and proven device simulator. It can be used for planar and non-planar structures in two or three dimensions (including some of the MOSFET-like isolation devices). It has been demonstrated that the enhanced drift-diffusion description of carrier transport in MINIMOS is sufficient to calculate the I-V characteristics of MOSFETs with channel length down to $0.1\mu m$ [14]. MINIMOS is being enhanced by models for the injection of carriers into the gate oxide and trapping of these carriers in the oxide [15]. This enables us to calculate self-consistently the degradation of MOSFETs due to hot carriers.

2.4 GALENE

Special device structures are simulated with the general purpose simulator GALENE [18]. Arbitrarily shaped devices can be investigated. A band-to-band tunneling model

[11] is implemented which facilitates the simulation of degradation in bipolar devices [12]. Since the solution algorithm of GALENE can be customized from the input deck level, devices like SOI structures that normally face severe convergence problems with other simulators can be tackled.

3.   System Architecture

3.1   Overview

Figure 2 shows the SATURN system architecture. The system consists of tool layer, data layer and framework layer. These layers are available for all applications.

The hierarchical database contains all design and response parameters of a SATURN application. Its structure is an exact copy of the program flow of a design thread across several program levels as well as all stored variants. The application thread is not restricted to a sequence of program calls, but can be branched at almost any point. The system assigns a table (relation) to each program step and creates a table entry for each calculated variant. Each program step is supplied with its own, user-defined control file. In addition to the usual program-specific directives, the control file contains the input parameters serving to characterize the simulation variants. The corresponding table contains the parameter values of all simulation variants, which differ in at least one parameter value. Characteristic results of a simulation process are also available as response parameters in table form. They can be easily related to design parameters either in table-oriented or graphical form.

All data exchanged between the individual simulation programs are stored in UNIX files in a compact form. Due to strict naming conventions, each file within the system corresponds unambigiously to one entry in the hierarchical database.

All simulators and utilities operating on interchange files are connected to the system only via UNIX files to make them largely independent of system-specific interface requirements and to integrate simulators without having to change the code.

The framework level performs all operations in connection with the generation of input files and the extraction of output parameters. In addition, it allows checking the consistency between input and output tables across all simulation levels. Each access to hierarchically connected tables checks and inforces the consistency of the database. All framework functions are stand-alone programs. They can be called interactively as well as from within a shell script and are thus available to all applications.

As mentioned above, SATURN was developed with the intention of allowing simulation experiments to be performed. They can be related directly to real experiments and measurements. For instance, the splitting of a lot at a specific point within a process flow corresponds to a branching point in the simulation design thread. From here, two or more simulation processes must be performed in parallel. If splittings are performed at several points, there are possibly a lot of parallel simulation branches to be calculated. The result is a tree structure with its root at the top (fig. 3). This tree structure can be mapped directly to a table where the split parameters appear on the left side and the finally determined characteristic response quantities on the right (tab. 1).

The idea behind the concept followed by SATURN is to store such a variant tree consistently in order to be capable of extracting the dependence of the response parameters on the "experiment" parameters at any time and even across several simulation levels. The results on one simulation level can be compared with each other without any restrictions. Design and response parameters are available in the

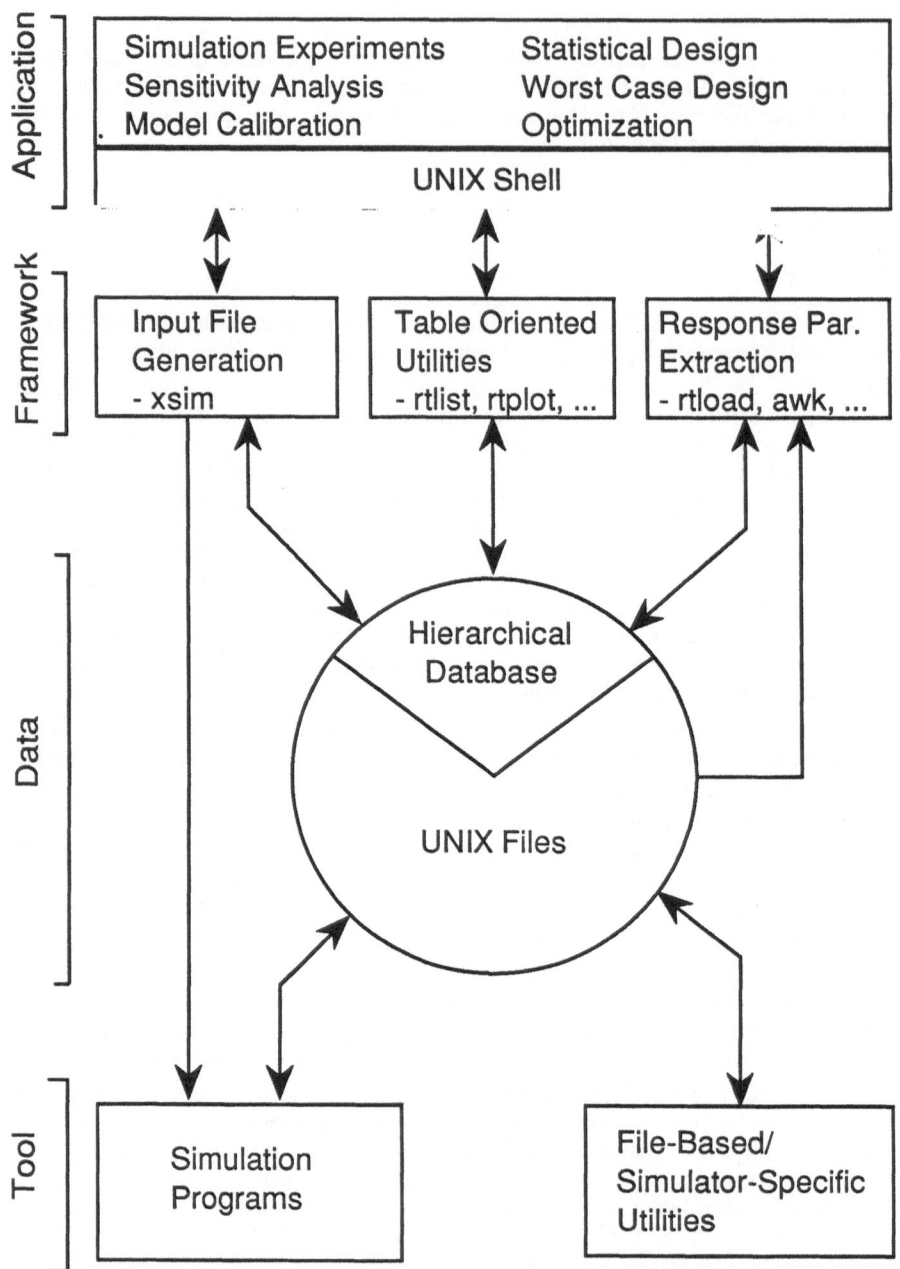

Figure 2: SATURN System Architecture

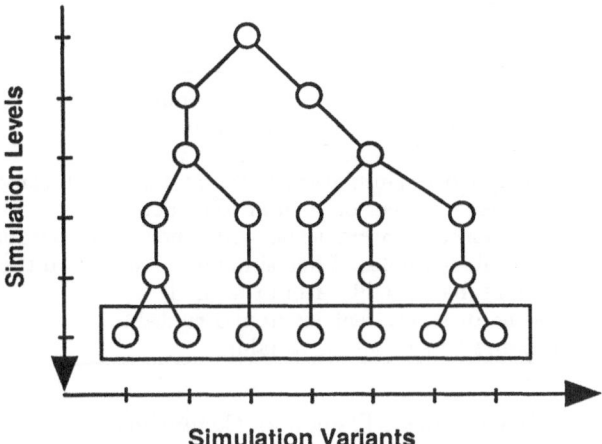

Simulation Variants

Figure 3: Multi-Level Variant Tree

Var	Dose	L_{gate}	U_{bulk}	V_{thr}
1	2.5e13	0.7	-1	0.98
2	2.5e13	0.7	0	0.71
3	2.5e13	0.9	-1	0.79
4	3.5e13	0.7	-1	0.83
5	3.5e13	0.9	-1	0.96
6	3.5e13	1.2	-1	1.05
7	3.5e13	1.2	0	0.87

Table 1: Table Representation of a Multi Level Variant Tree

form of one single table which can be further processed by SATURN utilities and external programs.

3.2 User Interface

The UNIX shell as well as the SATURN framework commands make up the user interface of the SATURN system. The shell programming language is used for supervising and controlling simulation programs as well as for automating simulation tasks. The values of input parameters of a design thread are available as shell variables and can be easily changed, which allows the whole design thread being rerun with different input parameters. The response parameters are extracted and are also made available on the shell level. Therefore simple control tasks like search of a minimum or maximum of a response value are easily handled by using standard UNIX tools like awk.

More complicated applications, like multi-dimensional optimization, require dedicated programs, which are also activated on the shell level. They use the shell as "simulation engine" to obtain the results belonging to a specific set of parameters. In addition, they can make use of the internal database to handle several runs of a design thread.

Command	Description
xsim	Determine simulation variant, create input file and call simulator
rtinit	Define prototype file and initialize local runtable
rtmod	Modify prototype file and update local runtable
rtjoin	Join runtables over simulation hierarchy and check consistency
rtlist	Filter and list variables of local or joined runtables
rtload	Extract response parameter and write them to a runtable
rtset	Copy runtable variables of one simulation variant to the shell
rtunset	Remove runtable variables from the shell
rtrm	Remove one or several entries from a runtable
rtplot	Plot response versus design variables

Table 2: Basic Framework Commands

The commands of the SATURN framework are listed in table 2.

3.3   Input File Generation

An important system feature of the framework is the generation of the simulator's input file, which contains all neccessary information for one program or simulation step within a given design thread.

Input file generation is independent of the attached simulator. SATURN requires only that one step is supplied with one input file and does not care about what this file is used for.

The input file is derived from a "prototype" or template file, which has to be created by the user. Usually, libraries of prototype files exist for each process or product which the user can choose from. A prototype file may contain all kinds of application-specific variables and makes use of macro facilities, which allow parts of the input file to be included or excluded depending on design variables. Local variables can be derived from design parameters by arithmetic expressions [20].

A variable part within a prototype file is marked in the following way:

```
    . . .
    diffusion temp=%{temp1-900:d5}
    . . .
```

A variable is defined by its name, a default value as well as the format of the character representation (cf. below).

The values of variables are set on the shell level as follows:

```
    set temp1 = 920
```

The prototye file needs only to be edited if new variables are introduced. The SATURN command **xsim** which activates the individual simulators evaluates the variables and replaces them with their corresponding values. If a variable value is not defined as a shell variable, the default value is used. Variables can be replaced by numerical values, by character strings or by names of files that contain interchange data of previous simulation steps.

Previously constant parts of a control file can be replaced by new variables at any time, and thus be changed from outside. The previous constant value must be specified as default value of the new variable to update the existing runtable entries.

```
before:  implant boron energy=25
after:   implant boron energy=%{energy-25:d3}
```

Variants which were previously calculated with an energy of 25 keV are extended by a new variable with a default value of "25". The **xsim** command recognizes, whether the specified prototype file is modified, and internally calls the **rtmod** command to update the runtable.

The control program converts the value of each variable with a user-defined format into an ASCII character string. This is required, for example, to recognise 25.00 and 2.5e+1 as identical. By specifying the field length, the user decides how many significant digits are stored.

The variable configuration is a character string consisting of the character representations of all parameters. To determine the simulation variant, the variable configuration is compared with all previous configurations (variants) stored in the runtable. If a configuration has already been stored, its variant number is taken over unless the user explicitly specifies another number. The built-in **make** mechanism checks whether the corresponding result file is newer than all input interchange files of this program step. If this is the case, the result of this simulation step is regarded as consistent and the simulation program need not be called any more. The existing result is used as input for the next simulation steps.

If one input file is newer than the result file, **xsim** takes the prototype file and replaces all its variables by their current values, creates the control file and, finally, starts the simulation program.

3.4 Connecting Runtables Across a Design Thread

The hierarchical database allows the parameter dependences to be traced across all simulation steps required. This is achieved by a special type of link variable, which has to be defined inside of a prototype file exactly in the same way as the design parameters described in the previous section. The linkage, once established, is static and cannot be changed during the design process.

The name of this reference variable reflects the name of the previous simulation step, which supplies an interchange file:

```
MOS CHANNEL=%{chan_pos2:R10}
```

Its value is simply the file name, which consists of the variant number and the file type:

```
set chan_pos2 = n005.pa
```

The example given above references the fifth simulation variant of the prototype file **chan** for simulator POSEIDON (**pos2**). The file type of the interchange file is marked by file extension **pa**. This kind of shell variables are automatically created when a simulation step is finished. Since the next simulation step uses the same naming convention for accessing its input files, the file names of interchange data are passed between program steps automatically.

3.5   Response Parameter Extraction

Response parameters like device characteristics are easily extracted from ASCII files using the **rtload** command. The extraction process may be applied to a set of simulation variants. Therefore, the results of several simulation variants of a design thread are collected in one single table. In addition, other quantities can be derived directly from the extracted variables.

Keeping the output table separate from the corresponding input table enables **rtload** to extract more than one block of response data per simulation variant. This is important for device characteristics data, because most device simulators work more efficiently when several bias points are calculated in a sequence.

The output table is linked with all input tables via the reference variables described above. Thus, it contains the complete amount of information required by an application.

The response parameters are extracted with the help of **awk** procedures. In cases where the number of response parameters is fixed, these procedures can be created by a program generator from marked ASCII files. This allows extraction of quantities which are not taken into account in the predefined procedures.

3.6   Interchange Data

Most of the integration effort was spent on allowing the simulation program to access interchange data supplied and expected by SATURN. Due to the development in this field, a clear standard was not available at that time (1987). Therefore a common format for all interchange data within the process and device simulation level of the planar design thread had to be introduced using a reasonable approach.

An interchange data file contains a sequence of several blocks, each of which consists of a key and a data part. The key part uniquely determines the type of the data object that follows. An application program is not forced to read the file sequentially but can read a specific block by specifying the corresponding keys. This behaviour allows new blocks to be added to a file without changing programs that do not expect the new data block. In addition, compatibility to existing programs as well as program maintenance are improved.

This common format enables the graphical postprocessor to access all available distribution data in a fairly flexible way.

However, this format is limited to finite-difference based distributions. No effort was spent to extend it to finite-element data structures, required by non-planar process simulation programs. Instead specialized conversion programs are provided for interfacing them to MINIMOS and GALENE.

3.7   Integration of Simulation Programs

As can be seen from fig. 2, simulation programs are coupled to the SATURN framework by UNIX files only. They read a single input or control file and create one or more output files. One of them contains interchange data processed by the next step of a design thread.

The framework command **xsim** forms the interface between the application level (shell) and a specific simulator. In addition to creating the simulator's input file, described above, **xsim** starts the simulation step. In this context, a simulation step can be made of either a single executable program or a shell script, which in turn calls a sequence of executable programs. In all cases, xsim supplies one single control file, which has to describe the simulation step entirely.

Interchange data, required as input, must be specified within the control file by the name of the corresponding SATURN file.

In general, every step is supplied with two different kinds of file-based data. The first type comprises installation files, like default model parameter files or syntax definition files. They are not changed between successive program calls and therefore stored in the same installation directory as the executable file. The second type contains all files which are specific for one simulation variant. The input file, generated by the system, as well as all output files, generated by the simulator itself, belong to this "run-specific" type. These file names have a common prefix, which links them to the corresponding entry within the hierarchical database.

xsim passes installation pathname and prefix as argument to the simulator call. They are processed by a special interface routine, which can be applied when source code is available. OPEN and CLOSE calls are replaced by subroutine calls, which expand a specified file type to the full file name either installation- or run-specific. Interchange input files or installation files are opened in readonly mode. Therefore simulation variants derived from the same prototype file are able to run simultaneously, because their output file names are unique. This feature has been especially useful for operating systems, which do not offer a hierarchical file system.

Without source code the simulation program is "wrapped" in a shell script, which intercepts the system parameters and takes over the task to connect the system files to the files required by the specific simulator. All system parameters are also available as special shell variables, created by xsim. They are marked as "hidden" variables and are not part of the variable configuration. They are used to insert SATURN filenames within prototype files. In addition, the shell script may call special programs for converting interchange data between different formats.

4.   Demonstration Examples

4.1   Example 1

A simple example is chosen in order to demonstrate the basic framework functions. The shell script given in fig. 4 facilitates the calculation of the threshold voltage of a MOS transistor for different channel length and bulk voltages. In this case, the channel implant dose is the design parameter and the threshold voltage U_thr is the response parameter. Channel length lgate and bulk voltage ubulk define a two dimensional space, in which the response parameter is calculated.

The shell script representing the application thread contains three simulator calls and is invoked as command VtvsL. It prompts the user to enter the dose of the channel implant. Afterwards, the command xpos2 reads the prototype file chan.pos2, specified by the -p option, replaces the variable dose by its current value and, finally, calls POSEIDON, which calculates the channel doping profile. PHOEBE (xpho1) creates the complete MOS structure as superposition of the channel profile and the source/drain profile calculated by MIMAS I. Finally, MINIMOS calculates the threshold voltage for each specified bulk voltage and channel length.

The rtload command extracts the response parameter U_thr from all eighteen MINIMOS simulation variants created by this script and writes them to the output table log1, specified by the -u switch. Option -C- makes rtload determine the different values of the parameter ubulk specified by the environment variable BY_VARIABLES.

The list of bulk voltages is available as shell variable _ubulk and is evaluated by the rtplot command later on. In addition, the references to input tables are resolved recursively. The table log1.j is created, which contains the design and response parameters of this application.

The rtunset command removes all input parameters of the whole application thread from the shell (option -j specifies the joined runtable of all simulation steps). Afterwards, variable dose is set again to its user-supplied value serving as filter value for the succeeding rtplot command.

Rtplot reads the table log1.j and plots variable U_thr versus lgate (options -X and -Y). For every bulk voltage, the (lgate,U_thr) points are sorted in ascending order with respect to lgate (-S a). The window boundaries are adjusted automatically (-W). VtvL and OUTPUT are files, which contain title and axis labels as well as directives to replot the whole picture in POSTSCRIPT format.

Fig. 5 shows the graphical representation created by the rtplot command. The design parameter dose appears on the lower left corner of the figure and thus connects the electrical characteristic with the process level.

4.2   Example 2

A one parameter optimization is given as second example.

The shell script listed in fig. 6 determines the gate voltage of maximum bulk current. The well-known golden section algorithm is used. The script is called as GetMaxIbulk <pdose> <fname>. MIMAS I calculates the doping profile for the specified LDD-implant dose pdose. The shell variable vector contains a maximum of four gate voltages and bulk currents which are required to determine the next gate voltage. Two gate voltages are supplied as initial values. All other elements of vector are marked as undefined. The command GetNextStep checks whether enough bulk currents are given, determines the gate voltage for the next MINIMOS call and returns its index within variable vector. The while loop is terminated if the relative change in bulk current is smaller than the specified accuracy (5%). The results are written to the file fname. The functions GetNextStep, GetIbulk and GetIdrain are implemented as awk functions.

This script is used inside an optimization script which minimizes the ratio of the bulk and drain current of a $0.6 \mu m$ MOSFET by changing the LDD-implant dose. The same method is used for finding the minimum. The MINIMOS call (xmin) is replaced by the script call. The result is shown in fig. 7.

```
#
 echo -n "Enter Channel Implant Dose "
 set dose   = $<
 set save   = $dose
#
 xpos2 -p chan
 foreach lgate ( 0.5 0.6 0.7 0.9 1.2 2.5 )
    xpho1
    foreach ubulk ( -3.0 -1.5 0. )
       xmin4
    end
 end
#
 setenv BY_VARIABLES 'ubulk: U_bulk=%g'
 rtload  -u log1 -X log -C-
#
 rtunset -j
 set dose = $save
 rtplot -u log1.j -X lgate -Y U_thr -WS a VtvL OUTPUT
```

Figure 4: Script to Calculate Threshold Voltage

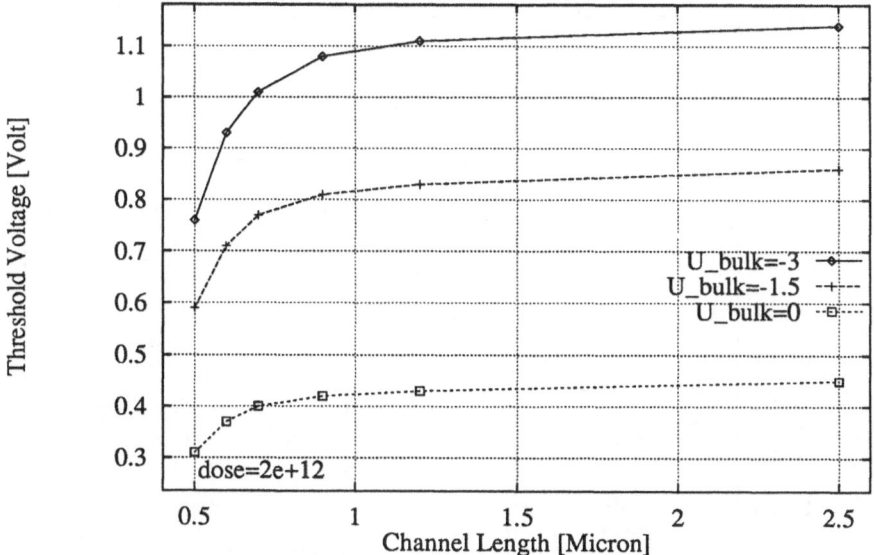

Figure 5: Threshold Voltage for Different Channel Lengths and Bulk Voltages

```
#                                    create doping profile
 set pdose = $1
 xmim1 -p ldd                        # calc. source profile
 xpho1 -p chan                       # get MOSFET structure
#                                    init. U_gate+I_bulk
 set vector = ( 2.5 4.0 u u u u u u )
 set idrain = ( 0 0 0 0 )
#
 @ maxiter = 10                      # limit iteration number
#
 while ( $maxiter>0 )
   @ maxiter--
   set vector = ( 'GetNextStep $vector[1-8] 0.05 1.0 5.0 0.2' )
   set i      = $vector[9]           # index of next U_gate
   @ k = $i + 4                      # index of corr. I_bulk
   if ( $vector[10] == 1 ) break     # converged ...
   set ugate  = $vector[$i]          # set gate voltage
   xmin4 -p ibulk -X log             # calc. I_bulk, I_drain
   set vector[$k] = 'GetIbulk'
   set idrain[$i] = 'GetIdrain'
 end
#
 echo "U_gate $vector[$i] : $idrain[$i] $vector[$k]" > $2
```

Figure 6: Script to Obtain the Maximum Bulk Current

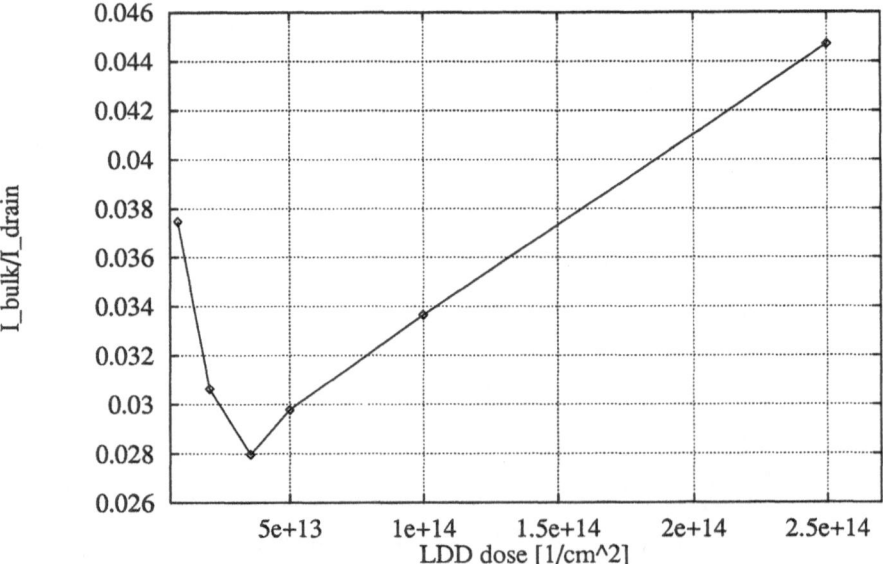

Figure 7: Minimum I_b/I_d as Function of LDD-Implant Dose

5. Conclusion

The approach of the SATURN TCAD system presented in this paper allows process and device engineers to solve their optimization problems in a thorough and systematic way. They are able to concentrate on physical and technological aspects. All tasks concerning the consistent book-keeping of simulation results like naming files and tracing parameter dependences across an arbitrary flow of programs are managed by the SATURN TCAD system itself. After successful calibration of the physical models, the user is able to run simulation variants without changing the procedure.

Experts can predefine a complete design thread, bring in their expert knowledge to write prototype files and hand over the whole project to technical assistants, who are able to do the work without knowledge of internal functions of the tools.

The hierarchical database allows simulations to be done in an incremental way. New simulation variants can be easily added at virtually every point of the variant tree. A design thread can be interrupted at any time and restarted with the same command. The design thread automatically starts from the last result available. This is especially important for application tools which are coded as separate programs. They can use the TCAD shell as simulation engine, which is controlled by design parameters only. They need not know how the design thread is structured and which design parameter is passed to which simulation step. Nevertheless, no computer time is wasted because SATURN recognizes already calculated simulation variants and skips the execution of the simulator. After running several variants of a design thread for a sensitivity analysis, e.g., the response and design parameters are available in one single table.

The SATURN TCAD system uses standard UNIX facilities wherever possible. Simulators are started on network servers by using the UNIX remote shell mechanism while UNIX files, containing interchange information, are accessed via the Network File System (NFS). Design and response parameters are available as ordinary shell variables. The UNIX utility awk extracts response data from output files and evaluates mathematical expressions. Every UNIX command and every shell directive can be used for writing application scripts.

Compared to other TCAD environments available, SATURN belongs to a generation of systems which restrict themselves to a command-driven approach to meet the basic requirements. Now, next generation TCAD systems are being developed both commercially [24, 25] and at academic institutions [22, 23], which make use of the emerging standards for data exchange [21], graphical user interfaces and visualisation. If these systems are able to meet the high requirements on user-friendliness, functionality and efficiency, they are expected to become de facto industry standards.

6. Acknowledgments

The author would like to acknowledge the contributions of many colleagues at the Central Research and Development Department of SIEMENS. In addition, the author appreciates managerial encouragement from H.Jacobs.

References

[1] C.P.Ho et al., *Stanford Electronics Laboratory Technical Report*, 1984.

[2] S.M.Hu, J.Appl.Phys. **57**, 1069(1985).

[3] M.Orlowski, Appl.Phys.Lett. **53**, 1323(1988).

[4] G.Hobler, S.Selberherr, IEEE Trans. Comp. Aided Des. **7**, 174(1988).

[5] F.Lau et al., Appl.Phys. **A49**, 671(1989).

[6] M.Paffrath, W.Jacobs, W.Klein, E.Rank, K.Steger, U.Weinert, U.Wever, *Concepts and Algorithms in Process Simulation* (to appear in Surveys on Mathematics for Industry, Springer Verlag).

[7] M.Paffrath, K.Steger, *Method of temporary coordinate domains for moving boundary value problems.* (to appear in IEEE Trans. Computer Aided Des.).

[8] E.Rank, U.Weinert, IEEE Trans. Computer Aided Des. **9**(5), 543–550 (1990).

[9] U.Feldmann, U.A.Wever, Q.Zheng, R. Schultz, H.Wriedt, Int. Jour. of Electr. a. Comm., **46**(4),274-285(1992).

[10] H.Brand, K.Dienstl, G.Punz *Der Prozeßsimulator POSEIDON.* Proc. NuTech 90, Numerische Simulation für Technologieentwicklung, Garmisch-Partenkirchen, September (1990).

[11] W.Bergner et al., Proc. ESSDERC 1992, 695(1992).

[12] W.Bergner et al., *Influence of Oxide-Damage on Degradation-Effects in Bipolar-Transistors*, Proc. SISDEP **5**,(1993).

[13] W.Hänsch, S.Selberherr, IEEE Trans. Electr. Dev. **ED-34**, 1074(1987).

[14] W.Hänsch, H.Jacobs, IEEE Electr. Dev. Lett. **EDL-10**, 285(1989).

[15] A.v.Schwerin, W.Bergner, H.Jacobs, IEDM Techn. Dig., 543(1992).

[16] G.Oldham et al., IEEE Trans.Electr.Dev.**27**, 1455(1980).

[17] W.L.Engl, R.Laur, H.K.Dirks, *MEDUSA - A Simulator for Modular Circuits*, IEEE Trans. CAD, April 1982.

[18] W.L.Engl et al., IEDM Tech.Dig., 543(1992).

[19] H.Schmidinger, *"CNIX Shell"*, SIEMENS AG, ZT ZTI DES 241, 1987, unpublished.

[20] B. Ludwig *"SUSE – Der SALVE Präprozessor"*, SIEMENS AG, ZFE BT ACM 24, 1992, unpublished.

[21] S.Duvall, IEEE Trans. **CAD-7**,489-500(1988).

[22] E.W.Scheckler et al., IEEE Trans. on **CAD-11**,(7),911-920(1990).

[23] H.Pimingstorfer et al., Proc. SISDEP **4**,409-416(1991).

[24] Silvaco International, *The Virtual Wafer Fab*, The Simulation Standard 3, 1992.

[25] Technology Modeling Associates, *STUDIO*, 1992.

The STORM Technology CAD System

J. Lorenz, C. Hill[†], H. Jaouen[‡], C. Lombardi[*], C. Lyden[+], K. De Meyer[#],

J. Pelka[§], A. Poncet[$], M. Rudan[%], and S. Solmi[&]

Fraunhofer-Institut für Integrierte Schaltungen, Bauelementetechnologie,
Artilleriestraße 12, D–91052 Erlangen, GERMANY
[†]GEC Marconi Material Technology, Caswell, UNITED KINGDOM
[‡]SGS Thomson, Grenoble, FRANCE
[*]SGS Thomson, Agrate Brianza, ITALY
[+]NMRC, Cork, IRELAND
[#]IMEC, Leuven, BELGIUM
[§]FhG-ISiT, Berlin, GERMANY
[$]CNET, Meylan, FRANCE
[%]DEIS, University of Bologna, Bologna, ITALY
[&]CNR-LAMEL, Bologna, ITALY

Abstract

In this paper an outline of the STORM TCAD system is given. STORM is
a program system for the two-dimensional simulation of semiconductor fabrication process sequences and the optimization of the electrical behavior of the
devices fabricated. It has been developed within an ESPRIT project by a consortium of European companies and research institutes. In this presentation,
the software structure of STORM is described, followed by a discussion of the
physical models developed. Finally, some application examples are given. A
more detailed description of the industrial evaluation of STORM is given in a
dedicated paper elsewhere [1].

1. Introduction

During the last few years, process modeling has become an indispensible tool for
the optimization of semiconductor fabrication processes and devices. Whereas first
two-dimensional process simulation programs such as ICECREM II [2] and SUPRA
[3] where limited to the simulation of the dopant distributions only, the application
to submicron technologies requires an integral and comprehensive treatment of the
changes of dopant distributions and device topography: E.g., ions are frequently
implanted through non-ideal mask windows, resulting in dopant distributions which
strongly depend on details of preceeding lithography and etching steps. The diffusion
of dopant atoms depends via the generation and recombination of point defects at the
silicon surface on the presence and growth of overlayers. The formation of shallow
junctions by outdiffusion from implanted polysilicon layers depends on the diffusion
of the dopants both in the polysilicon grain and along the grain boundaries, with
the size of the polysilicon grains depending on preceeding deposition, implantation,
and high temperature steps. Furthermore, it is necessary to finally calculate the
electrical behaviour of the devices fabricated. The subsequent optimization of the

process step sequence and the devices fabricated should as far as possible be performed automatically, with as little user interaction as possible.

Within the STORM software system, all these issues are being addressed. STORM consists of a universal two-dimensional process simulation program to which a two-dimensional device simulator is closely coupled. This simulation system is embedded in an advanced environment for the automatical optimization of electrical and physical target functions. To achieve this ambitious goal, the STORM consortium which consists of the European semiconductor companies GEC and ST and the European research institutes CNET, FhG (IIS-B and ISiT), IMEC, LAMEL, NMRC, and the University of Bologna started from various software tools which had been developed at these institutions before, including the data base DAMSEL (CNET), the graphical user interface IDAS (ST), the two-dimensional process simulators COMPOSITE (FhG) and TITAN (CNET), and the two-dimensional device simulator HFIELDS (University of Bologna). The specific strengths of these tools were combined and further enhanced to form the integral STORM simulation package. Furthermore, for the simulation of optical lithography which is one of the process steps where three-dimensional effects are most important, in the STORM system also results from true three-dimensional simulation using the SOLID lithography simulator (FhG-ISiT) can be used.

2.   STORM Program System and Data Structures

STORM (Software for Technology Optimization in Research and Manufacturing) is a software environment providing the user with a unique interface to access and manage all the necessary tools for developing and analyzing advanced semiconductor structures. Its main features are:

1. Functionality: Leading-edge modules for the two-dimensional simulation of semiconductor fabrication process steps have been integrated into STORM, and a state-of-the-art two-dimensional device simulator has been closely linked to the STORM system.

2. Modularity: The integration among the different software tools has been performed in such a way to guarantee a flexible management of the STORM modules. Moreover, new modules can be easily inserted, keeping full compatibility with the existing ones.

3. Efficiency: In a large environment the easiness of the connection among the different modules is mandatory and it has been carried out using the software library DAMSEL [4], a Data Base Management System which allows a simple control of complex data structures.

4. Interactivity: The communication from the user to the program is mostly direct, step by step, saving time and preventing from mistakes. Alternatively, a batch mode may be used.

5. Friendliness: The STORM interface language is deeply user oriented and mostly common to all tools. Further than that, a HELP-ON-LINE feature makes the approach to the program easier, limiting the necessity to consult the user guide.

6. Graphics capabilities: Great efforts have been made in order to obtain very powerful graphic features, to display input and output data. For the sake of efficiency, standardization and portability, the standard graphic library GKS has been used.

2.1   The STORM Program System and User Interface

The STORM system consists of the integrated STORM project code and some external programs which are interfaced to STORM. The structure of the system is outlined in Fig. 1. The various modules of the integrated STORM project code communicate with each other via a data management library which is based on the DAMSEL software developed by CNET [4]. For user interaction, the STORM system provides a graphical user interface which is based on the IDAS software developed by SGS-Thomson [5].

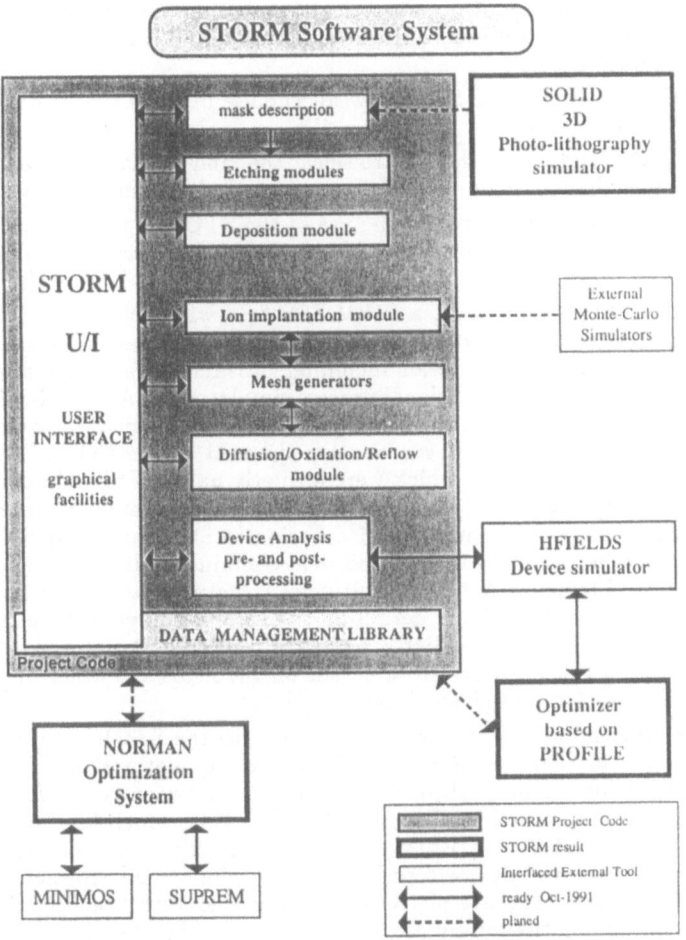

Figure 1: The STORM software system

STORM has been developed providing full modularity of its structure. Each STORM module is called a FIELD. At any time one and only one FIELD is active, and the name of that FIELD is always displayed on the base window. Having activated STORM, the user automatically enters the MAIN field, which may also be activated again at any time of a STORM run. The MAIN field includes various utility programs, like the activation of any other FIELD, running the device simulator HFIELDS, creating a doping file for MINIMOS, executing a command file, on-line help, display of contents of commands, current setting, or the last messages, and terminating the STORM session. DEVICE commands may among others be used to open, load, copy, modify, or close a device, or to select the device output variables which should be plotted. Whereas more than one device may be present in the STORM data base, only the one device activate at a time will be addressed by all other STORM commands. GRAPHIC WINDOW commands may be used to address the graphic windows of STORM.

The STRUCTURE field is used to initialize devices, to perform user-defined modifications of the devices, e.g. using analytical formula for the growth of a bird's beak, or to load device structures which are result from numerical simulations carried out in the PROCESS field.

The GAME and the ATMOS field allow for the use of these mesh generators for the creation of device meshes.

The SOLVINPU field allows the user to select the kind of analysis, the biases and the physical models that must be taken into account in the numerical device simulation.

The OUTPUT field is dedicated to the use of the STORM graphics section in batch mode. Most commands in this field address graphics windows. This field allows to perform 1D, 2D and 3D plots.

Within the STORM framework, a PROCESS field which consists of advanced simulation modules for all important process steps is integrated. The modules for the simulation of diffusion and oxidation including glass reflow have been developed based on modules from the two-dimensional process simulator TITAN of CNET [6], whereas the modules for the simulation of layer deposition including metallization, for etching, and for ion implantation have been developed based on the COMPOSITE program from FhG-IIS and FhG-ISiT (which were formerly named FhG-AIS and FhG-IMT). Among others, the PROCESS field has been complemented by the integration of an advanced module for the simulation of diffusion in and from polysilicon, developed by GMMT. Furthermore, the PROCESS field contains modules for the generation of adaptive Finite-Element meshes in multilayer structures, and for pre- and postprocessing for device analysis. More details on the various components of the PROCESS field are given in the sections 3 to 6, whereas the data structures of STORM are described later in this chapter. More details on the fields and other properties of the STORM user interface are given elsewhere [8].

Various external programs are closely interfaced to the STORM project code. The most important of these programs are the device simulator HFIELDS [9] from the University of Bologna, the NORMAN optimization software from IMEC [10], and the 3D lithography simulator SOLID from FhG-ISiT [11], which is briefly described in section 3.2. In STORM, HFIELDS is used for the simulation of the devices fabricated, starting from the device topographies and dopant distributions simulated by the PROCESS field. The use of NORMAN together with the PROCESS field and HFIELDS allows for the automatic optimization of the electrical behaviour of the devices fabricated, after desired target functions and the process parameters which should be chosen in an optimum way have been specified. More details on the capabilities of STORM in terms of device optimization are given in section 7 of this paper.

In addition to the close link with HFIELDS, NORMAN, AND SOLID, which are vital parts of the STORM system, also other external programs like the Monte-Carlo simulator CRYSTAL from the University of Surrey, MINIMOS, and SUPREM are interfaced to STORM.

2.2 STORM data structures

The data base DAMSEL[4] is used for the internal data structures of most of the STORM modules and for the communication within the whole program system. In DAMSEL, the data structures are hierarchically organized into BOXES which contain informations of similar kind. E.g., the simulation area consists of various subdomains which are called FACES and which are stored in a box named FACES. Each FACE is defined by referring to the identifiers of the LINES which surround the face in an anticlockwise manner. All LINES are stored in a box named LINES. Furthermore, for each FACE also a reference to an entry in the box REF_GEOM is given, where, among others, the name of the material is stored. This also links a FACE to its physical properties, e.g. diffusion model and parameters. The LINES stored in the box LINES are defined by referring to their start and end point which are stored in the box BOUNDS, and by giving the polygon points which lie between the beginning and the end of the line. Furthermore, again a reference to the box REF_GEOM is made, where, e.g., interface types are stored. In the box BOUNDS, the coordinates of points at which lines begin or end are stored, together with identifiers for the corresponding entries in the box REF_GEOM. Comparable hierarchies are used to store Finite-Element meshes and solutions.

All the information allocated to the various boxes is stored in a large vector, the so-called CASE, and is accessed by the corresponding DAMSEL routines. Details are given elsewhere [4]. Via these DAMSEL routines, the different modules integrated in STORM access the global data.

3. Topography

In the STORM software system, modules for the two-dimensional simulation of CVD layer deposition, etching, and metallization, and for the three-dimensional simulation of optical lithography are implemented. Whereas for the simulation of optical lithography the cell-removal algorithm is employed, in the other topography steps mentioned the string algorithm is used.

3.1 CVD Layer Deposition

3.1.1 CVD Models

In STORM, three different approaches may be used for the simulation of CVD layer deposition processes. The first approach simply consists of creating a new layer with a flat horizontal surface on top of the deposition area specified by the user before. In the second approach an isotropic deposition is performed, using the string algorithm.

For the simulation of nonconformal CVD processes a novel model is used within STORM [12]. Similar models have been developed in parallel by other groups [13]. The model is based on the assumption that the mean free path of the reactive species is large compared with the feature sizes in question, and that the particles move isotropically far from the wafer surface. This condition is fulfilled for LPCVD processes. Furthermore, it is assumed that a reactive particle which hits the surface has a probability less that unity for being absorbed. This probability is called the sticking coefficient s. With a probability 1-s the particle is reemitted and can hit the surface at another position or go back into the reactor volume. In the model it is assumed that the angular distribution of the reemitted particles follows a cosine law. The lower the sticking coefficient the higher is the number of reemissions for a particle, resulting in a better conformity of the deposited layer than for higher values of the sticking coefficient.

In the algorithm implemented, for each surface segment the equilibrium between incoming, outgoing, and incorporated particles is calculated. The condition for the equilibrium for segment μ is

$$A_\mu + G_\mu - D_\mu = R_\mu \tag{1}$$

where A_μ is the total flux from all other surface segments to segment μ, G_μ is the flux from the reactor gas space to segment μ, D_μ is the desorption from the segment and R_μ is the portion which is absorbed at the surface.

Taking into account all surface segments, the portions A, D, and R forms the so-called transfer matrix T, which describes the redistribution of the reactive species on the surface, whereas G is the flux to the surface from the reactor volume. In this way an inhomogeneous linear equation system is defined by

$$T * C = G \tag{2}$$

where C is the surface concentration of the reactive particles.

For two segments $\overline{AB}$ and $\overline{CD}$ which have direct view to each other, the transfer coefficient, that means the matrixelement $t_{\mu\nu}$ is

$$t_{\mu\nu} = \frac{|\ \overline{AC} - \overline{AD} - \overline{BC} - \overline{BD}\ |}{2 \cdot \overline{AB}} \qquad ; \qquad \mu \neq \nu \tag{3}$$

If there is no direct view from segment μ to segment ν, then the matrix element is 0. The diagonal elements of the matrix contains the surface reaction and desorption fraction. The element for segment μ is

$$t_{\mu\mu} = -\frac{2.0}{1.0 - sc} * L_\mu \tag{4}$$

where sc is the sticking coefficient and L_μ is the length of segment μ. The source term which describes the flux portion from the reactor volume to segment μ is

$$G_\mu = -[\cos(\alpha_L) - \cos(\alpha_R)] * L_\mu \tag{5}$$

α_L and α_R are the angles at the left and right side of the sector where the flux to the segment is not shadowed by other segments. They depend on the segment orientation and the geometry (shadowing). L_μ is again the length of segment μ. As for the desorption a cosine distribution is assumed as flux characteristic.

In STORM, model parameters are available for 1D CVD deposition of polysilicon, Si_3N_4, high- and low temperature oxide (LTO). Whereas the deposition of the other materials mentioned shows a good conformity, in case of LTO the step coverage is rather poor. This is caused by a quite high sticking coefficient. More detailed investigations have shown that two components contribute to the LTO deposition process, a highly conformal one and a highly unconformal one. There values and also their relative contribution depend on the process conditons and are given elsewhere in the literature [14].

3.1.2   Implementation of the CVD model

The simulation of a CVD step consists of the following main parts:

1. Adaptive length discretization of the surface polygon, including refinements at sharp corners and replacement of relevant convex corners by infinitesimal circle segments.

2. Calculation of the view angles of each pair of surface segments and of each segment with free space.

3. Collection of the transfer matrix and solution of the linear system which is in general neither sparse nor symmetric.

4. Update of the boundary polygon using a modified string algorithm.

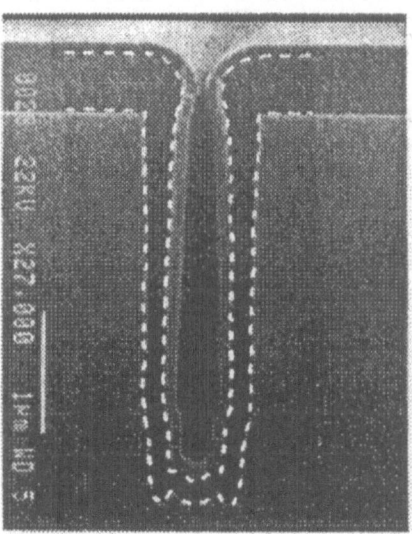 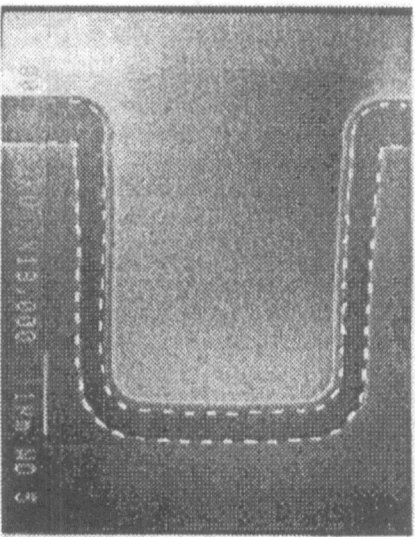

Figure 2: Comparison of STORM simulations with SEM micrographs for LTO deposition

3.1.3   Example for CVD simulation

In Fig. 2, SEM micrographs of LTO layers deposited at a temperature of 570° and a pressure of 13 Pa in trenches of different aspect ratios are compared with simulations using default STORM parameters.

3.2   Lithography

For lithography simulation, the SOLID program developed by FhG-ISiT was included in the STORM software system. SOLID provides full-scale three-dimensional simulation of photolithographic process steps in IC production. Physical effects caused by the illumination system of the projection optics of the stepper are completely covered, as well as effects caused by the resist film itself and the layer stack underneath. Several resist types can be considered including positive and negative type Novolac resists or chemically amplified resists. The latter ones are becoming increasingly attractive for application in deep UV regime, i.e. wavelength of 248 nm and below.

A lithography process can be subdivided into three major substeps: optical imaging, exposure, and development of resist material. The imaging process has to be simulated by the calculation of two dimensional irradiance distributions caused by partially coherent illuminated two-dimensional mask features. Included in this feature is the correct handling of any type of phase-shifting mask, of spatial filtering techniques for resolution improvement, and of modern illumination concepts like SHRINC or CQUEST. Moreover, aberrations of the projection lens can be considered up to the 9th order.

The exposure process is modeled by calculation of light propagation within the layer stack. A post-exposure bake process can be simulated as well. Development simulation is performed in two dimensions for selected cross-sections or in three dimensions, fully covering three dimensional exposure and development aspects, which are of major importance in modern lithography processes.

The Fig.3 demonstrates the capabilities of SOLID. The figure shows a resist pattern generated by a chrome-less phase-shifting mask using i-line lithography and fully coherent illumination. The pattern was printed into a negative type resist. Feature size is aproximately 0.7 microns [15]. More details on SOLID are given elsewhere [16, 11, 17].

3.3   Etching

Pattern transfer from a resist mask into the underlying layer can be simulated using the dry etching capablities of STORM. Whereas the lithography simulator SOLID was interfaced to STORM, the 2d dry-etching simulator ADEPT was completely integrated. Because of the poor knowledge about the chemical and physical processes at the wafer surface in combination with the wide variety of process and equipment variations, dry etching simulation is not so far developed compared to lithography simulation. The etching simulation is restricted to the basic processes like sputtering, ion induced etching and chemical etching, its combinations and some interactions between these contributions. However, quite complex process variations can be modeled as presented in Fig.4. It shows a non-ideal spacer etch process. Selectivity is poor and chemical etching is not negligible. Therefore, the etch process has not stopped at the underlying silicon and the spacer was slightly reduced in width. More details on ADEPT are give elsewhere [18, 19, 20].

3.4   Metallization

For the simulation of metallization steps, in STORM an approach similar to that one used in SAMPLE [21] is implemented and extended to additional metallization equipment. The evaporation and sputtering souces which can be simulated by STORM are the unidirectional, dual, hemisperical, planetary, and conical source, the evaporation with a spherical calotte, and a sputtering system with disk-shaped electrodes. From the angular distribution of the particles emitted from these sources and from the shadowing conditions on the silicon wafer local growthrates are calculated. The string algorithm is used to calculate the updated topographies from these rates.

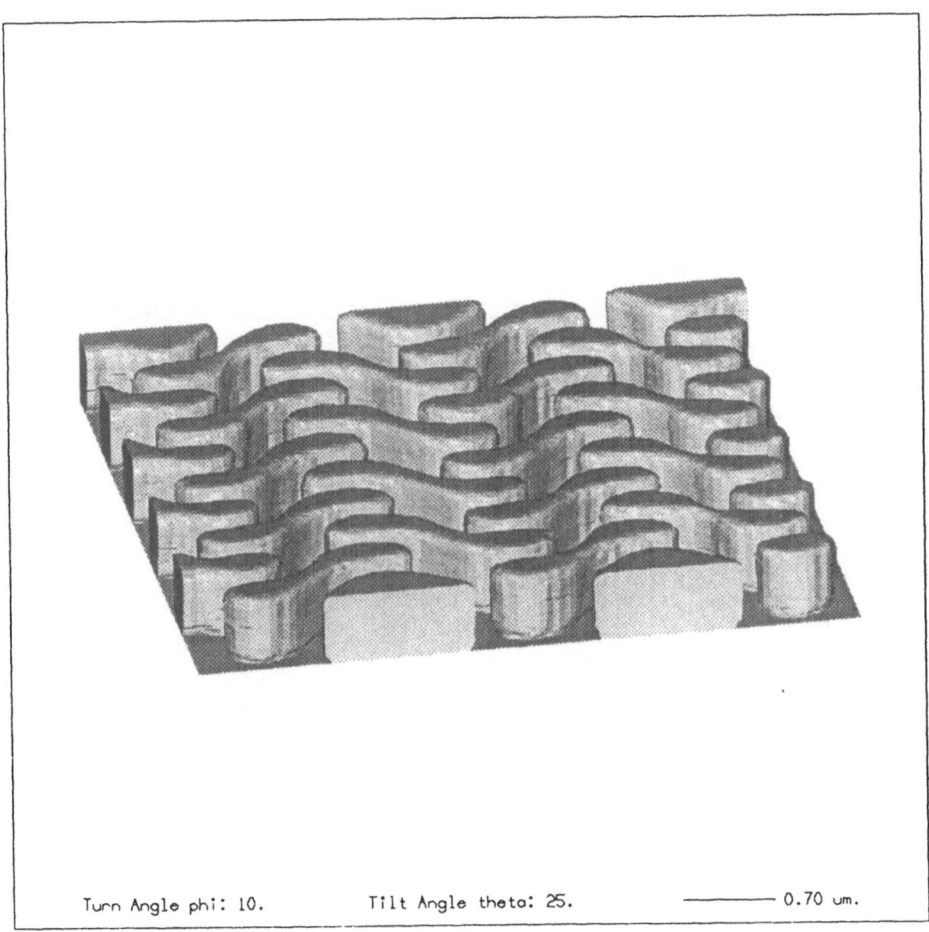

Figure 3: SOLID simulation of resist pattern after i-line lithography step

4. Ion Implantation

For the simulation of ion implantation, leading-edge analytical models have been implemented into STORM. This includes as well the accurate description of vertical dopant profiles in multilayer structures, including the effects of residual channeling, as the inclusion of the depth-dependent lateral shapes of implanted dopant profiles. Furthermore, STORM includes an interface to the crystalline Monte-Carlo simulation program CRYSTAL of the University of Surrey.

4.1 Vertical Dopant Distributions

4.1.1 Vertical Dopant Distributions in Single Layers

For the description of vertical dopant distributions in semi-infinite materials, the set of proven models which has been used in the 2D process simulator COMPOSITE [7]

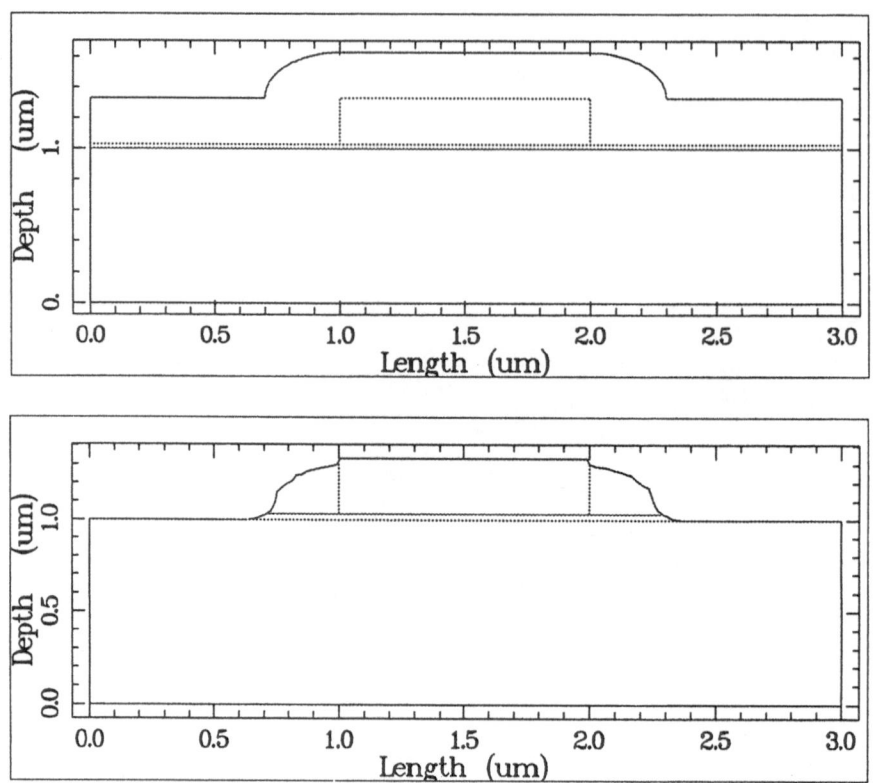

Figure 4: STORM-ADEPT simulation of non-ideal spacer etch process: Initial structure (upper figure) and result after etching (lower figure)

has been further enhanced and now also includes the dose-dependency of the residual channeling. Depending on the ion- target combination in question, different Pearson distributions are being used for the simulation of the dopant profiles: Crystalline targets are best described using Pearson IV distributions, whereas in amorphous targets the values of skewness and kurtosis require the use of either Pearson VI or Pearson I distributions or, in a limiting case, also Pearson III or Pearson V [22]. For the description of the dose-dependency of residual channeling which is caused by the amorphization occuring during high-dose implantations, dose-dependent range parameters which have been extracted by IMEC in the STORM project are used for B and P implantations. For B and BF_2 implants the weighted sum of two Pearson distributions may alternatively be used [23].

4.1.2   Multilayer Models

For the simulation of ion implantation into multilayer structures, in STORM an enhancement of the Numerical Range Scaling model [24] is implemented. In the Numerical Range Scaling model, the dopant profile in a multilayer structure is obtained by using the profiles $C_i(x)$ of the semiinfinite layers i. In all layers except the topmost one, the dopant profile is shifted to take into account the different stopping powers of the layers. Based on investigations on the backscattering behavior of implanted ions, a novel physical model has been developed and implemented into STORM which also accurately describes the modification of the width of the implanted dopant profile in a silicon layer due the different stopping power of a masking layer. In case of a masking layer 1 on top of layer 2, the differences in the stopping powers lead to a projected range straggling ΔR_{p2}^* in layer 2:

$$\Delta R_{p2}^* = \Delta R_{p2} + t \left[\frac{\Delta R_{p1}}{R_{p1}} \cdot \frac{R_{p2} + \Delta R_{p2}}{R_{p1} + \Delta R_{p1}} - \frac{\Delta R_{p2}}{R_{p1}} \right] \tag{6}$$

This model yields excellent agreement with experiment and Monte-Carlo simulations in all cases investigated [25].

4.2 Two-dimensional Dopant Profiles

For the simulation of ion implantation in two dimensions using analytical expressions, in most simulation programs a lateral convolution with Gaussian distributions is employed. However, lateral point response functions after ion implantation are in general neither Gaussian nor, even more important, independent of the depth. Therefore, in STORM a much more advanced model is used. The major aspects of the model are:

1. For each point in the lateral convolution the vertical distribution is calculated using the Enhanced Numerical Range Scaling model including residual channeling, which has been outlined in the preceeding section. For this distribution the mask thicknesses on top of the convolution point are used, not those on top of the mesh point where the dopant calculation has to be calculated.

2. Depending on the values of the kurtosis valid for the implantation conditions, Pearson II, Pearson VII or, in a limiting case, also Gaussian distributions are used for the lateral distributions needed in the convolution integral.

3. The parameters of the lateral distribution depend on depth. A mixed parabolic/ exponential approach for the lateral range straggling yields very good agreement with Monte-Carlo data [27]:

$$\Delta R_{pl}(x) = R_{pl} \cdot \begin{cases} \sqrt{1 + a + b \cdot z - a \cdot z^2} & : \quad z \leq 0 \\ \exp(\, A + B \cdot z + C \cdot z^2 \,) & : \quad z > 0 \end{cases} \tag{7}$$

with the normalized depth $z = (x - R_p)/\Delta R_p$.

In STORM, the parameters a and b are deduced from range moments which have been calculated for all amorphous targets in question using the Boltzman transport program RAMM [26]:

$$a = \frac{X^2 Y^2 - 1 - XY^2 \cdot \gamma}{1 - \beta - \gamma^2} \quad ; \quad b = \frac{XY^2(1 - \beta) + \gamma(X^2 Y^2 - 1)}{1 - \beta + \gamma^2} \tag{8}$$

Here, β and γ are the (vertical) skewness and kurtosis, respectively, and XY^2 and $X^2 Y^2$ are mixed range moments. The parameters A, B, and C are deduced from a and b by expansion of the exponential for small z:

$$A = ln(\, 1 + a \,) \quad ; \quad B = \frac{b}{1 + a} \quad ; \quad C = \frac{-2a(1 + a) - b^2}{2(1 + a^2)} \tag{9}$$

However, in silicon technology one frequently encounters non-vertical mask edges. Whereas the depth-dependency of the lateral spread outlined above influences the lateral shape of the dopant profiles, the increase or decrease of the lateral spread caused by the implantation through non-vertical mask edges or overlayers even modifies the effective channel length: E.g. implantation through nitride layers may reduce the lateral spread in the silicon, because of the smaller lateral spread in the nitride. This means that a model for the depth-dependent lateral spread is only useful if it includes multilayer structures.

In STORM, such a model has been implemented. First, for each layer a depth-dependent lateral spread is assumed as it would be if all the masks would consist of the same material as the layer in question. Then, for all the layers except the topmost one a shift of the lateral range straggling is calculated which depends on the thickness of the masking layers and the depth-dependent lateral straggling within single layers at that depth. This shift is added to the single-layer value of the lateral range straggling.

This model gives good agreement for the implantation of B and As into multilayer structures. It is described in detail and evaluated elsewhere [28].

In Fig. 5, the result of a STORM simulation using this model is compared with Monte-Carlo simulations using TRIM. Boron has been implanted at an energy of 100 keV through a nitride layer of 0.1 μm thickness near a vertical edge of an impenetrable mask. The left side of the figure shows the lateral range straggling versus depth, whereas the right side shows the two-dimensional dopant profiles. Very good agreement between the analytical model and the Monte-Carlo simulation is obtained.

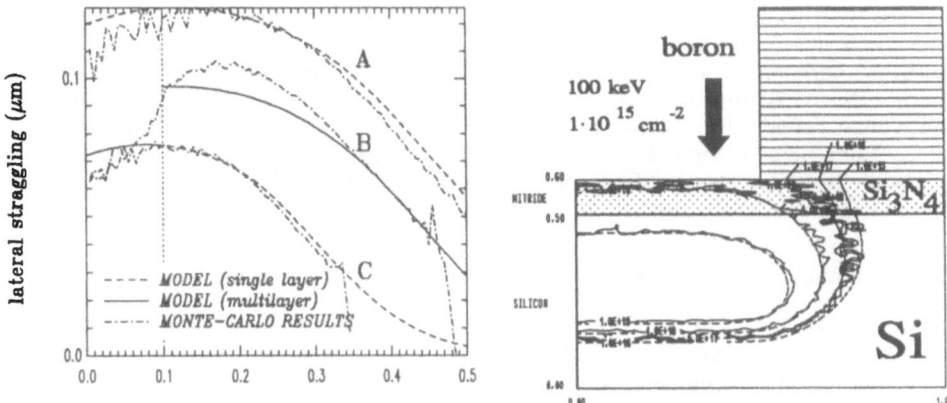

Figure 5: Comparison between Monte-Carlo simulation (drawn line) and analytical model for the implantation of 100 keV boron through 0.1 μm nitride near a mask edge. Left: depth-dependent lateral range straggling: a) in silicon, b) in a two-layer structure consisting of 0.1μm nitride on top of silicon, c) in nitride; right: two-dimensional dopant distribution

5. Diffusion

STORM has a comprehensive suite of models and parameters for dopant diffusion in the most important VLSI materials, many of them novel and developed by the STORM consortium. They are all embodied in the diffusion module of the code, for which CNET had the main integration responsibility. The major models are briefly described below. In addition to these, the consortium generated a large database of diffused experimental profiles, much of which was used to validate the models, in bulk silicon (IMEC, ST, LAMEL, GMMT), in silicides (IMEC, NMRC, CNET), and in polysilicon (GMMT).

5.1   Diffusion in Bulk Silicon

General modeling of diffusion in single crystal silicon must incorporate a wide range of effects and process variables, as can be seen from the 95 page review by Fahey et al. [29]. The main mechanisms which perturb diffusion in silicon from the classical dilute diffusion behaviour with diffusion coefficient D depending only on temperature are: dopant-dopant interaction (e.g. precipitation); dopant-point defect interactions (e.g. pairing and break-up of phosphorus-vacancy pairs); changes in point-defect concentrations due to both local equilibrium effects (e.g. increase in charged point defect concentrations with changes in Fermi level) and to remote non- equilibrium effects (e.g. silicon interstitial injection by oxidising surfaces). Enhanced and new models for all these are implemented in STORM, as summarized below.

5.1.1   Point Defect Interactions

The diffusion of dopant atoms in silicon must be described in terms of their interactions with point defects. In STORM, the effects of point defects can be modelled
at various levels, to allow the user to select a proper compromize between required
accuracy and computation time for each application. The simplest option is a constant diffusion coefficient; the next level is the Fair-Tsai model, suitable for equilibrium point defects. For conditions where there are significant non-equilibrium effects,
or complex point defect- dopant interactions, a novel model, MMG, based on the
Mathiot-Pfister approach [30, 31] is available in STORM. The main assumptions of
the MMG model are:

1. Dopant diffusion is assisted by both vacancies and self-interstitials;

2. Isolated substitutional dopant atoms are immobile. The diffusion takes place
   only through the diffusion of pairs, i.e. E centers (dopant-vacancy pairs) and
   interstitialcies (dopant-self-interstitial pairs);

3. Both the point defects and dopant-point defect pairs exist in several charge
   states. The appropriate charge states depend on the dopant considered and are
   chosen in agreement with what is known from basic point-defect studies in Si;

4. The dopant-point defect pairs are in local equilibrium with the isolated point
   defects and substitutional dopant atoms;

5. Local equilibrium is not assumed between the vacancies and self- interstitials.
   It is simply assumed that they can annihilate each other by bimolecular recombination. This bimolecular recombination takes place not only between the free
   defects, but also through reactions involving the dopant- point defect pairs. This
   leads to a strong recombination enhancement at high dopant concentrations;

6. Boron and arsenic form clusters at high concentrations;

7. Above $2 - 3 \cdot 10^{20} cm^{-3}$, the formation of percolation clusters of dopant atoms
   enhances effective diffusivity of both vacancies and dopants;

8. Above the solubility limit, boron precipitates. These precipitates do not diffuse,
   but modify boron diffusivity in acting as a source of dopants. The SOLMI model
   outlined below describes this phenomena.

With the assumptions made above, the total flux of dopants is the sum of both
mechanisms, the E center flux and the dopant interstitialcy flux.

5.1.2   Initial Conditions for Point Defects

In STORM, the model described above has been fitted for dopant profiles obtained
after predeposition steps. In this case, the point defects can be considered in thermal
equilibrium in the whole bulk at the beginning of the step. As defects are normalized
to the thermal equilibrium during calculation, defect concentrations are initialized to
the value 1.

In case of implanted profiles, this assumption is not valid. For long anneals, e.g. 30
min at 950°C, the above approach is an acceptable approximation, since the implantation induced defects anneal out in the first few minutes. For anneals at times shorter

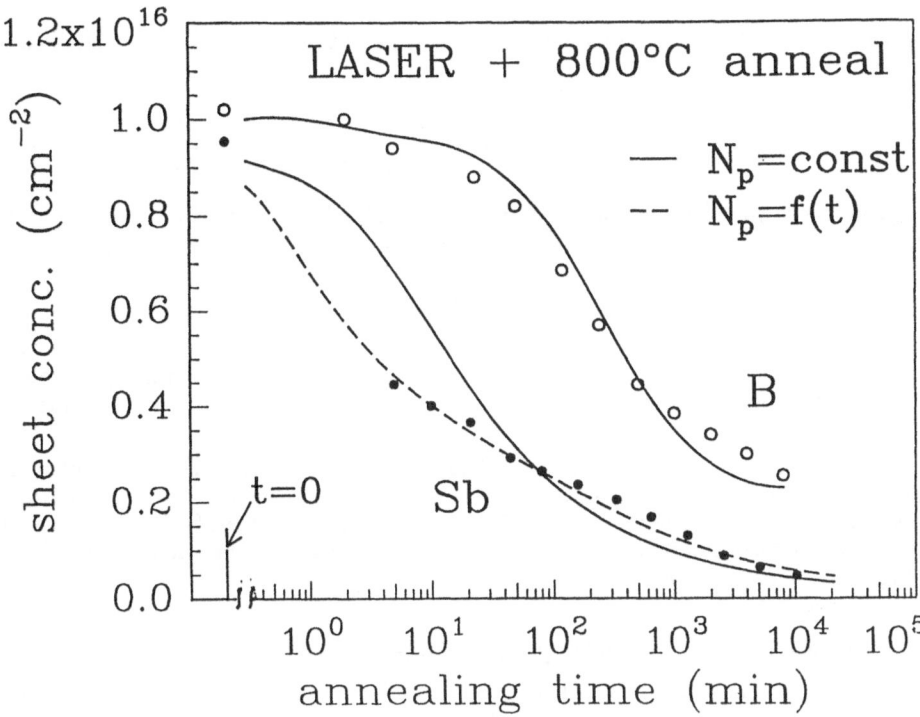

Figure 6: Sheet carrier concentration vs. annealing time at 800°C for Sb and B implanted specimens with dose $1 \cdot 10^{16} cm^{-3}$ and laser annealed before the isothermal treatment

than this, significant implant-induced non- equilibrium defect populations are present for most of the anneal time and influence dopant diffusion. Defects generated by ion implantation are in general represented either as an initial amount of vacancies and interstitials located in the bulk at the same place as the dopants, or as dislocations at the boundary of the amorphised silicon, or as amounts of defects which decrease with time independently from diffusion and recombination phenomena. The algorithm implemented in STORM is robust enough for all these cases. The influence of defects on dopant diffusion was shown even at temperatures as low as 850°C.

5.1.3 Dopant-Dopant Interactions

The precipitation kinetics of boron and antimony in silicon at high concentrations are accurately modelled by the LAMEL SOLMI model incorporated into the STORM code by CNET; the model also simulates very high concentration As precipitation. In STORM, the model has been combined with the MMG model described above. In order to take the precipitation into account, an additional term P(x,t) which acts as a source or sink for the dopant has been added to the diffusion equation. P(x,t) is the amount of dopant going to or leaving the precipitates in the interval dt:

$$P(x,t) = 4\pi \cdot D(C) \left[C(x,t) - C_{sol} \right] \cdot r_p(x,t) \, N_p(x,t) \qquad (10)$$

This source term depends on the dopant solubility C_{sol}, and the number N_p and radius r_p of precipitates. C_{sol} is a function of temperature and co-diffusing dopant concentration only. N_p remains constant with time at the initial critical nuclei density $N_p(x,0)$:

$$N_p(x,0) = K_n exp\left[-\frac{16\pi}{3}\left(\frac{\sigma}{kT}\right)^3 \frac{1}{[C_p ln(C(x,0)/C_{sol})]}\right] \qquad (11)$$

r_p grows from the initial nucleus size $r_p(x,0)$ to $r_p(x,t)$ by a diffusion limited growth rate:

$$\frac{dr_p(x,t)}{dt} = \frac{D[C(t) - C_{sol}]}{C_p r_p(x,t)} \qquad (12)$$

Here, C_p is the dopant concentration in the precipitate, σ is the surface free energy between precipitate and matrix, and K_n, the only fitting parameter, is related to the density of available nucleation sites.

In Fig. 6, a comparison of the simulated and measured electrical conductivity of B and Sb oversaturated silicon as a function of the anneal time is shown. In the simulations, both a constant precipitate density N_p and a precipitate density $N_p(t)$ decreasing with time are considered. Good agreement between experiment and simulation is obtained.

5.2   Diffusion in and from Polycrystalline Materials

As in some other simulators, STORM can approximate diffusion in polycrystalline materials by using a (high) constant diffusion coefficient, and for some applications, this is a fast and satisfactory approach. However, in many cases, the microstructure of the material must be taken into account, and in STORM this option is available through the incorporation of the GMMT physically-based polysilicon model. This model has also been used by IMEC, with adjusted parameters, to simulate dopant diffusion in and from CoSi$_2$.

5.2.1   Diffusion in and from Polysilicon

The polysilicon model incorporated into the STORM code is the most sophisticated 2D model currently available. The basic model, described elsewhere [33, 34, 35], uses a local homogenisation approach to solve simultaneously the grain growth and the dopant diffusion in grain boundaries, in the grain interior and along interfaces, and the segregation at grain boundaries and at interfaces. It has been successful in predicting a wide range of experimentally observed profiles. The influence of doping concentration and temperature on grain growth, dopant diffusion and dopant segregation is accurately modelled. The principles of the model are summarized in Fig. 7 and Fig. 8.

The model described above has been extended for true simulation of non-planar microstructures by including a modification of the Mei model [36] for grain growth with a term related to the local curvature at each location inside the polysilicon layer, determined from the mathematical divergence of the columnar orientation vector function stored for each point. The effect of the curvature is to enhance growth rates at convex corners and to reduce it at convace corners, through a change in the total free energy per unit volume.

Another extension of the model consists in the inclusion of epitaxial alignment, using the basic mechanism proposed earlier [34], to define a nucleation time for slits to

Non-Planar Polysilicon Diffusion Model

Local Homogenisation

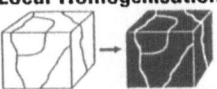

$\rho(\underline{x})$ density of grain boundaries

$\hat{\underline{\xi}}(\underline{x})$ grain boundary orientation

Single grain boundary

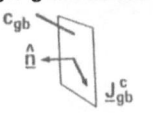

basic flux $\underline{J}_{gb} = D_{gb} \nabla c_{gb}$

constrained to 2D grain boundary plane

$\underline{J}_{gb}^c = \underline{J}_{gb} - (\hat{\underline{n}} \cdot \underline{J}_{gb}) \, \hat{\underline{n}}$

Grain boundary network

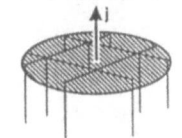

sum over 3D homogenisation volume - flux tensor F

$\underline{J} = \underline{F} \, \underline{J}_{gb}$

F_{ij} = i component of the projected gb density per unit area normal to the j direction

In principal axes (ξ, ν, x) F_{ij}= diag (2/L, 1/L, 1/L)

Figure 7: Principle of the non-planar polysilicon diffusion model

Polysilicon Dopant Diffusion Equations

Variables	c dopant concentration in grain
	w bulk gb homogenised dopant concentraton
	$w = \rho \, c_{gb}$
	L local grain size $\rho = 2/L$

gb network $\dfrac{dw}{dt} = \nabla_i F_{ij} D_{gb} \, \nabla_j \dfrac{w}{\rho} + G$

Grain interior $\dfrac{dc}{dt} = \nabla_i D \, (c) \, \nabla_i c - G$

Segregation $G = \rho \, q(\dfrac{N_{gb} - w}{N_{gb}} c - K \, \dfrac{N_g - c}{N_g g} \dfrac{w}{\rho})$

Segregation velocity $q = \dfrac{1}{\alpha} \dfrac{dL}{dt} + \dfrac{D(c)}{L/2}$

Figure 8: Polysilicon dopant diffusion equations

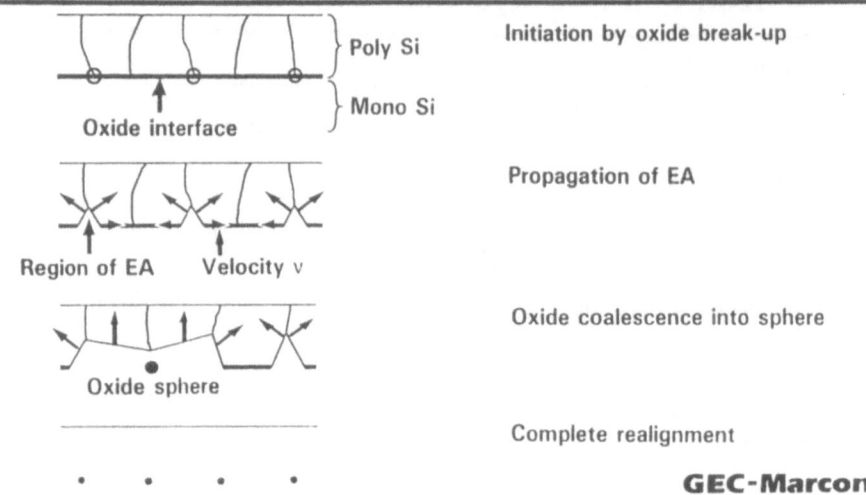

Figure 9: Epitaxial alignment (EA) of polysilicon layer

appear in the interfacial oxide which then allow propagation of the epitaxial regrowth front at a rate determined by grain size, temperature and doping, as summarized in Fig. 9.

The dopant diffusion is modelled by treating the internal grain boundaries by the local homogenisation approach, as in the standard model, but by modeling diffusion along external interfaces explicitly, so that when, for example, the interfacial oxide disrupts, the interfacial diffusion is switched off. A comparison of simulated and experimental dopant profiles for arsenic implant and diffusion in polysilicon at $1100°C$ is shown in Fig. 10. In this case, complete epitaxial alignment has occurred, as shown by the sloping As profile in the polysilicon. The arsenic peak remains at the original interface position, and an excellent agreement between simulation and experiment is obtained.

5.2.2   Diffusion in and from Silicide

In STORM, the diffusion of boron in $CoSi_2$ can be predicted using the poly model described above with modified parameters. The main features are correctly simulated: The interfacial peaks increase in greater portion than the implanted dose due to the oversaturation of the grain interior, and the outdiffused profile becomes insensitive to dose above a threshold concentration of about $2 \cdot 10^{20}$ cm^{-3}.

5.3   Two-dimensional Diffusion

STORM, as a 2D simulator, can simulate all diffusion effects observed in 1D in two dimensions. Because of its physically based models, it also has the capability to model effects only observable in 2D structures; for example, the influence of laterally adjacent regions on each others point defect concentrations (2D defect diffusion) and the influence of laterally adjacent regions on each others temperature (2D thermal diffusion).

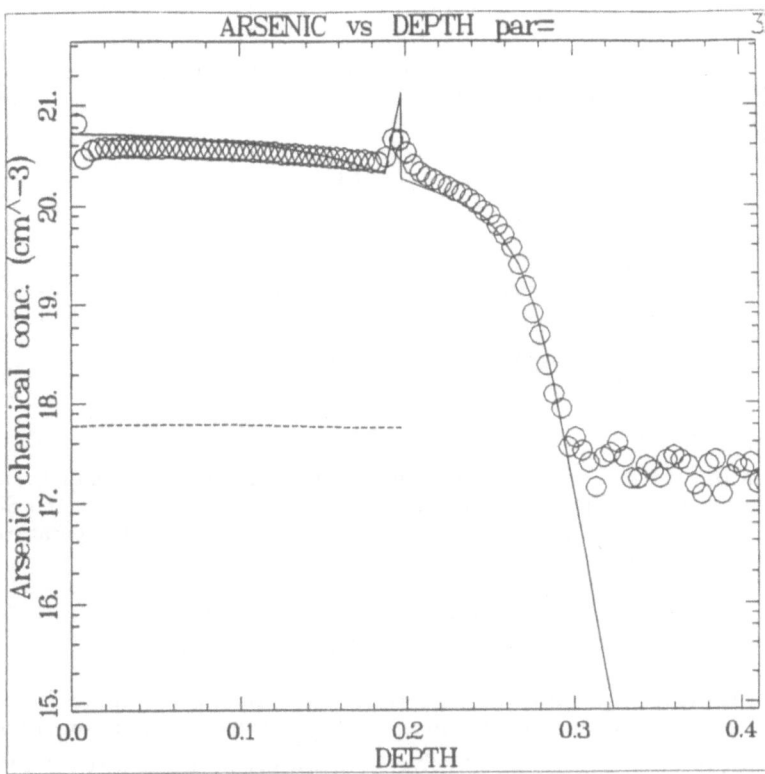

Figure 10: Arsenic profile for 0.2μm polysilicon layer on top of crystalline silicon substrate, implanted with 10^{16}cm^{-2} As at an energy of 40 keV, and annealed for 30 sec at 1100°. Circles represent SIMS result, drawn line the STORM simulation using the full polysilicon model

In 2D dopant diffusion, the additional parameters over 1D simulation are the influence of 2d geometry on the lateral distribution of defects, and the influence of different surfaces and their geometry on point defect generation and recombination rates.

2D thermal diffusion becomes important where the surface conditions or geometry of the wafer influence the local heating, as in the case of optically heated RTA. These effects can be simulated in STORM, using the GMMT models fully described elsewhere [32].

6. Oxidation and Planarization

The numerical algorithms implemented in STORM for the simulation of thermal oxidation includes the diffusion and reaction of the oxidizing species, the growth of a new oxide layer at the interface between silicon and SiO$_2$, and the displacement of layers resulting from dilatation and surface tension. In case of glass reflow or mechanical stress during thermal annealing in inert ambient, only the layer displacement is calculated. Furthermore, the coupling with the diffusion of dopant atoms and point defects is included.

6.1 Diffusion and Reaction of Oxidizing Species

For the diffusion and reaction of the oxidizing species a straight-forward generalization of the well known Deal & Grove model [37] is used, which assumes stationary diffusion with an effective diffusivity D_{eff} in the oxide

$$div\ (D_{eff} \cdot grad\ C) = 0 \tag{13}$$

and the boundary condition

$$D_{eff} \cdot \frac{\partial C}{\partial n} = k_f \cdot (C - C^*) \quad \text{on the free surface} \tag{14}$$

$$D_{eff} \cdot \frac{\partial C}{\partial n} = k_s \cdot C \qquad\qquad \text{on the Si/SiO}_2 \text{ interface} \tag{15}$$

$$\frac{\partial C}{\partial n} = 0 \qquad\qquad\quad \text{on the SiO}_2\text{/Si}_3\text{N}_4 \text{ interface} \tag{16}$$

Here, C^* is the surface concentration resulting from Henry's law, and k_f and k_s are the transfer coefficient at the free surface and the reaction rate at the Si/SiO$_2$ interface, respectively. Diffusion and reaction coefficient depend on temperature, HCl concentration and stress:

$$D_{eff} = D_{eff,0} \cdot D_{HCl} \cdot exp\ (-(V_{d0} + P \cdot V_{dp}/kT) \tag{17}$$

$$k_s = k_{s0} \cdot k_{HCl} \cdot exp\ (-(V_{k0} + \sigma_{nn} \cdot V_{kp}/kT) \tag{18}$$

Here, P is the hydrostatic pressure, and σ_{nn} is the normal stress along the Si/SiO$_2$ interface. The model parameters used in eqs. (14-18) have been taken from literature.

The diffusion-reaction equations are solved numerically by using a standard finite-element method with piecewise linear shape functions and numerical quadrature rules.

6.2 Oxide growth near Si/SiO$_2$ interfaces

During a time step Δt, an infinitesimal new oxide layer with thickness δ is grown:

$$\delta = \Delta t \cdot C \cdot k_s / N_1 \tag{19}$$

For the calculation of the movement of the interface between Si and SiO$_2$, two different approaches are being used. In the first one, the interface between Si and SiO$_2$ is displaced by the incremental additional oxide thickness δ. Afterwards, suitable equations of motion are solved to express the mechanical reaction of the silicon. In the second approach, the interface is displaced by $0.44 \cdot \delta$ to express the consumption of the silicon, and the dilatation of the layers is considered afterwards. In this latter case, a mesh is built in the newly formed oxide in each time step. This also decreases the numerical noise otherwise resulting from unstructured triangular meshes. In contrast to the first approach this one also correctly includes that part of the stress which results from 1D oxidation, whereas the first approach includes 2D extra stress only. Both principles are shown in Fig. 11.

6.3 Displacement resulting from surface tension and dilatation forces

Following the approaches described above, the oxidation kinetics either leads to Dirichlet boundary conditions on velocities at the interface between rigid Si and SiO$_2$,

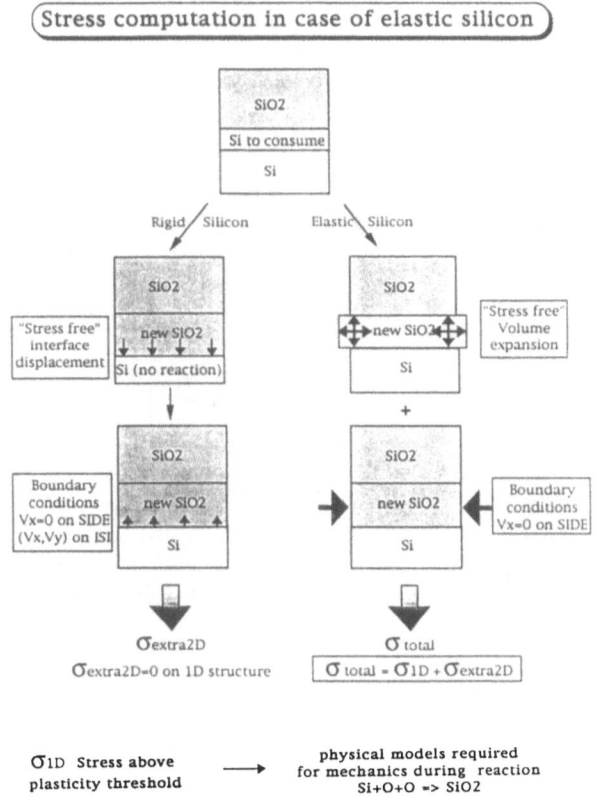

Figure 11: Alternative approaches for the simulation of thermal oxidation

or internal dilatation force in the newly formed oxide resulting from the reaction with elastic silicon. In both cases, the initial displacement of the interface between Si and SiO_2 is obtained, while the other boundaries are left unchanged.

In a second step, all non-rigid layers (SiO_2, Si_3N_4, polysilicon,..) are moved, using either a purely elastic incremental model with maxwellian stress relaxation or a completely visco-elastic model which differs from the classical Stokes equations insofar as it includes a total derivative of the stress deviator. In both cases it is assumed that pressure and normal stress are not changed during the first stage of motion.

6.3.1   Incremental elastic model with Maxwellian stress relaxation

In the elastic option of the STORM oxidation module, the oxide velocity $\vec{V} = (V_x, V_y)$ in the different elastic materials is depending on the dilatation $\vec{D}$ by [38, 39]

$$G\Delta\vec{V} + (\lambda + G)\nabla(\nabla \cdot \vec{V}) = \vec{D} \tag{20}$$

The Lame' coefficients G and λ are related to the Young modulus E and the Poisson ratio ν via the classical expressions (plain strain):

$$G = \frac{E}{2\,(1+\nu)} \tag{21}$$

$$\lambda = \frac{\nu E}{(1+\nu)\,(1-2\nu)} \tag{22}$$

The components of the stress tensor σ are given by

$$\sigma_{xx} = \lambda \left[\frac{1-\nu}{\nu} \frac{\partial V_x}{\partial x} + \frac{\partial V_y}{\partial y} \right] \tag{23}$$

$$\sigma_{yy} = \lambda \left[\frac{1-\nu}{\nu} \frac{\partial V_y}{\partial y} + \frac{\partial V_x}{\partial x} \right] \tag{24}$$

$$\sigma_{xy} = G \left[\frac{\partial V_x}{\partial y} + \frac{\partial V_y}{\partial x} \right] \tag{25}$$

$$\tag{26}$$

Shear stress and pressure are deduced from the stress tensor as follows:

$$\sigma_{shear}^2 = (\sigma_{xx} + \sigma_{yy})^2 + \sigma_{xy}^2 \tag{27}$$

$$P = -\frac{\sigma_{xx} + \sigma_{yy}}{2} = -\frac{\lambda}{2\nu} \nabla \cdot \vec{V} \tag{28}$$

and the normal stress along any boundary or interface depends on the normal component V_n and the tangential component V_τ of the velocity via

$$\sigma_{nn} = \lambda \left[\frac{1-\nu}{\nu} \frac{\partial V_n}{\partial n} + \frac{\partial V_\tau}{\partial \tau} \right] \tag{29}$$

The boundary conditions are that $\vec{V}$ is zero in rigid materials, σ_{nn} is zero on the free surface, and V_x is zero at the left and right sides of the structure. For the interfaces between SiO_2 and silicon,

$$\vec{V} = -0.56 \cdot \vec{n} \cdot C \cdot k_s / N_1 \tag{30}$$

Here, N_1 is the number of oxidizing molecules incorporated into a unit volume of the oxide layer.

In STORM, the elastic model outlined here is used as the default one for silicon nitride, and also recommended for oxide layers at temperatures below 1000°C.

The link between viscosity and elasticity coefficients in silicon dioxide is achieved by introducing a relaxation time τ, which is applied to cumulative stresses in the expressions of diffusivity and reaction rate. The Maxwell viscoelastic model is assumed for the shear modulus G and the viscosity μ:

$$\mu = \tau \cdot G \tag{31}$$

In STORM, thermal dilatation is accounted for during thermal ramps (under oxidizing or inert ambients) by simulating the different expansion between bulk silicon and the other materials. Thermal stresses are computed in all materials defined as elastic or visco-elastic. The internal forces which result from temperature variations lead to an additional term in the elasticity or visco-elastic equations; from first principles, the internal pressure is given as $\alpha_{th} K \delta T$, where α_{th} is the volume expansion coefficient, K is the bulk modulus, and δT denotes the temperature variation.

6.3.2 Visco-elastic Model

Alternatively, a viscoelastic model may be used in STORM. In this model, the components of the strain tensor ε depend on the components of the stress tensor σ via the stress deviator σ', the hydrodynamic pressure p, the shear stress modulus G, the viscosity μ and the components D_x and D_y of the dilatation force:

$$\varepsilon_{xx} = 1/2G \cdot d\sigma'_{xx}/dt + \sigma'_{xx}/2\mu \tag{32}$$

$$\varepsilon_{xy} = 1/2G \cdot d\sigma'_{xy}/dt + \sigma'_{xy}/2\mu \tag{33}$$

$$\varepsilon_{yy} = 1/2G \cdot d\sigma'_{yy}/dt + \sigma'_{yy}/2\mu \tag{34}$$

$$\sigma'_{xx} = \sigma_{xx} - p \tag{35}$$

$$\sigma'_{yy} = \sigma_{yy} - p \tag{36}$$

$$\sigma'_{xy} = \sigma_{xy} \tag{37}$$

$$\frac{\partial \sigma_{xx}}{\partial x} + \frac{\partial \sigma_{xy}}{\partial y} = D_x \tag{38}$$

$$\frac{\partial \sigma_{xy}}{\partial x} + \frac{\partial \sigma_{yy}}{\partial y} = D_y \tag{39}$$

In addition to the boundary condition (30) of the elastic model, the surface tension at the free surface is inversely proportional to the local curvature radius R:

$$P_s = \gamma/R \tag{40}$$

This condition has negligible influence in case of thermal oxidation, but is the driving force for glass reflow.

In this model, the pressure is computed as a piecewise constant solution and is used a a Lagrange multiplier for expressing the uncompressibility of the materials at a fixed temperature. A modified version of the well-known Uzawa algorithm is used to combine this uncompressibility, the elastic term in eqs. (32-34), and the different dilatation forces.

6.3.3 Stress-reduced Viscosity

In STORM, stress reduced viscosity has been expressed according to the work by Suturdja and Oldham [40]:

$$\mu = \mu_0 \cdot exp(\ E_\mu/kT\ ) \cdot \mu/sinh(\eta) \tag{41}$$

with

$$\eta = V_\mu \cdot \sigma_{shear}/kT \tag{42}$$

Here, μ is the local viscosity, σ_{shear} is the shear stress, and the parameters V_μ, μ_0, and E_μ are taken from literature [41].

6.3.4 Viscosity of BPSG Glasses

The viscosity parameters referred to above are valid for thermal oxidation. For planarization, materials with much lower viscosity are used, such as phospho-silicate and boro-phospho-silicate glasses. For BPSG, the pre- exponential term μ_0 has been expressed versus boron concentration for different deposition processes, whereas E_μ shows no clear dependence on process conditions.

6.3.5 Finite Element Simulation of Dilatation Forces

Dilatation forces result either from the transformation of Si into SiO_2 or from thermal dilatation. They can be expressed as gradient of a scalar function d. This fits well to the Finite Element approach to solve the oxidation problem. If some silicon is not treated as a rigid body, the full stress tensor needs to be evaluated, including 1D stress occuring during full wafer oxidation. In this case, the scalar function d in the new oxide layer is

$$d = (\lambda + 2\ G)\ (\ 1 - \rho)/\rho \tag{43}$$

Otherwise, only extra stress resulting from local 2D effects is simulated.

Furthermore, at each time step the thermal dilatation is simulated, depending on the dilatation coefficient α_{th} of each material:

$$d = K\ (\ T_t\ -\ T_{t-\Delta t}\ )\alpha_{th}\Delta t \tag{44}$$

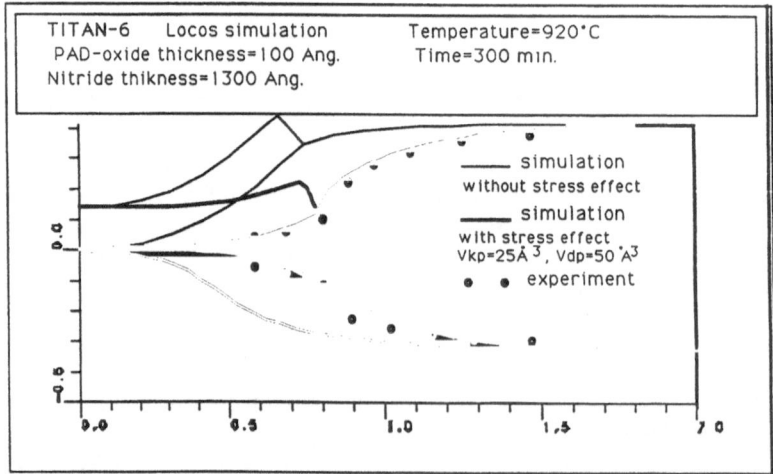

Figure 12: Comparison between experiment and STORM/TITAN simulations of local oxidation at 920°C for 300 min

6.4   Examples for the Simulation of Oxidation

In the following figures some examples for the flexibility and the capabilities of the STORM oxidation module are given. In fig. 12, a comparison between a STORM/TITAN simulation of a LOCOS oxidation at 920° for 300 min is shown. The stress-dependent simulation yields very good agreement with experiment. Fig. 13 shows the initial structure and the process results after 20, 40 and 60 min oxidation of a SOI structure at 900°C, whereas Fig. 14 shows the initial structure and the result for glass reflow for 2 min at 1100° C. Fig. 15 shows the distribution of the hydrostatic pressure P, eq. (28), in silicon, oxide, poly, and nitride after oxidation of a poly-buffered LOCOS at 900 °C.

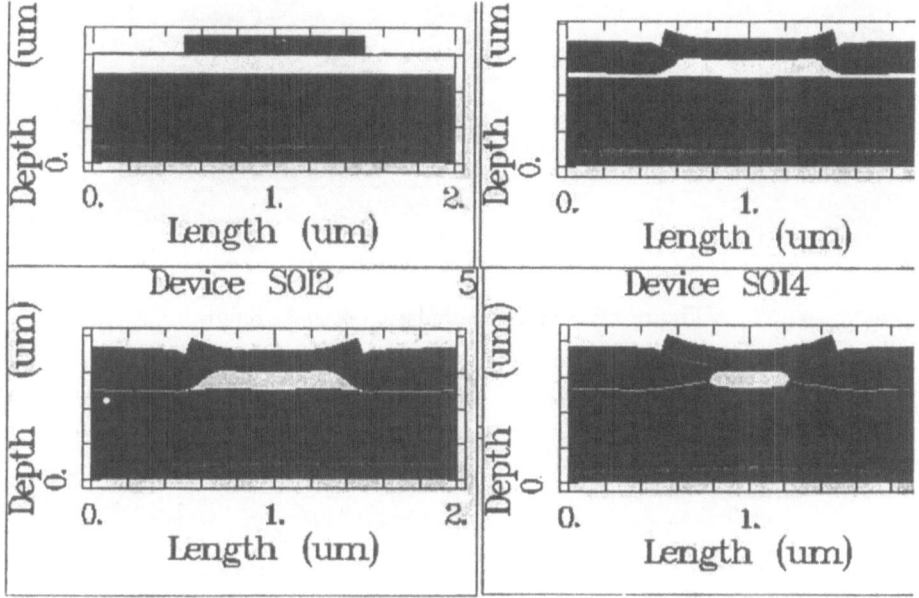

Figure 13: STORM simulation of the oxidation of an SOI structure

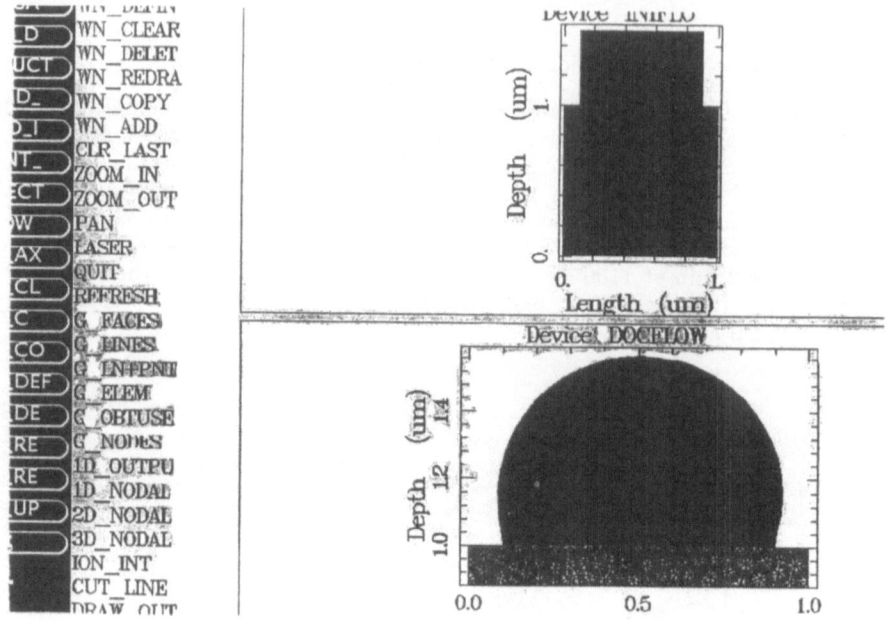

Figure 14: STORM simulation of glass reflow

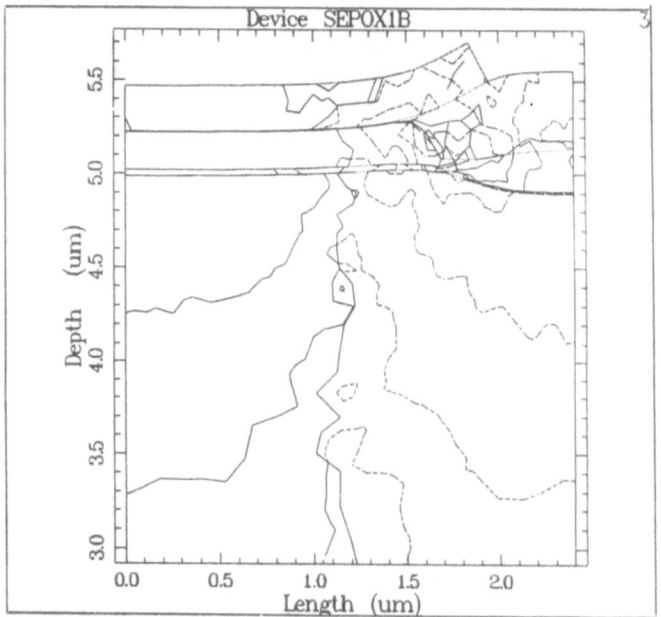

Figure 15: STORM simulation of oxide growth and hydrostatic pressure P in a poly-buffered LOCOS structure after oxidation at 900°C: Two lines per decade are given for the compressive pressure (drawn lines) and the tensile pressure (dashed lines)

7.   Device Optimization

For device optimization, two approaches have been implemented in STORM: the Response Surface Method and the Sensitivity Analysis. In case of an optimization which is based on the sensitivity analysis, at each point of the coupled process and device simulation not only the process results or the electrical properties of the devices, respectively, are calculated, but also their derivatives versus some process or device parameters which were defined before as the parameters to be optimized. In case of the Response Surface Method, a set of numerical experiments is being performed with different values of the free process and/or device parameters where the optimum values have to be found. A suitable interpolation is performed between the results of these simulations. New simulation experiments are being performed until the final optimum set of process or device parameters has been found. The sensitivity analysis has some advantages in the neighborhood of the solution, whereas the Response Surface Method is more suitable to study the global behavior of the target functions and to start the optimization if a good estimate of the solution is not known. In Fig.16, the flow of information during automatic device optimization is outlined. Fig. 17 shows the principles of the sensitivity analysis method and of the response surface method.

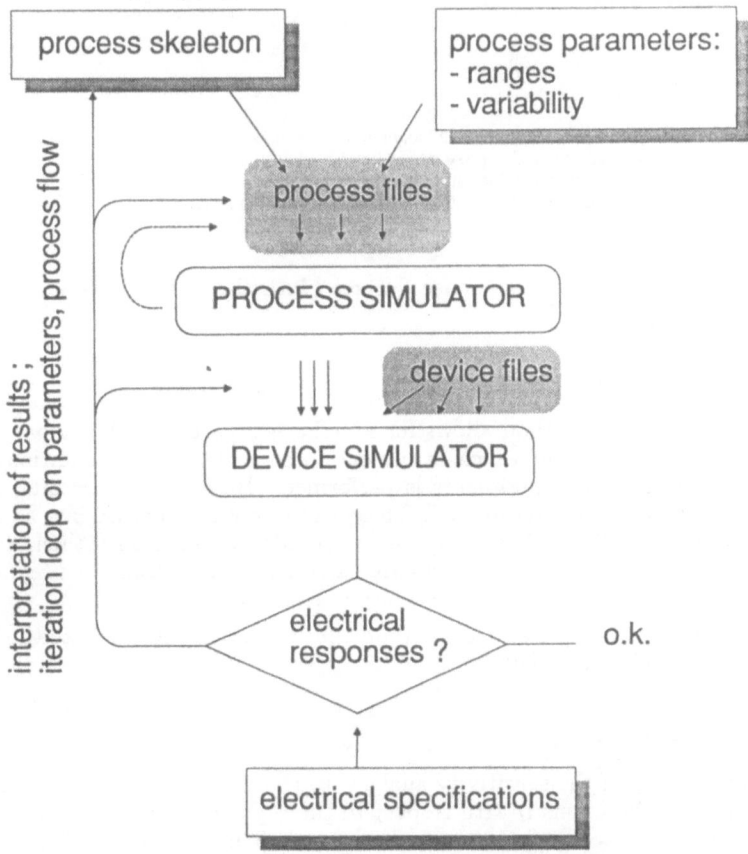

Figure 16: Flowchart for device optimization using simulation tools

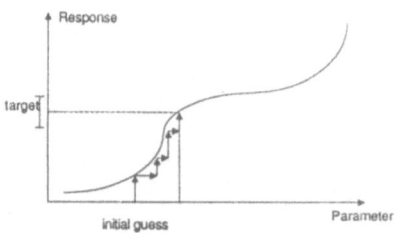

GRADIENT: finds a 'solution' + reveals local sensitivity, but
could be trapped in a local minimum and only provides a
'looking glass' magnified view around the solution

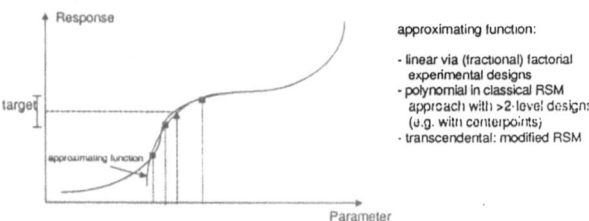

RSM-based method: finds a 'solution' via carefully planned
experiments and approximating function determination.
It also provides information over wider range of parameter
variation (and as such it can also work in a system with
a higher degree of freedom: process windows !)

Figure 17: Principles of the Response Surface Method and of the Gradient Method
(Sensitivity Analysis)

7.1   Response Surface Method

The Response Surface Method allows for an efficient search for the global optimum
of an optimization problem which is subject to some boundary conditions. A set
of numerical simulation experiments is performed. Based on the results obtained,
"Response Surfaces" are interpolated. These can easily take constraints as shown in
Fig.18 into consideration. Depending on the results obtained until then, additional
simulation experiments are defined and are carried out. This leads to a refinement of
the response surface, until finally the solution is obtained.

More details of the response surface method and its implementation into the STORM
system are given elsewhere [10].

7.2   Sensitivity Analysis

For the application of the sensitivity analysis to the device optimization, the derivatives of all target functions f_i with respect to the input parameters w_j for which the
optimum values are to be chosen must be known. The calculation of this Jacobian
matrix requires considerable modifications to the kernels of the simulators to be used.
In general, both during process and device simulation the implicit solution of large
systems of partial differential equations is required. However, for the calculation of

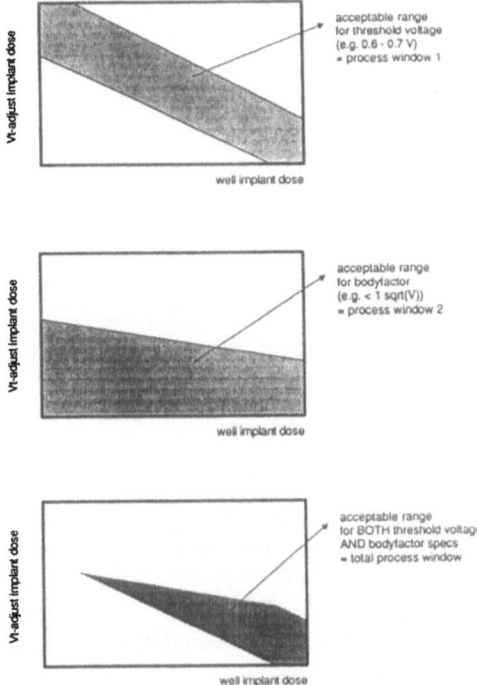

Figure 18: Response Surface Method and Process Window

the Jacobian matrix only terms which are first order in the perturbed input variable have to be considered. In consequence, the Jacobian matrix $\partial f_i / \partial w_j$ depends only on the unperturbed solution and some additional partial derivatives which have to be calculated additionally during the calculation of the unperturbed solution, but requires no additional solutions of the original large linear systems. For this reason, the direct calculation of the sensitivity by tracing the necessary gradients during the process- and device simulation is considerably more CPU efficient than numerical differentiation of the final results versus the input parameters in question.

In STORM, the following elements of the sensitivity analysis are implemented:

1. Calculation of the partial derivatives of results of process steps with respect to the initial conditions before the process step (equal to the result of prior process steps) and with respect to the parameters of the process step (e.g. implantation dose) and of the models used (e.g. implantation range parameters)

2. The calculation of the derivatives of the electrical parameters with respect to device geometry and dopant distributions has been implemented in the device simulator HFIELDS [9]. This enhanced program has been coupled with the optimizer PROFILE profile. More details on this approach are given elsewhere [44].

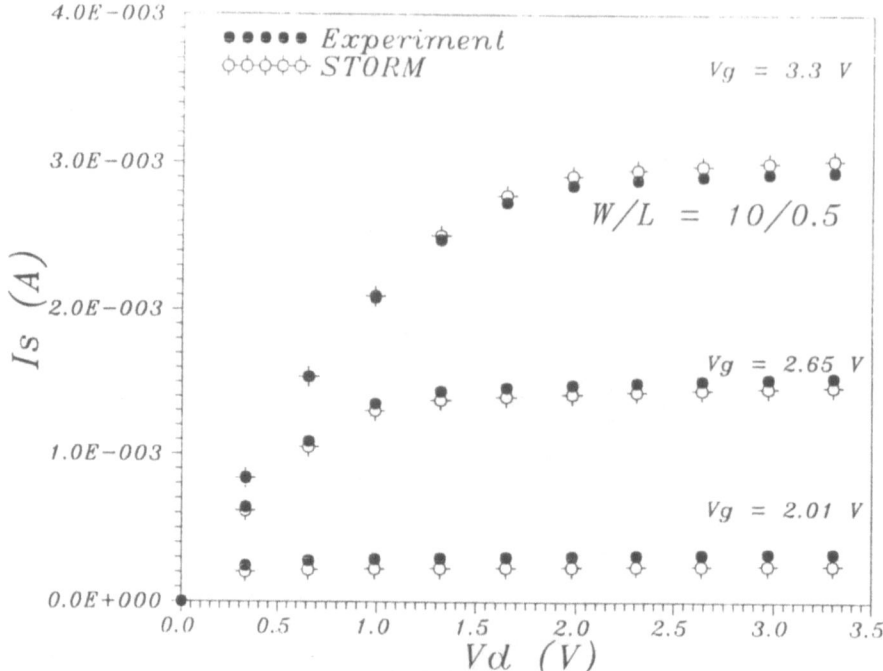

Figure 19: Comparison between STORM simulations and experimental data for a short-channel NMOS device

8. Validation

During the STORM project, a lot of evaluation work was performed on single process steps and full device fabrication sequences, both at the research institutes and at the semiconductor companies involved in the project. In general, a very good agreement between simulated STORM results and measurements was obtained. Some of the results are discussed elsewhere [1, 42]. Fig. 19 and Fig. 20 show some results from an industrial evaluation of STORM on $0.5\mu m$ technologies.

9. Conclusion

In this paper, an overview of the STORM software system for two-dimensional process simulation and automatical device optimization has been given. STORM has a modular program structure and a very user-friendly user interface. The STORM software systems includes leading-edge program modules for the simulation of all process steps important in main-stream silicon technology, and a two-dimensional state-of-the-art device simulation program. The accuracy of its models and the capabilities of its optimization features have been evaluated and demonstrated both in industrial and research benchmarks.

10. Acknowledgements

This work has been funded in part by ESPRIT within the project 2197 "STORM". Some recent improvements of the physical models described in this paper were part of ADEQUAT (JESSI project BT1B) and were funded as ESPRIT project 7236. Work

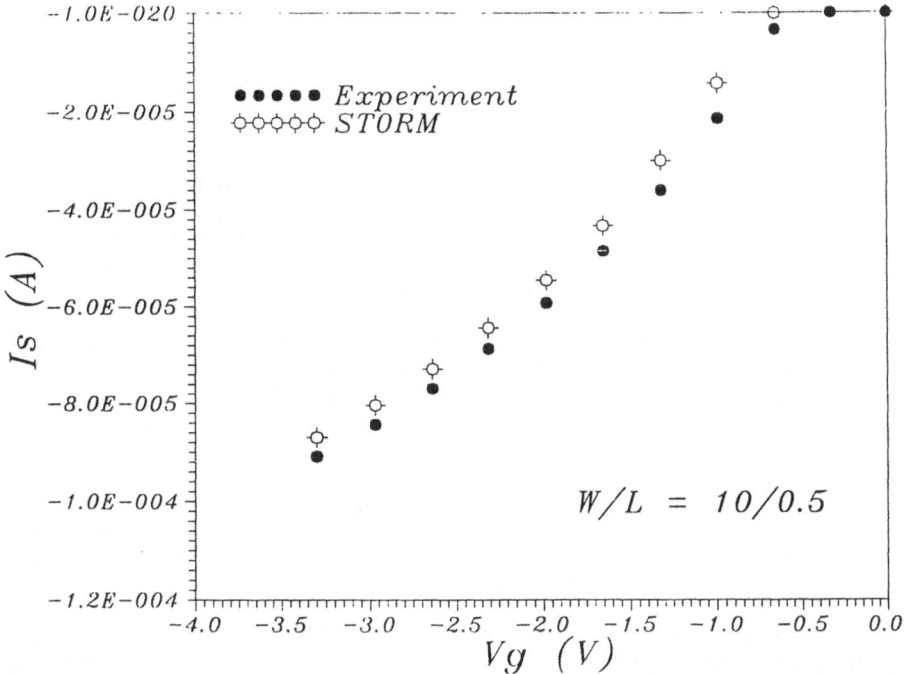

Figure 20: Comparison between STORM simulations and experimental date for a short-channel PMOS device

carried out at CMMT was on behalf of, and financially supported by, GEC Plessey Semiconductors. The authors would like to acknowledge the important contributions of several colleagues from the each of the authors' sites, especially G. Baccarani, F. Baruffaldi, R. Booth, G.C. dal Brun, B. Carniti, R. Cartuyvels, P.D. Cole, C. Corbex, G. Crean, L. Dupas, W. Eichhammer, A. Gerodolle, R. Guerrieri, W. Henke, P. Ciampolini, S.K. Jones, A. De Keersgieter, S. List, K. Maex, S. Martin, D. Mathiot, T. Pedron, P. Pearson, M. Rohan, M. Schäfer, W. Schoenmaker, R.J. Wierzbicki, and H. Wille.

References

[1] S.K. Jones et.al.,*STORM: A European Platform for Sub-Micron Technology Simulation and Optimization*, in: Proc. ESSDERC '93, 1993

[2] H. Ryssel, K. Haberger, K. Hoffmann, G. Prinke, R. Dümcke, A. Sachs, *Simulation of Doping Processes*, IEEE Trans. Electron Devices ED-27 (8), 1484 (1980)

[3] D. Chin, M.R. Kump, H.G. Lee, R.W. Dutton, *Process design using two-dimensional process and device simulators*, IEEE Trans. Electron Devices ED-29 (2),336 (1982)

[4] C. Corbex, A. Gerodolle, S. Martin, A. Poncet, *Data Structuring for Process and Device Simulations*, IEEE Trans. Computer-Aided Design CAD-7 (4), 489 (1988)

[5] C. Lombardi, M. Vanzi, E. Torri, *IDAS: An interactive device analysis environment*, in: Proc. NASECODE IV (ed. J.J.H. Miller), p. 384, Boole Press, Dublin (1985)

[6] A. Gerodolle, C. Corbex, A. Poncet, T. Pedron, S. Martin, *TITAN 5, a two-dimensional process and device simulator*, in: Software Tools for Process, Device, and Circuit Modeling (Lecture Notes of the Short Course and Digest of the Software Forum held in association with the NASECODE VI conference) (ed. W. Crans), p. 56, Boole Presss, Dublin (1989)

[7] J. Lorenz, J. Pelka, H. Ryssel, P. Pichler, *Programs for VLSI Process Simulation*, in: Software Tools for Process, Device, and Circuit Modeling (Lecture Notes of the Short Course and Digest of the Software Forum held in association with the NASECODE VI conference) (ed. W. Crans), p. 179, Boole Presss, Dublin (1989)

[8] STORM User's Guide, Version 3.3, March 1993

[9] G. Baccarani, R. Guerrieri, P. Ciampolini, M. Rudan, *HFIELDS: a highly flexible 2-D semiconductor-device analysis program*, in: Proc. NASECODE IV (ed. J.J.H. Miller), p. 3, Boole Press, Dublin (1985)

[10] R. Cartuyvels, R. Booth, L. Dupas, K. De Meyer, *Process Technology Optimization Using An Integrated Process and Device Simulation Sequencing System*, in: Proc. ESSDERC '92 (eds. H.E. Maes, R.P. Mertens, R.J. Van Overstraeten), p. 503, Elsevier, Amsterdam (1992)

[11] W. Henke, D. Mewes, M. Weiss, G. Czech, R. Schiessl-Hoyler, *Simulation of Defects in 3-Dimensional Resist Profiles in Optical Lithography*, Microelectronic Engineering 13, 497 (1991)

[12] H. Wille, E. Burte, H. Ryssel, *Simulation of the step coverage for chemical vapor deposited silicon dioxide*, J. Appl. Phys. 71, 3532 (1992)

[13] M.M. IslamRaja, M.A. Cappelli, J.P. McVittie, K.C. Saraswat, J. Appl. Phys. 70 (11), 1991

[14] H. Wille, E.P. Burte, *A Dual Sticking Coefficient Chemical Vapor Deposition Model*, in: Proc. ESSDERC '92 (ed. H.E. Maes, R.P. Mertens, R.J. Van Overstraeten), p. 503, Elsevier, Amsterdam (1992)

[15] H. Watanabe et al.; IEDM Technical Digest, p.821 (1990)

[16] W. Henke, G, Czech, *Simulation of Lithographic Images and Resist Profiles*, Microelectric Engineering 11, 629 (1990)

[17] R. Pforr, R. Jonckheere, W. Henke, K. Ronse, P. Laenen, K.-H. Baik, L. Van den hove, *New Resolution Enhancing Mask for Projection Lithography Based on In-situ Off-axis Illumination*, Presented at the SPIE Conf. on Microlithography, San Jose, CA (1993)

[18] W. Pilz, J. Pelka, P. Banks, *Profile Evolution in the Multi-Level Technique*, Microelectronic Engineering 11, 521 (1990)

[19] J. Pelka, *Simulation of ion-enhanced dry-etch processes*, SPIE Vol. 1392, 55 (1990)

[20] K. Boernig, *Modeling a collisional, capacitive sheath for surface modification applications in radio-frequency discharges*, submitted for publicatation in J. Appl. Phys.

[21] C. Sung, *Simulation and Modeling of Evaporated Deposition Profiles*, SAMPLE Report No. SAMD-4, Memorandum No. UCB/ERL M81/8, University of California, Berkeley, (1981)

[22] H. Ryssel, L. Gong, J. Lorenz, *Improvements in Simulation of 2D Implantation Profiles*, in: Proc. 1989 International Symphosium on VLSI Technology, Systems and Applications, p. 102, Taipeh, Taiwan, May 17-19, 1989

[23] A.F. Tasch, H. Shin, C. Park, *An Improved Approach to Accurately Model Shallow B and BF$_2$ Implants in Silicon*, J. Electrochem. Soc. 136 (3), 810 (1989)

[24] H. Ryssel, J. Lorenz, K. Hoffmann, *Models for the Implantation into Multilayer Targets*, Appl. Phys. A41, 201 (1986)

[25] R.J. Wierzbicki, J.P. Biersack, A. Barthel, J. Lorenz, H. Ryssel, *Reflection Approach for the Analytical Description of Light Ion Implanted into Bilayer Structures*, in: Proc. COSIRES 1992 (ed.J.P. Biersack), Berlin (1992)

[26] J. Lorenz, W. Krüger, A. Barthel, *Simulation of the Lateral Spread of Implanted Ions: Theory*, in: Proc. NASECODE VI (ed. J.J.H. Miller), p. 513, Boole Press, Dublin (1989)

[27] J. Lorenz, R.J. Wierzbicki, *Efficient Multidimensional Simulation of Ion Implantation into Multilayer Structures*, in: Proc. of the 1993 International Workshop on VLSI Process and Device Modeling (VPAD 1993), p. 84, Nara, Japan, May 14/15, 1993

[28] R.J. Wierzbicki, J. Lorenz, H. Ryssel, *Advanced Analytical Models for the Multidimensional Simulation of Ion Implantation*, to be published

[29] P.M. Fahey, P.B. Griffin, J.D. Plummer, Rev. Mod. Phys. 61 (2), 289-384 (1989)

[30] D. Mathiot, J.C. Pfister, J. Appl. Phys. 55, 3518 (1984)

[31] D. Mathiot, S. Martin, J. Appl. Phys. 70, 3071 (1991)

[32] C. Hill, S. Jones, D. Boys, in: Reduced Thermal Processing for VLSI (ed. R.A. Levy), pp. 143-180, Plenum Press, New York (1989)

[33] A.G. O'Neill, C. Hill, J. King, C. Please, *A new model for the diffusion of arsenic in polycrystalline silicon*, J. Appl. Phys. 64 (1), 167 (1988)

[34] C. Hill, S.K. Jones, *Modelling Dopant Diffusion in and from Polysilicon*, Mat. Res. Symp. Proc. 182, 129 (1990)

[35] A. Gerodolle, S.K. Jones, *Integration in the 2D Multi- Layer Simulator TITAN of a Advanced Model for Dopant Diffusion in Polysilicon*, in: Simulation of Semiconductor Devives and Processes Vol. 4 (eds. W. Fichtner, D. Aemmer), p. 381, Hartung-Gorre Verlag, Konstanz (1991)

[36] L. Mei, R.W. Dutton, *A Process Simulation Model for Multilayer Structures Involving Polycrystalline Silicon*, IEEE Trans. Electron Devices ED-29 (11), 1726 (1982)

[37] B.E. Deal, A.S. Grove, J. Appl. Phys. 36 (12), (1965)

[38] A. Poncet, IEEE Trans. Computer-Aided Design CAD-4 (1), (1988)

[39] P.G. Ciarlet, The Finite Element Method for Elliptic Problems, Studies in Mathematics and its Applications Vol. 4, North Holland (1978)

[40] P. Suturdja et al., IEEE Trans. Electron Devices ED-36 (11), (1989)

[41] D.J. Chin, PdD Thesis, Stanford University, (1983)

[42] S.K. Jones, A. Gerodolle, C. Lombardi, M. Schäfer, C. Hill, *Complete Bipolar Simulation Using STORM*, in: Proc. IEDM '92

[43] G.J.L. Ouwerling, F. van Rijs, B.F.P. Jansen, W. Crans, *Inverse Modeling with the PROFILE optimization driver*, in: Software Tools for Process, Device, and Circuit Modeling (Lecture Notes of the Short Course and Digest of the Software Forum held in association with the NASECODE VI conference) (ed. W. Crans), p. 78, Boole Press, Dublin (1989)

[44] M. Rudan, M.C. Vecchi, A. Gnudi, *Integrated Tools for Device Optimization*, IEICE Trans. Electron E75-C (2), 216 (1992)

The Viennese Integrated System for Technology CAD Applications

S. Halama, F. Fasching, C. Fischer, H. Kosina, E. Leitner, Ch. Pichler,
H. Pimingstorfer, H. Puchner, G. Rieger, G. Schrom, T. Simlinger, M. Stiftinger,
H. Stippel, E. Strasser, W. Tuppa, K. Wimmer, and S. Selberherr

Institute for Microelectronics, TU Vienna,
Gußhausstraße 27–29, A–1040 Wien, AUSTRIA

Abstract

In order to meet the requirements of advanced process and device design, a new generation of TCAD frameworks is emerging. These are based on a data level providing a common data interchange format. Such a format must be suitable for building simulation databases, and needs to be accompanied by supporting tools and by a procedural interface with multi-language bindings for data storage and retrieval by application programs. The complexity and scope of a rigorous TCAD framework requires special efforts to create a system which is both transparent to the user and comprehensible to the programmer. A consistent architecture and strict adherence to general software engineering guidelines can contribute significantly to the solution of this problem. We discuss general requirements and architectural issues of the data level, the user interface and the task level environment, and present their implementation in VISTA, the *Viennese Integrated System for Technology CAD Applications*.

1. Introduction

The *Viennese Integrated System for Technology CAD Applications* is an integration and development framework for process and device simulation tools. VISTA consists of a data level part which provides a common library for accessing and manipulating simulation data[1], a set of utilities for visualization and high-level data manipulation, and user interface and interactive shell[2] which integrate all services (including the simulation tools) on the "task level".

In the following sections, we will investigate the role of the data level, the user interface and the task level environment within a TCAD scenario first from a general perspective, and then present the VISTA implementation.

General requirements for the data representation, user interface, task level environment, and related services will be discussed in Section 2. Then, after a review of existing implementations and general software architecture guidelines, we will motivate the major ideas and choices for data level, user interface, and task level of VISTA. The structure and implementation details of data level, user interface, and task level will be described in Sections 4, 5, and 6, respectively.

2.   Requirements

2.1   The TCAD Scenario

The term "user interface" suggests to start with a closer look at the human aspects of engineering. With the introduction of frameworks into the TCAD field, the simple "programmer creates application for users" model of the TCAD software situation is no longer sufficient. The modern TCAD scenario looks more like the model depicted in Fig. 1, which will be used in the remainder of this paper (Fig. 1 already implies the most often used framework architecture, where applications are controlled by some sort of integrating task level shell and share common libraries).

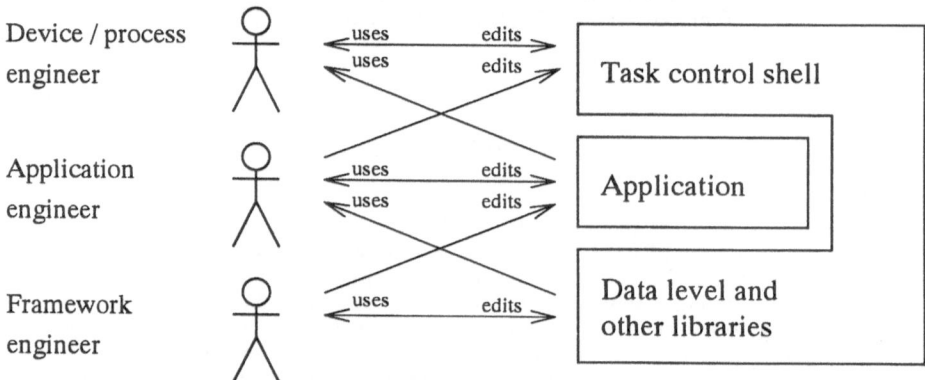

Figure 1: Process, application, and framework engineers interact with the TCAD system by both using (arrows pointing left) and modifying (arrows pointing right) the components of the system.

The *process* or *device engineer* uses simulation tools, or, more generally speaking, TCAD applications, to simulate a given process or device. For this purpose the engineer edits task level information (like manufacturing parameters, the process flow, or optimization goals) and explicitly or implicitly uses applications (simulators) which are provided and edited by *application engineers*. The application engineers, in turn, make use of the framework's global libraries and similar facilities to add new process and device simulation functionality and to maintain existing simulators. The process/device engineers make use of framework services (like data level access or the task level environment), too, but in contrast to the application egineers, they do not know about the libraries and details of "programming" with the framework.

The framework is edited and extended by *framework engineers* who also take care of generic (framework-related) applications.

2.2   The Engineers' Perspective

All three categories of engineers interact in some way with the user interface and task level part of the framework, either in a "user role" or in a "programmer role". Hence, one of the major difficulties in choosing a strategy for a user interface and task level implementation (in contrast to the data level or other internal libraries which are "just" visible from the programmer's point of view) is the vast variety of interests and perspectives which must be considered. Depending on the "engineering role" and the user's technical background and needs, different demands may arise:

- For casual users who seldomly need to use simulation tools (for instance to track down bugs in manufacturing), ease of use, robustness, and continuity of the user interface properties are the most important features. The user interface should employ familiar visual elements where they are available.

- For device or process engineers who use TCAD tools more often, flexibility on the task level is the most crucial issue. It should be easy to define new, complex simulation tasks without having to bother with the internal workings of the TCAD system. Within the task level environment, the details of simulation sequences should not be too simulator-dependent and should not, as is often the case in UNIX-shell based solutions, depend on the operating system at all.

- For specialists in physical modeling or numerical techniques, the most important features are openness of the system as well as good aid for making modifications and extensions. The user interface should support full access to the simulation tools and models. Furthermore, it should be possible to integrate existing tools into a homogeneous user interface without having to redesign the tools.

- From the framework and application engineer's point of view, the use of a high level of abstraction is desirable, as it usually reduces the effort for using and maintaining the system.

- For software support groups, besides the points listed above, maintainability and portability are very important. This includes the use of open portable subsystems since the entire system will be less portable than its least portable subsystem.

2.3   The Applications' Perspective

Seen from the applications' point of view, there are numerous requirements which the data level of a TCAD system must satisfy. Firstly, there has to be a persistent simulation database where simulation problem descriptions, histories and results are stored. A clear, procedural interface provides access to the simulation data and conveys all physical and nonphysical information used by the application. The interface must contain language bindings for those programming languages which are commonly used to develop TCAD tools. Moreover, the interface must be sufficiently operating system- and machine-independent to achieve easy portability to different platforms. Integrability with external TCAD tools must be ensured by providing access to simulation data on different levels of functionality and abstraction (this includes well-defined low-level interfaces).

The procedural interface must be characterized by its ease of use, and an orthogonality to minimize the effort involved in the creation of new tools. The interface should be able to automatically perform conversions of coordinate systems, physical units, simulation grids, etc., so that the application engineer can concentrate on the actual task of the application. Fast random access to simulation data and compact database sizes are crucial for three-dimensional simulation, so these issues cannot be neglected when designing a procedural interface and data representation. Since some TCAD applications may want to use their own internal data structures, the interface has to adapt easily to application-specific data structures.

Once these demands are satisfied, simulators may be run as standalone applications coupled by a common data format. However, a full TCAD integration imposes further requirements upon the data level, under the assumption, that simply "wrapping" the simulator is not a desirable integration method.

Since many different tools with a highly complex sematical background have to interact in a TCAD framework, it is indispensable that the data level has to express semantic rules to ensure the "understandability" of common simulation data for all tools integrated in the environment. Whether all of this semantics can and should be implemented in software and hence be reflected by framework services is a tough question, both technically and economically.

2.4   Maintenance and Comprehensibility

The size of a typical "classical" single process or device simulation program lies in the range of one to two megabytes of source code. Fig. 2 compares a few prominent examples of single simulation tools with VISTA (as an example of a TCAD framework) which currently requires nine megabytes of code in different implementation languages (predominantly C), not including simulators.

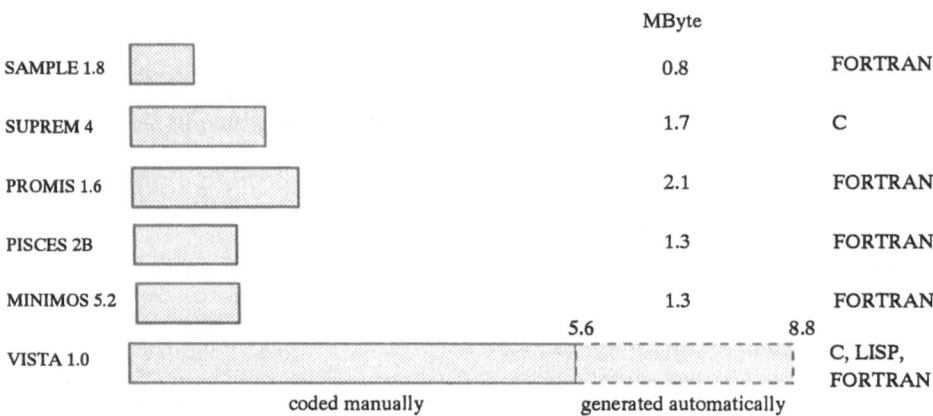

Figure 2: Comparison of source code sizes

Because of the remarkable size of the code, it is obvious that special care is required to ensure the consistency and maintainability of the framework, and that pure software issues become much more relevant than in the case of single simulation tools, regardless of their sophistication. In fact, in the case of a single simulation tool, underestimating or neglecting software issues might lead to an improper and unflexible implementation, although the tool will probably still work. In the case of a whole TCAD framework however, the resulting implementation (if ever achieved) would be completely unusable.

Additionally, it is indispensable that the basic structure and the details of the system can be learned and understood with only a moderate expenditure of time and effort. Therefore, a major demand is that the design and implementation of all components adhere to a few, simple, and mutually consistent basic concepts. We will see that this demand has a severe impact on the potential use of existing solutions from which the framework can be built.

3.   Architecture

3.1   Existing Approaches

Several workstation-based systems can be found which address the issue of multi-tool integration into a unified user interface, mostly based on the X Window system. PRIDE[3], based on SunView, exhibits a user interface and task level architecture which is strongly influenced by the preprocessing — computation — postprocessing task model of TCAD. SIMPL-IPX[4], based directly on Xlib, features a central interactive graphical editor which has menu-oriented facilities for running simulators. In both cases, implicit or explicit assumptions about the design cycle have an impact on the (top-down) *design* of the software and restrict design tasks which can be performed or implemented. A more flexible and extension-oriented user interface architecture has been accomplished in PROSE[5], which is mainly due to the use of the generic Tcl interpreter[6] and Tk toolkit[7].

Other well-known simulation frameworks are an integrated system for statistical VLSI design from Hitachi[8], the MECCA system from AT&T[9], or the SATURN system from SIEMENS[10].

However, only few of these frameworks feature a data level for simulation data access. Most of the existing TCAD environments use data converters to couple simulators using different data formats. Doing this not only causes the number of converters needed to rise quadratically with the number of simulators present, it also prevents the user from taking advantage of the services provided by a TCAD-oriented data level. Using a data level, simulators can be split up into separate tools of well-defined functionality, allowing tool developers to concentrate on their particular task.

Early implementations of data levels, like the DAMSEL system from CNS/CNET [11], feature two-dimensional geometries and simple data structures for easy usage by existing simulators. Among data levels designed for TCAD environments there are the CDB/HCDB from CMU [12], and the BPIF implementation from UC Berkeley [13]. Another data level built on PIF featuring object-orientedness is the PIF/Gestalt system from MIT [14].

A recent approach is the SWR 1.0 specification[15][16] (issued by the *Semiconductor Wafer Representation* technical subcommittee of the *CAD Framework Initiative* (CFI), an international standardization committee for electronic CAD) which defines an object-oriented application interface for TCAD data access and suggests the use of a client-server framework architecture. The intriguing idea of this standard definition is to separate the physical modeling completely from tedious tasks such as grid generation, interpolation, or geometry handling by providing these functions as a black-box server which is accessed by the simulation clients via a procedural interface. This method is very well-suited for, e.g., the simulation of topography formation, however, it can be detrimental to applications with high data throughput or applications which exhibit performance advantages thanks to a tight coupling between physical models and numerical techniques. Furthermore, the sole definition of a rather high-level interface — which implicitly requires a very large amount of functionality — makes it difficult to implement this standard in a rigorous way. This is due to the absence of intermediate-functionality definitions and goals, there is no layering that would provide natural milestones for implementation and verification.

3.2   General Guidelines

It seems that the inherent semantic complexity and diversity of information flow between user and TCAD system, which is presumably related to the rich physical background inherent to process and device simulation, can only be represented by corresponding specialized user interface elements. Furthermore, the rapid development of advanced process and device simulation tools calls for a data level and user interface which can easily accommodate new demands without necessarily having to change the underlying concepts.

Unfortunately, there is no publicly available monolithic user interface toolkit which is flexible enough to meet the changing requirements, while simultaneously providing the specialized functionality to support TCAD information flow efficiently.

The requirements described in section 2 are not easy to fulfill within today's software environments. However, some general rules for data level, user interface, and task level can be derived:

- *Bottom-up* — As the very top TCAD problem and application is hard to narrow down (there is no "generic design task" in TCAD), a *bottom-up design* is favorable. This implies that the definition of higher-level functionality and behavior is shifted towards the end of the development phase.

- *Layering* — Where possible, implementation should be done in distinct layers of increasing functionality and abstraction.

- *Separation and orthogonality* — All of the framework code should be kept independent from the (rapidly changing) TCAD tools themselves.

- *Consistency* — Where possible, the generalization of existing concepts within the framework should be favoured over the introduction of new ones.

- *Interprete design tasks* — The need for defining and using design task macros in a flexible (non-taxative) manner suggests to ask for full programming capabilities, the task level programs being executed by an *interpreter*.

On the data level, a certain architectural transparency is desirable to be able to accomodate and implement possible framework architectures, like client-server, master-slave, parity, ... Although for some, especially geometry-oriented applications a client-server architecture might be advantageous, it is highly questionable, if a true client-server architecture using the network will exhibit the required performance for more general simulation requirements, since large data amounts like grids, attributes or solver stiffness matrices (especially in three-dimensional applications) have to be communicated between client and server. Although mapping client memory into the server substantially improves performance, this approach is neither portable, nor does it work over the network.

A well-balanced and consequent layering of the functionality and semantics of the data level implementation is most important as this is indispensable for re-using the implementation, when, e.g. an RPC-based client-server interface is introduced between two layers, or an object-oriented system is imposed (presumably on the top layer).

An early confinement to a specific architecture would result in an inflexible data level and thus lead to a framework that cannot be adapted to (unpredictable) environmental needs of a simulation site. A firm precondition, however, is the multiprocessing ability of the application interface, to enable parallel simulator runs using the same data set as well as ensuring clusterwide access to the data. The data level of a TCAD environment must be able to manage and archive simulation sequences in order to ensure the reproducability of the results and easy backtracking through the simulation history. The environment has to provide facilities for intersite data exchange, message passing between applications, and error reporting, handling and recovery, which have to be consistent with the data-level implementation.

3.3   The VISTA Design

For the data level, we have decided to start with the *Profile Interchange Format* (PIF), as initially proposed by S. Duvall[17] and to extend and modify it to meet the requirements stated earlier. From the above considerations, it is clear that a binary implementation and a procedural interface with different levels of functionality (described in Section 4) are required. Since there is no public and efficient implementation available, we had the opportunity to implement the application interface from scratch.

For the user interface, the X Toolkit[18] already offers an ideal method for achieving a very flexible and consistent architecture when the required specialized parts of the user interface are implemented as so-called *widgets*. From this set of building blocks, all higher-level functions and applications can be built. This coincides with the proposed bottom-up concept and, due to the object-orientedness of the X Toolkit and widget set, is very well suited for future extensions. A widget set has to be choosen from which the required specific widgets for TCAD purposes can be subclassed.

For the task level environment, a proper choice is in general non-trivial, but becomes almost obvious, when the proposed architectural guidelines and requirements are considered. A UNIX- (or any other operating system) shell based solution does not fulfill the portability requirement, whereas the use of an integrating master application (like an interactive device editor) alone does not offer programming language features.

A portable interpreter appears to be the only appropriate solution that meets all demands. Hence we have chosen to build on XLISP[19], a public domain LISP interpreter, which is available in source code. It is coded in highly portable and comprehensible C code, fulfills all software-oriented requirements and provides full programming capabilities. It can be extended for TCAD purposes by adding C-coded primitives or by loading LISP code at run time.

There are some remarkable implementations of task level environments in related fields, which confirm the feasibility of using LISP as extension language:

The well-known *GNU Emacs*[20] text editor features a LISP extension language which is used to implement special editing modes and to provide text browser style interfaces to a number of unix applications. The computationally expensive parts are still coded in C, so that LISP mainly ties together high-level primitives.

The generic CAD system *AutoCAD* uses the Scheme-like extension language interpreter *AutoLISP*[21] which allows direct access to the data level. It is an invaluable tool for customization and for implementation of a multitude of specific applications.

Winterp[22] (*"Widget Interp*reter") is an application development environment, based on the public-domain XLISP interpreter. It provides interfaces to the X11 Toolkit Intrinsics and to the OSF/Motif widget set and is distributed with the public-domain MIT X11 distribution. Unfortunately, this potential candidate for building a TCAD task level environment and user interface upon it lacks two requirements: It does not readily accomodate additional C-code layers between the X Toolkit (OSF/Motif widgets) and XLISP interpreter, which inhibits the introduction of higher-level user interface layers which need to be shared among C applications, and the object-oriented interface in LISP can not easily be extended to be used by C applications in a *homogeneous* way.

3.3.1   Interaction of Framework Components

The user interface has to allow easy access to *all* services provided with the framework, such as the data level implementation (including high-level data manipulation and interactive editing of device structures), visualization, or the error system.

Additionally, a good link between the TCAD extension language interpreter, which integrates all system components on the task level and represents the "main program" of the TCAD system, and the user interface is required in a way that the existing interpreter is simultaneously used for all interpreted user interface parts.

But there are also other software components which need to be accessible from within the extension language environment, so that a *generic method* for linking C-coded functions to XLISP is highly desirable. External simulator executables need to be started and provided with appropriate input and their termination needs to be recognized to trigger subsequent simulation steps.

4.   Data Level Implementation

The data level is the *backbone* of the whole framework. The data level of the VISTA system was designed to meet most of the above requirements. It features:

- A layered procedural interface for applications to store and retrieve all TCAD relevant data,

- Language bindings to FORTRAN, C and LISP,

- A common ASCII interchange format (*Profile Interchange Format*, PIF),

- A compact binary inter-tool and storage format (*PIF Logical Binaries*, PLBs),

- Parallel access to PLBs,

- The ability to build databases of PLBs into *PIF Binary Files* (PBFs),

- Database utilities to manage PBFs,

- Networking capabilities.

The procedural interface to the database services is called *PIF Application Interface* (PAI, [23]) and makes extensive use of automatic code generation to achieve platform independence and generate the individual language interfaces. It is described in subsection 4.2.

4.1   VISTA's PIF Implementation

The ASCII version of the PIF is used as an intersite data exchange format. The binary form[23] is used as database storage format of the data level. Fig. 3 shows the logical PIF structure with corresponding object relationships. Note that the majority of the simulation information is carried in the grey shaded **geometry**, **grid** and **attribute** constructs, while the **objectGroup** and **meta** objects are important extensions for conveying TCAD-related data. Both the **geometry** and the **grid** constructs are built out of primitive geometric objects (points, lines, faces and solids). The **geometry** construct additionally holds a simulator's point of view of a simulation geometry through **segmentList** and **boundaryList** constructs.

The **attribute** construct is used to attach *any* kind of information to an object. The **attributeType** subconstruct describes the meaning of an attribute. Thus the PIF attribution mechanism is the most flexible means in attaching information to geometries and grids, since they can express anything ranging from a simple descriptive string to a vector field defined on a tensor product grid. With this unified concept there is no separation between fields and attributes necessary, which is another milestone to a clearly structured architecture allowing a simple implementation. Fig. 4 shows a **materialType** attribute defined over a segment and Fig. 5 shows an **electricField** attribute defined over a three-dimensional grid.

In contrast to other approaches, attributes types are semantically standardized to prevent incompatibilities in tool communication (e.g. one tool writing a "Potential" attribute, and a second tool trying to read an "ElectricPotential" attribute), although each tool is free to define its own local attribute types.

Due to its generality and flexibility, a PBF may hold an unlimited number of PLBs, and one PLB (conforming to one ASCII PIF) in turn may hold an infinite number of objects. So, a PBF may contain anything from just one PLB with a few comments, to hundreds of PLBs, each holding several geometries, attributes, grids and process flow descriptions. The maximum size of a PBF is limited by the adressing capability of the PAI, which in turn is affected by the machine word length. On a 32-bit machine the PAI can address one gigabyte (some bits are reserved) which is therefore the maximum PBF size, supposed the operating systems file size limitation is higher. Since the PAI is capable of opening up to 16 PLBs, an application has a maximum of 16 gigabyte of data available. Typically, a single PBF holds one or two PLBs containing a geometry, attributes and grids of a single tool run. Through the special **link** construct objects in other PLBs or even other PBFs may be referenced.

The binary format is closely related to the ASCII format inasmuch as the hierarchical structure of the ASCII PIF is preserved in the binary form through the use of LISP-like constructor nodes. However, to improve performance and data compactness, several additional features have been implemented, such as a symbol hash table for fast object access by name and a compressed array storage format for large arrays which typically occur in TCAD applications for attributes on grids.

It is important to note that, although the structure of the PAI is derived from the PIF syntax, the PAI itself is independent from the underlying database, and thus could be interfaced (probably with losses in performance and compactness) to other databases, since the TCAD application sees just the PAI procedural interface and has to know little about PIF. Thus, multiple different implementations of the low-level application interface routines are possible, because the applications have just to rely on the specification of the PIF application interface services.

The decision to use PIF was made in conjunction with the decision to adopt LISP as the VISTA task level extension language: PIF uses a LISP-like syntax and LISP as the

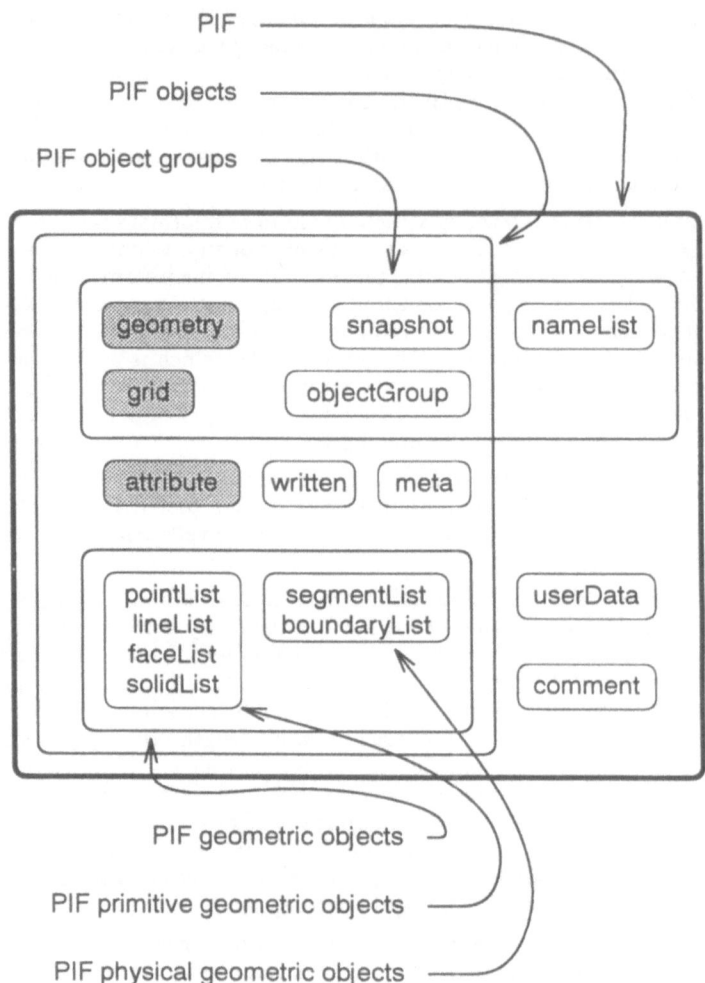

Figure 3: The logical PIF structure.

```
(attribute geometry_attribute
    (attributeType "MaterialType")
    (nameList (ref my_segments (valueList 1)))
    (valueType asciiString)
    (valueList "Silicon")
)
```

Figure 4: Attribute defined on a segment.

```
(attribute grid_attribute
   (attributeType "ElectricField")
   (nameList (ref my_grid))
   (valueType (vector 3 real))
   (valueList          1.2 3.4 6.5
       4.4 3.5 4.7
       .....)
)
```

Figure 5: Attribute defined on a grid.

task level language provides seamless and homogeneous fusion of data and task level concepts. With this unique combination it is equally possible to modify simulation data in the database directly as LISP data as well as store LISP expressions (e.g. task level programs) in the PBF. Thus a process flow representation can be directly embedded in the TCAD data level; there is no artificial separation, and homogeneous data storage, retrieval and maintenance services are available for both semiconductor wafer and process flow representations.

4.2   Implementation of the PIF Application Interface

The PAI is split into seven layers with strict interfaces between each other. The different layers are shown in Fig. 6.

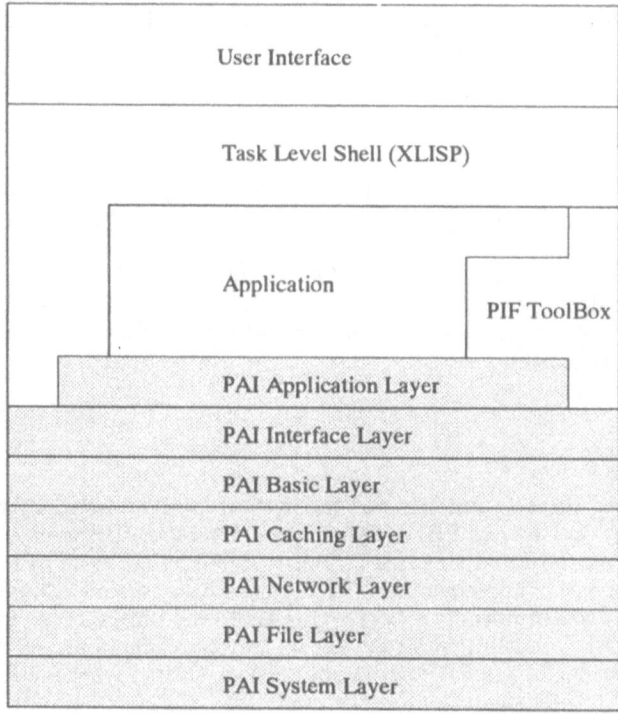

Figure 6: Layout of the PIF application interface.

Each layer calls only functions in the underlying layer. This mechanism leads to separate modules with distinct functionality as used by individual tools. Each layer is responsible for a unique storage concept of the whole PBF, with increasing complexity towards the upper layers. The application interface works on PBFs (intertool format); for data exchange with other hosts there is the PIF ASCII form (intersite format). To convert PIF files between these two formats there is the *PIF binary file manager* (see section 4.4), implemented as a separate PIF tool on top of the PAI.

The PAI is able to handle simulation data in three geometric and infinite nongeometric dimensions. Thus it is possible to read and write distributed attributes ranging from scalar to N-order tensor values on one- to three-dimensional grids. All PIF objects can be selectively and directly accessed with the PAI, either by handle or by name. The PAI will read only the necessary parts of a PBF into a cache avoiding performance drawbacks of most file-based systems.

4.2.1   Error Handling

Errors detected in the PAI are signaled to the global VISTA error handling system, which allows the user to specify different error handlers for each type of error. In addition to program-signaled errors, the error system handles system faults and program exit too. The default error handler prints out the function, the line number and source file name of the function, where the error occured.

New error handlers can be registered by each application to handle error conditions in a program specific way. For example, the caching layer installs its own exit handler on initialization to panic-close all open PBFs through the error system if a memory fault or address violation occurs.

4.2.2   System Layer

This lowest layer of the PAI is the link to the operating system and defines simple access routines to the file input and output services. In **ANSI C** only the buffered file I/O is defined and standarized, but buffering is not needed by the PAI since this is done in the caching layer above. If the unbuffered UNIX style file I/O exists in a specific operating system, this is used instead. This is the only layer which has system dependent functions and implements also basic functions for network access (TCP/IP and DECnet).

4.2.3   File Layer

The standarized file I/O functions of the system layer are used by the file layer to handle the physical I/O of PBFs. It guarantees that a PBF is only opened by one application at a time for writing (file locking). Avoiding multiple write accesses to one PBF allows an easier implementation of the data base, since the physical file cannot change during access (unless it is closed); multiple read-only accesses are allowed. The locking of a PBF is not implemented through system functions. It works through a mark in the header of the PBF and a special lock status, where multiple accesses of the same file at the same time are detected. The file layer also allows the creation of temporary PBFs for intermediate storage of simulation data. Temporary files are stored in PBFs without a physical name and deleted automatically upon closing.

4.2.4   Network Layer

The functional interface exhibited by this optional layer is equivalent to the one of the file layer for access to PBFs, but allows instead accesses to PBFs over the network. In order to minimize network traffic, the functions of the file layer are used for local and temporary PBFs. The network databases are accessed through a database server as shown in Fig. 7, which opens, reads, writes and closes PBFs.

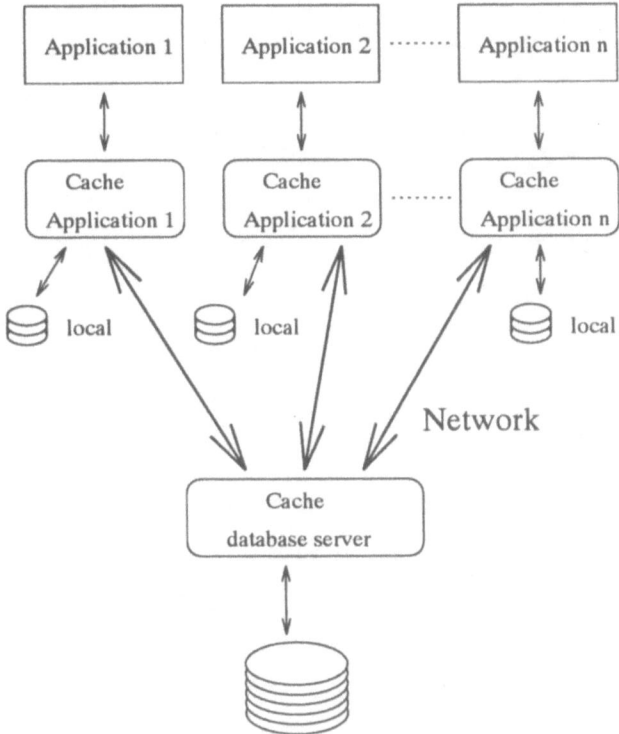

Figure 7: Local and network storage.

The client uses the database server for all file I/O functions on the network PBFs, but all database operations are done locally with the help of the basic and interface layers. For fast access to the data, the server holds some data blocks of the files in a local cache similar to the caching layer. This cache is shared by all clients and is not cleared upon closing a PBF, so that a following reopen and usage of the same file, even by a different application, is fast due to its remaining in the server cache. All write operations are delayed and buffered through a cache to maximize performance. The runtime option of unbuffered write operations ensures consistency of the PBF during update operations, and allows to examine a PBF while a tool is running and writing to it, which is an invaluable help in debugging simulators.

Another aspect of the network layer is the capability of message passing. It allows the application to contact other programs (e.g. the XLISP interpreter on the task level) over the network. Fig. 8 shows an example network configuration with tools and database servers interacting over the network.

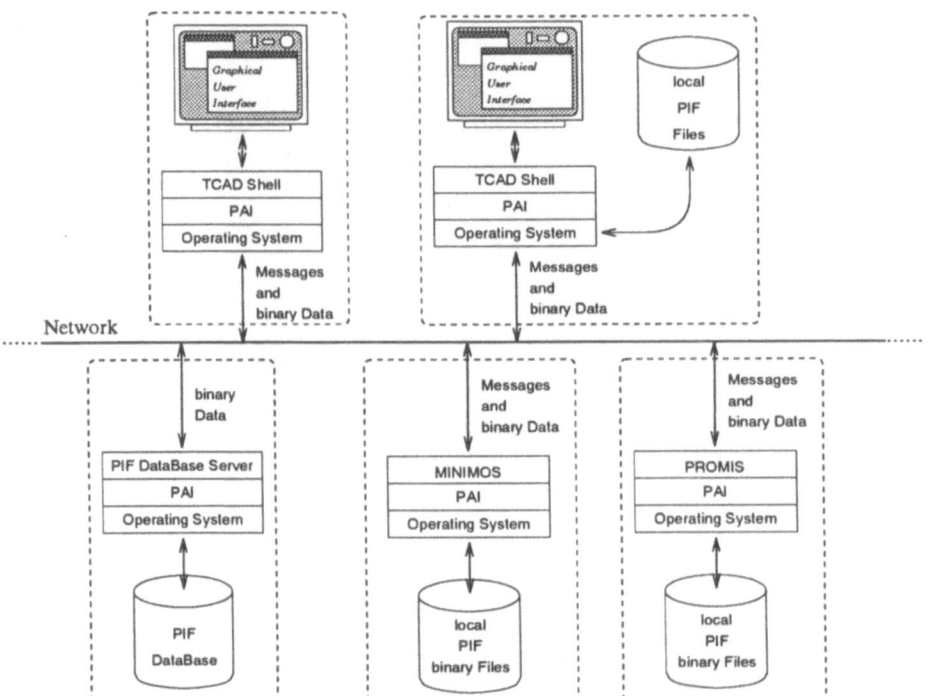

Figure 8: Examples of PAI networking capabilities.

4.2.5   Caching Layer

This layer buffers I/O data to minimize disc and network accesses. Depending on the application, the size of this buffer can vary from a few hundred kilobytes to several megabytes. The advantage of the cache is that data requested by read operations frequently can be found in the cache, while write operations can be delayed until closing of the file, depending on the page size and total cache sizes and on the page replacement algorithm. The caching layer is designed in such a way that the page-replacement algorithm can be substituted with a different one like LRU or random replacement of memory pages [24]. Currently, an algorithm implementing a combination of these two methods is used. The memory pages are usually as big as or — for better performance — even bigger than operating system cache pages. This layer also allocates file space for all types of objects. To optimize cache hits, all small objects with a few allocation units in size are contiguously stored in one big chunk whereas large data pieces are always appended at the current end of the file. Since the above layers need the functionality to update data items, a `free()` operation is also implemented so that no space on the physical file is permanently wasted. From the above layers, the caching layer can be seen as a big `malloc()`/`free()` library with access functions that perform cached file access.

4.2.6   Basic Layer

This layer is the lowest to implement structured data nodes. Fig. 9 shows the general structure of such a basic layer node. The header word of the node determines its type and structure, i.e. the type and number of the generic and specific data slots. The former are common to each node type whereas the latter carry the actual data visible to upper layers. Thus the shaded fields in this figure are maintained and used by the basic layer. For the unshaded fields the basic layer just reserves space and provides access functions.

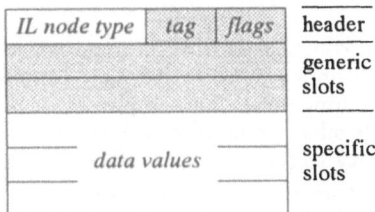

Figure 9: Implementation of a basic layer node.

The possible data types of the *specific* slots as determined by the *tag* field of the header word are:

Car   pointer to another node

Symbol   unique symbol name in the logical PIF file

Symref   reference to a symbol node

Char   character node

Byte   unsigned byte value (8 bit)

Short   short data value

Word   unsigned short data value (16 bit)

Long   long data value

LongWord unsigned long data value (32 bit)

Float   real data value

Double   double presision data value

LongDouble quad precision data value

All these types correspond to the **C** language types of the same size. To connect nodes together into a list or to implement arrays like strings (consequently stored as character arrays), the `flags` in the header word are used. The possible values are any combination of the following definitions, responsible for determining the *generic* slots of the node:

Cdr   the node has an implicit pointer to a successor node

Array   the node is an array (its size is stored as a separate entry)

Compressed the data of the node is compressed

With the basic layer a functional interface to a LISP-like information storage concept is implemented. The interface presents the notion of atoms (primitive data items like a number, character or string value) and constructor nodes (*CONS* nodes for list creation) to the upper layers, as described in [25]. One significant difference to a LISP interpreter's memory structure is that every basic layer node is implicitly a *CONS* node providing a *CDR* pointer as one slot in the data slots and carrying an atomic data value, i.e. the *CAR* pointer is redundant and therefore removed. The actual *CONS* node is implemented as a basic layer node with the atomic data value being a *CAR* pointer.

It should be noted that in contrast to LISP storage concepts, all nodes of the basic layer (and hence the PIF Application Interface) are originally kept on a file and are just cached through the caching layer. This implies, that all reference pointers stored in a basic layer node are file offset pointers and thus do *not* point to memory locations. It is the caching layer's duty to resolve those references correctly.

To illustrate the different storage concepts, let us consider the simple PIF expression `(ref P (valueList 1 2))`.

This construct represents a reference to the first two points of a `pointList` P. Fig. 10 shows how this construct would be stored in the XLISP interpreter with separate *CONS* nodes. The corresponding PBF structure, as it is handled in the basic layer, is shown in Fig. 14. The *CONS* nodes are fused with the *CAR* data values, to compact the data structures and minimize data access time, which is crucial on slower external storage media. Furthermore, this concept retains the principle extension language storage structures on the data level.

Using this concept of LISP-like information nodes the PLB is stored, whereas the PBF is built as a linked list of PLBs, shown in Fig. 11. Since this list is only searched when the file is opened, this is no performance drawback. The data area is shared by all PLBs in the PBF, but on write operations it is checked that no crosspointers into disjoint PLBs occur. To allow fast access to all symbols, these are stored in a hash table which is unique for each PLB so that there are no conflicts between different PLBs.

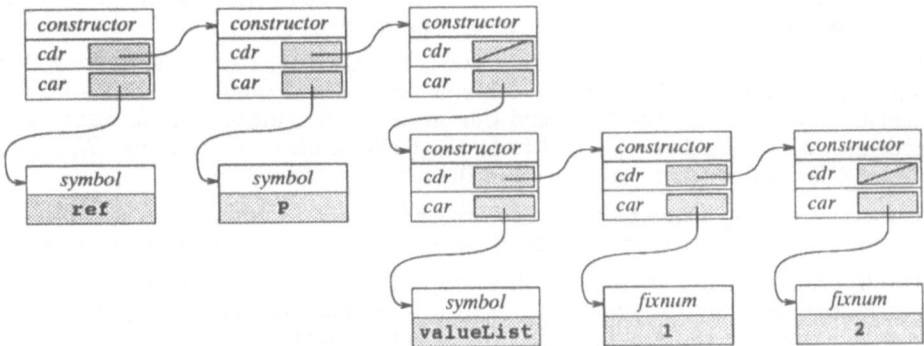

Figure 10: Example of a LISP internal data structure.

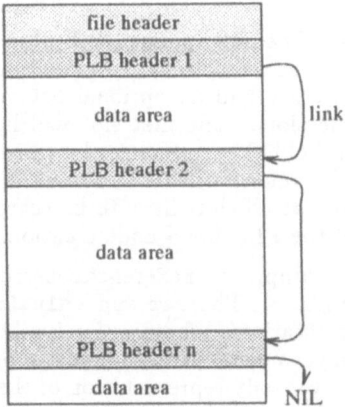

Figure 11: Layout of a PIF physical file with multiple logical files.

4.2.7   Interface Layer

This layer is the implementation of the PIF syntax, providing administration, access and inquire functions. To improve performance, it uses a structure of the basic layer array node to implement the interface layer nodes. A reserved field of the header word of a basic layer node Fig. 9 is used to store the type information of the interface layer node (labelled *IL node type*) as an integer value.

The **tag** field of the basic layer node contains the *CAR* type identifier, stating that this node contains *CAR* pointers to other nodes as data values. The **flags** field of the basic layer node has the *Array* and *Cdr* bits set, indicating that the interface layer node may have successors (pointed to by the *CDR* pointer) and multiple *CAR* data values (the number of which is stored in the *Size* field of the node).

The corresponding name of the object and all related information is stored in an automatically generated syntax table. This reduces the file size of a PLB significantly.

The *IL node type* identifies the PIF object type like *pointList*, *geometry* or *valueType* which are defined by the syntax. This field is automatically checked upon creation of a node through the syntax table. All access functions and the syntax table are generated automatically from a syntax description an example of which can be seen in Fig. 12.

```
(rule 'snapshot
      '(deriv
      LPAR SNAPSHOT OBJNAME
      (opt comment)
      (llist nameList)
      (llist attribute)
      RPAR))
```

Figure 12: Abstract syntax description of the PIF **snapshot** construct.

The *snapshot* construct is defined as a named PIF object, whose name can be used to search for the object. It has an optional comment, an optional list of references through the *nameList* construct and an optional list of *attributes*. Therefore, the node will have four specific slots. The first slot will hold the unique name of the *snapshot*. The second will hold the comment, the third the references and the last the *attribute* definition. This information is also used to limit the search depth when traversing the tree (in case not all slots have to be searched) on PLB inquiries, and to check the correctness of the PLB upon node creation.

The previously mentioned example of a **reference** construct represented with interface layer nodes is shown in Fig. 13. The **ref** and **valueList** constructs have specific interface layer node representations, whereas the symbol name and the **valueList** indices are genuine basic layer nodes, since they just represent primitive data values. Compared with the basic layer-only representation of the same reference construct in Fig. 14, a significant reduction in the number of required nodes and total storage size can be seen, resulting in faster data access.

Since the syntax defines many fields optional to allow a wide range of possible constructs, additional "language rules" are needed to define a well constructed PLB which can be understood by different simulators. Many of these rules are implemented by the application layer, others are described in the PIF CookBook (see [26]) which defines the semantic meaning of the PIF.

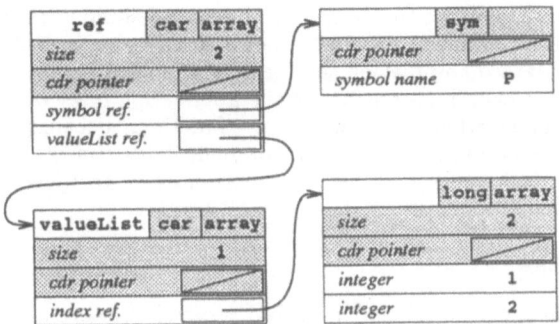

Figure 13: Example of an interface layer data structure.

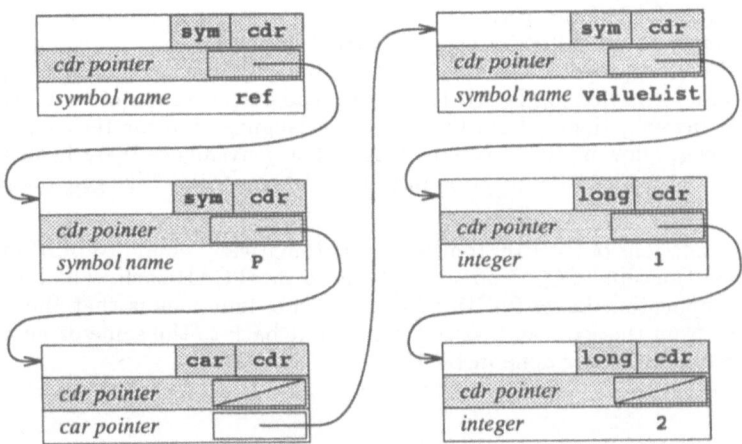

Figure 14: Example of a PIF binary file and basic layer data structure.

4.2.8   Application Layer and Language Binding

This layer implements some functionality common to simulators and utility programs. Its design is intended to be extendable in order to adapt the interface to new simulators or special demands. Many semantic rules and checks are implemented in the application layer, making the adaption of existing simulators to PIF easier, and ensuring interoperability in the VISTA framework. High-level functionality and automatically invoked data-manipulation services are provided to relief TCAD tools from tedious "everyday" work. The routines of the application layer implement geometry-manipulating as well as attribute-manipulating functions, because we think that both aspects are closely related in a TCAD environment. This fact is expressed in the uniform data representation of geometries and attributes on geometrical objects in PIF.

FORTRAN Interface

As the application layer is written in **C** and most simulators are written in **FORTRAN**, we have developed language bindings for most application layer functions and all inquiry functions to **FORTRAN**. This binding is strongly dependent on the two compilers, since there is no standard in parameter passing of strings in **FORTRAN** and the implementation of logical values (.TRUE. may be represented as the cardinal number 1 or -1). So we generate all binding functions automatically out of a formal description and additional information about the specific **FORTRAN** compiler. All string and logical variable conversions to **C** types are done automatically before the user-supplied **C** code is called. Adding a new binding or another compiler requires only few additions in the configuration files.

LISP Interface

The LISP interface of the PAI is not built on top of the application layer, since it makes no sense to use LISP for computationally intensive calculations on PIF data. The extension language of the task level is primarily used to generate input PLBs and control information for TCAD tools or to read output values of simulation results for further investigations. Thus the extension language interpreter connects to the interface layer, allowing full access to PBFs. For convenience there is an additional LISP library to support the creation of whole PIF constructs (like generated with the application layer).

High level functions of the PIF ToolBox are automatically bound to LISP by the Tool Abstraction Concept (TAC) and so available to the TCAD shell. The big difference between Application Layer functions and ToolBox functions is that the second get their input from the PLB and write their output back to the same or another PLB. No data manipulation is done in LISP.

4.3   Use of the PAI

The short code example in Fig. 15 shows the C calls to generate an example of a PIF data structure. `namelist` is the handle to the parent `nameList` object [26]. The two element array `points` holds the indices of the points on which the line is created. In the Application Layer code example Fig. 16 this part of information is generated by the function `palWriteLineList1`. In addition to the reference construct this function generates the whole `lineList` construct, as can be seen in Fig. 17.

```
{
    /* local variables */
    paiObject valuelist, ref;
    paiLong points[2];

    points[0] = 1;
    points[1] = 2;
    ref = pilCreateRef(
        namelist,                /* parent nameList construct */
        pointlist,               /* referenced pointList P */
        pilCREATE_NESTED);       /* create a new reference construct
*/
    valuelist = pilCreateValueList(
        ref,                     /* parent ref construct */
        pilDATA_INTEGER,         /* data type is integer */
        points,                  /* data points indices */
        0, 2,                    /* which values to write */
        pilCREATE_NESTED);       /* create a new valueList construct
*/
}
```

Figure 15: Interface layer code example.

```
{
    /* local variables */
    paiObject linelist;
    paiLong endindices[1];
    paiLong objdx[2];

    endindices[0] = 2;
    objdx[0] = 1;
    objdx[1] = 2;
    linelist = palWriteLineList1(
        parent,                  /* handle to PIF file */
        "myLine",                /* name of the lineList */
        1,                       /* number of lines */
        endindices,              /* endindices of the lines */
        2,                       /* number of used points */
        pointlist,               /* handle to referenced pointlist P */
        objdx,                   /* point indices */
        palCREATE_NEW)           /* create a new lineList */
}
```

Figure 16: Application layer code example.

```
(lineList "myLine"
        (nameList (ref P (valueList 1 2))))
```

Figure 17: PIF construct produced by example code.

4.4   PIF Binary File Manager

As mentioned above, the whole PAI works on a binary representation of the data for fast access. This type of data storage is optimized for architecture-dependent coupling of simulators in non human-readable form. For data exchange via eMail or FTP, or making PLBs human-readable, there is the ASCII PIF representation holding the same information. The PBFM (Fig. 18) is able to convert the binary to ASCII PIF and vice versa. Thus data exchange between machines with different byte ordering (little and big endian) and floating point formats (e.g. IEEE, VAX and IBM) is possible by converting PLBs to ASCII PIF and back to binary format on a machine with different architecture. The maintenance functions of the PBFM allow the user to list all PLBs of a PBF, delete any PLB within a PBF, repair a not cleanly closed PBF and check PIF ASCII files for lexical correctness.

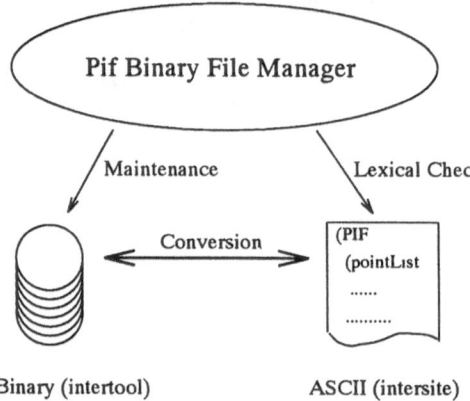

Figure 18: Using the PIF Binary File Manager.

4.5   Semantic Issues and High-Level Functionality

The PIF itself, be it its ASCII or binary form, only defines a syntax. It does not prescribe any interpretation of the stored TCAD data. This is one issue which accounts for the flexibility and general-purposeness of the PIF. On the other hand, many ambiguities arise from the possible multilateral description of the same physical problem in terms of PIF syntax. There are many ways to describe a geometry, ambiguities in recognizing a grid or an attribute, and general PIF semantics. These ambiguities arise from different coordinate systems, hierarchical or non-hierarchical geometry specifications and using one or many lists of primitive geometric objects, with or without references to other PLBs. Grids can be of unstructured or tensor product type, defined on a segment or the whole geometry and attributes can be defined on the grid, its points, lines, faces or solids. Moreover, the application interface has to know what to do with different units of measure, when to write and what to reference in a **snapshot**, **geometry** or **written** construct, where to define attributes, what attribute types to use, ...).

However, ambiguities and multilateral descriptions are a general problem, because the more general a syntax is and the more functionality a procedural interface has, the more semantic standardizations are needed to make applications work properly in a common environment.

In order to unambiguously interpret PIF data, there have to be both semantic constraints which applications have to adhere to (losing PIF flexibility), and ambiguity resolution mechanisms built into the application interface. However, only few additional semantic rules specified in the PIF CookBook [26] have to be obeyed. The PAI takes care of different coordinate systems through a transformation matrix applied to geometrical data, and accounts for different units of measure through a unit conversion system (e.g. point coordinates can be written in micrometers and read in inches, different spatial axes can have different units). It automatically resolves links to other PLBs and provides a multitude of inquiry functions for locating a certain PIF construct wherever it appears in the PLB.

The more severe semantic differences between simulators (e.g. a simulator working on an unstructured grid coupled to a simulator using a tensor product grid) are dealt with in the PIF ToolBox, comprised of generic PIF tools such as grid generators, interpolators, attribute and geometry manipulators using the PAI and preparing a PLB according to the semantic standards of the PIF CookBook [26]. However, these tools are controlled by the task level and belong to the tool rather than to the data level.

A particularly difficult problem is the support mechanism for the innumerable different grid types used today. A distinction between tensor product and unstructured grids has been made, because we didn't want to lose an orthogonal grid's unique features by decomposing it into rectangles/cuboids. Therefore the special orthoProduct construct was introduced, which significantly enhances the efficiency of storing tensor product grids while preserving its advantageous structure. However, since the number of different unstructured grid types increases steadily, a specification mechanism for dynamically adding new grid types just by providing a unique name, an interpolation and a decomposition function has been implemented. Using automatic code generation tools, these routines are linked into generic PIF ToolBox functions, thus adding support for the new grid element to the whole framework. Fig. 19 shows some example elements and how they are referenced in a PIF grid. Through using an attribute defined over the faceList it is possible to specify different element types in one and the same grid.

New element types are introduced by specifying the name, dimensionality, number of nodes and a decomposition and interpolation function in a element definition table. After recompiling the PAL, the new element type is known to applications through a unique constant identifier. But most applications need not explicitly take care of new element types: Reading attributes defined over a grid can be done without knowledge of the grid, since there is a generic interpolation function, which is automatically invoked when requesting an attribute value at a location (x,y). The generic interpolation routine knows the grid type the attribute is defined on, and correspondingly invokes the orthoProduct interpolation or determines the element in which the requested location lies, then invoking the element interpolation function defined for that element type.

Tensor product grids are supported through the orthoProduct construct. The grid has an origin point, may have different topological and topographical dimensions and each dimension may have a different base vector. Conforming to the PIF syntax, the number of supported dimensions is infinite. The example Fig. 20 shows a tensor product grid of topological dimension 2 lying in three-dimensional space. This capability is needed e.g. to describe distributed boundary conditions of a three-dimensional device.

The assembly of solver matrices is not supported by the PAI, since we believe that this task is very problem-specific and current networks don't exhibit the necessary

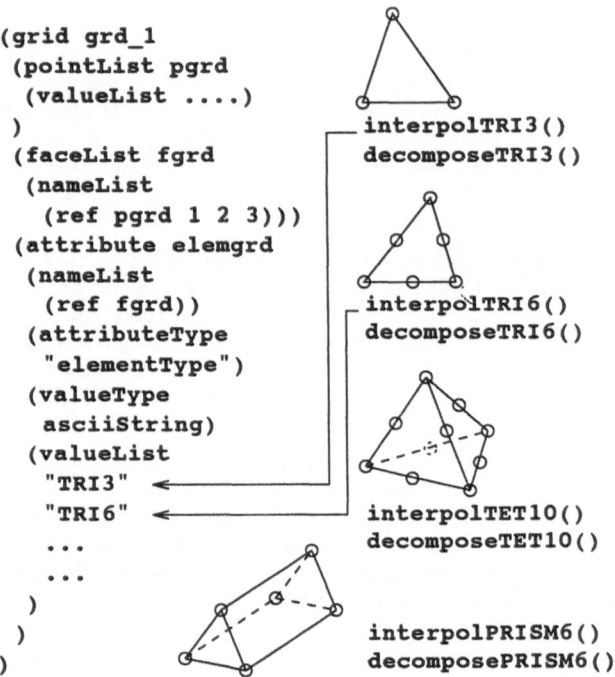

```
(grid grd_1
 (pointList pgrd
  (valueList ....)
 )
 (faceList fgrd
  (nameList
   (ref pgrd 1 2 3)))
 (attribute elemgrd
  (nameList
   (ref fgrd))
  (attributeType
   "elementType")
  (valueType
   asciiString)
  (valueList
   "TRI3"
   "TRI6"
   ...
   ...
  )
 )
)
```

interpolTRI3()
decomposeTRI3()

interpolTRI6()
decomposeTRI6()

interpolTET10()
decomposeTET10()

interpolPRISM6()
decomposePRISM6()

Figure 19: Support for unstructured grids.

```
(orthoProduct my_tensor_grid
   ; the 3D base point
   (origin
     (units "um")
     (valueType (point 3 real))
     (valueList 1.0 1.0 0.0))
   ; the two axis vectors
   (base
     (valueType (vector 3 real))
     (valueList 1.0 0.0 0.0
        0.0 1.0 0.0))
   ; the two axis specifications
   (axes
     (valueType real)
     (valueRange 0.0 5.0 0.23)
     (valueList 0.0 0.1 0.22 0.37 0.4))
)
```

Figure 20: Tensor product grid example.

performance to transfer these large amounts of data in an acceptable time frame to a solver server. The "know how" of a simulator is always contained in its physical models, the knowledge of which is essential in matrix assembly. A simulator using a standard matrix assembly method would lose much of it's advantages. This holds true for grid generation and the partial differential equation solver too.

4.6   Performance Evaluation

Besides the goals of classical intertool PIF implementations featuring object-oriented-ness (PIF/Gestalt, [14]) or suitability for TCAD environments (BPIF, [13]) our implementation stresses efficiency in terms of run-time performance and database compactness. Thus, writing and reading 10 000 points (in three-dimensional space) of a PIF `pointList` takes 0.51 and 0.66 seconds (real time) respectively on a DECstation 3100; the database written is 250 kB in size. Therefore, linking a TCAD application to the VISTA environment is not a performance issue. In contrast to a client-server approach, the administrative and communication overhead is negligible for any application consuming a few seconds of CPU time – the commonly used argument, that PIF is not practical because of its low run-time performance no longer holds true.

5.   Implementation of the VISTA User Interface

5.1   Structure

The structure of the VISTA user interface is shown in Fig. 21. The bottom layer is the X Toolkit[18], an object-oriented subroutine library, designed to simplify the development of X Window applications. The X Toolkit defines methods for creating and using widgets, which appear to the user as pop-up windows, scrollbars, text-editing areas, labels, buttons, etc. Basic functionality is provided by the generic *Athena widgets*, which are part of the MIT X11 distribution. We have decided to use this widget set rather than any other open standard, because a migration from these generic widgets to another widget set (like OSF/Motif, or Open Look) is significantly easier than vice versa.

A widget-wrapping layer has been put on top of these widgets in order to achieve some widget-set independence. All widgets are created and modified via specific functions rather than via the generic interface of the X Toolkit. This facilitates the potential migration of the entire user interface onto another X Toolkit-based platform.

In addition specialized VISTA widgets have been developed on top of the widget-wrapping layer for supporting TCAD-related information flow. The VISTA widgets are also created and accessed via specific functions, so that they can more easily be replaced by other widgets, should the need arise.

The top layer, the VUI (*VISTA User Interface*) library serves two purposes. It provides some often needed higher-level operations and it simultanously contains most of the policy which is shared among VISTA applications. In other words, the VUI library takes care that different parts of VISTA look alike and behave similar. Interactive applications (like visualization clients or the device editor) have their own VUI-based user interface, whereas applications requiring no user interaction (like simulators or converters) are provided with a front-end user interface which is executed by the XLISP interpreter.

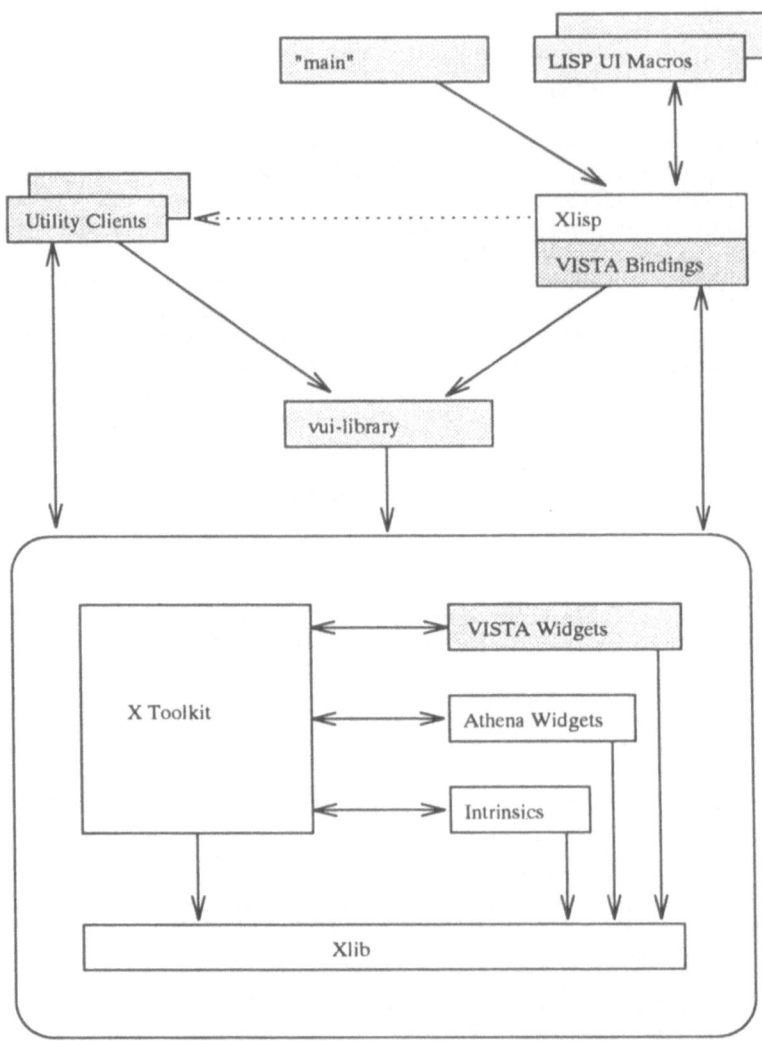

Figure 21: The structure of the VISTA user interface. Shaded boxes represent extensions to the public domain products XLISP and the MIT X Window system. The arrows indicate the sequence of function calls between different parts of the user interface.

5.2 The VISTA Widget Set

The *Athena* widgets were developed by MIT's project Athena (it is part of the X release). This widget set was not intended to be sufficient for all purposes and thus does not fulfill all the needs of a TCAD user interface, but it does provide the required generic functionality, it is highly portable, it is available on virtually every modern workstation platform, and it is easy to comprehend.

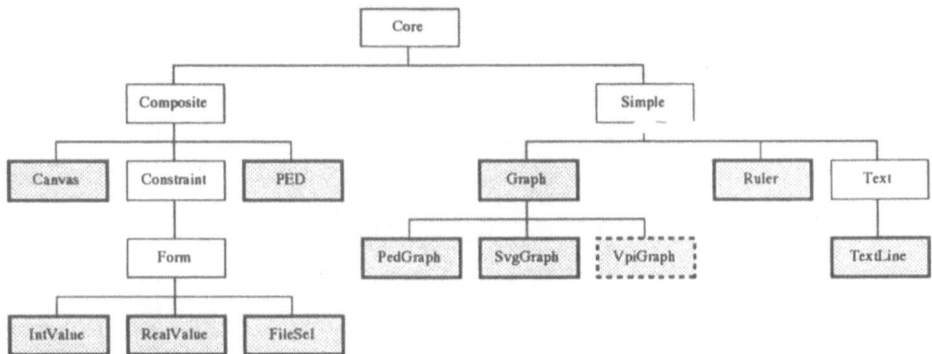

Figure 22: The widget set used by VISTA. The VISTA extensions are shaded, the Intrinsics and Athena Widgets used for subclassing are blank.

The VISTA widgets are subclassed from either X Toolkit Intrinsics or Athena widgets (Fig. 22). The *Canvas*, *PED*, *PedGraph*, and *Ruler* are parts of the interactive PIF Editor (PED), the *IntValue*, *RealValue*, *TextLine* and *FileSel* are widgets for the specification of integer, real, and string values, and files respectively, and the *SVGraph* widget is a widget for displaying simple vector graphics plots.

5.2.1 The PIF Editor

A data level implementation would be incomplete without an interactive graphical editor for manipulating the geometrical data (device geometries) stored in the binary PIF. The *PIF editor* (PED) is the front-end user interface for the interactive creation and modification of geometrical data in one, two and three spatial dimensions and of all attributes (like the material type) which define the device structures (see Fig. 23)

The PED makes use of the *Canvas*, *Ruler*, and *PedGraph* widgets and is implemented as a widget itself (see Fig. 23). This allows the use of multiple subwindows for editing one and the same device geometry, editing of several logical PIF files in one PED process and even using the PED as a component in "surrounding" applications. Thus, arbitrary additional menus or other widgets can be added without interfering with the PED itself.

The PED can work on all PIF files independent of the specific semantical contents. It is a generic tool for building a simulator input PIF file from scratch, for modifying existing device structures, and for visualizing geometric PIF information.

The PIF data is held in a memory-resident intermediate representation which is slightly extended with respect to the binary PIF to allow efficient processing of inter- active manipulations. At the end of the editing session or during intermediate save operations the PIF file is updated.

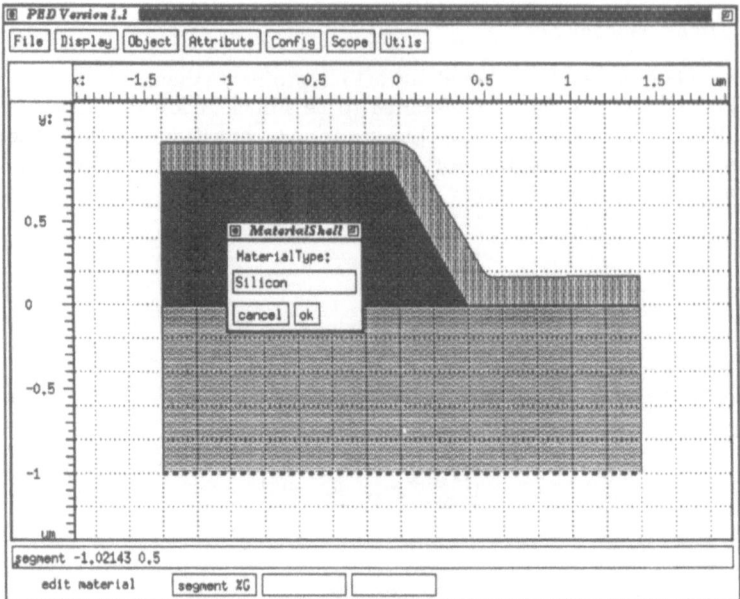

Figure 23: The PIF editor widget.

The hierarchical geometry structure is supported by "snapping" on existing lower-level geometrical objects and by automatically creating missing intermediate-level objects during input. Common techniques like background grid or coloring according to different physical or logical criteria are used to facilitate the comprehension and assimilation of the spatial information.

The top-level execution control of the PED is implemented as an extensible and configurable automaton which filters all user input and triggers appropriate actions. Through the default configuration for this automaton, all graphical functionality is accessible via mouse and keyboard input is used for all non-graphical data.

The implementation as a widget, together with the overall architecture of the user interface and task-level allows LISP-coded editing macros to be added easily. It should also be noted that the use of LISP as an extension language for an interactive geometry editor has already proven to be a very successful strategy for interactive geometrical CAD[21].

5.2.2   Vector Graphics Widget

The X Toolkit and Athena widget set do not provide "classical" two-dimensional vector graphics capabilities, which are a firm requirement for any CAD discipline. To support platform-independent vector graphics output we have implemented a minimum-functionality vector graphics widget (Fig. 24) which is built directly on the generic Xlib and X Toolkit.

The widget remembers all drawing commands and provides zoom and pan functions for the user, which henceforth, the programmer does not need to bother with. Callbacks can be utilized for example to digitize data points. This widget is used as interactive back-end of VISTA's visualization library.

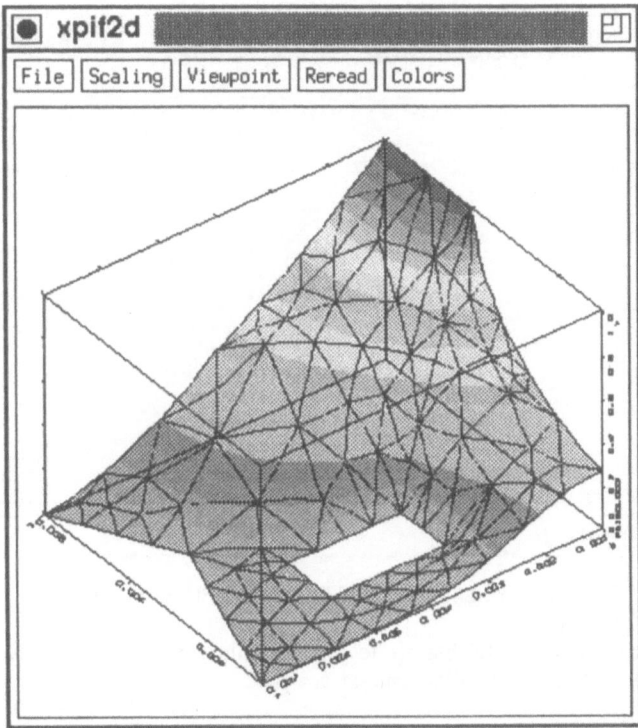

Figure 24: The vector graphics display widget is used for displaying the output of PIF-based visualization tools.

The contents of the widget (i.e. the "plot") can be converted to PostScript format, other converters can be added easily due to the limited set of drawing commands which is used.

5.2.3   File Selection

As it is not provided with the Athena widget set, we have implemented an advanced file selection widget (see Fig. 25), which allows operating system transparent specification of files (including a *GNU Emacs* like filename completion) using a string subwidget and operating system independent traversal of the directory tree and selection of existing files using list subwidgets.

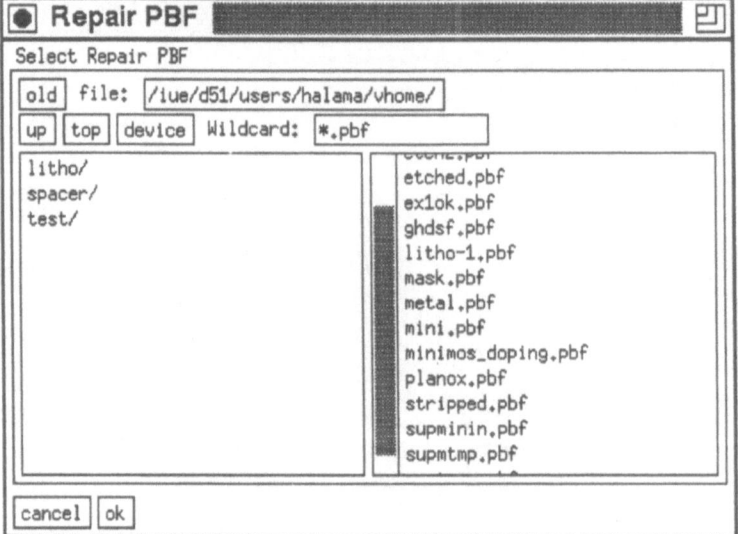

Figure 25: The VISTA file selection widget.

The selection of logical PIF files (one physical file can contain multiple logical PIF files) is implemented as so-called *widget bouquet.*

5.3   The VUI Library

5.3.1   Bouquets

The VUI library contains functions which create often-used combinations of several widgets in one step, arrange them and set up all required connections and callback functions. These widget *bouquets* behave as if they were single composite widgets and are indistinguishable from the user's point of view. This approach is similar to the OSF/Motif "Convenience Function" concept[27], and helps to maintain a unified appearance for different VISTA applications.

The following examples show some widget bouquets for performing TCAD-related subtasks:

The periodic table shown in Fig. 26 is implemented in C and is used by applications to let the user select a "pure element" material from a material database, which is

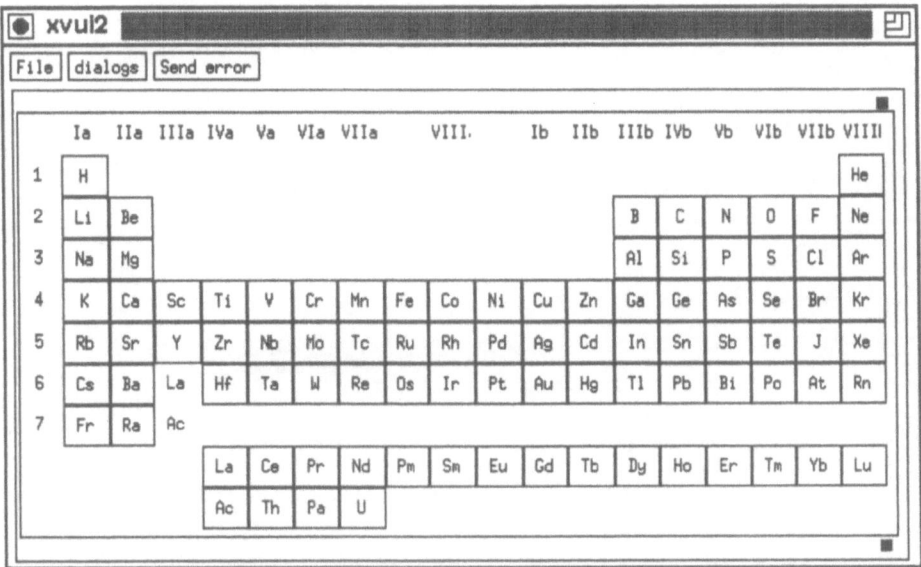

Figure 26: The periodic table bouquet lets the user select chemical elements as bulk or implantation material.

shared by all (fully integrated) simulation tools. It provides a familiar method for the identification and specification of single chemical elements. This bouquet is used, e.g., to ask the user for the bombarding ion species for the Monte Carlo simulation of ion implantation.

The symbolic PIF browser bouquet (see Fig. 27) is generated by a protoype LISP program. It is a generic intuitive facility on the task level for the selection of PIF objects and represents the hierarchical structure of the PIF file in iconic form. It can be used in any step which requires the specification of one or more PIF objects, like for example visualization (choice of attribute to be visualized), inquiry, or post-processing operations.

5.3.2   Tool Control Panels

Applications which don't require user interaction (like all classical simulation programs) can very easily be provided with a "supply parameters and run" user interface. The widget bouquets shown in Fig. 28 and Fig. 30 are tool control panels which are created from a formal specification of the tool and its parameters by an interface generator, which is implemented in LISP. This high-level user interface tool, in most cases, relieves the application engineer from the need to use X Toolkit programming to make new tool control panels.

6.   Implementation of the Task Level Shell

The applicability of XLISP for the execution of practical TCAD optimization tasks has already been demonstrated[28]. The programming language features offered by LISP are a powerful and efficient basis for carrying out complex task flows, like, e.g., nested optimization loops.

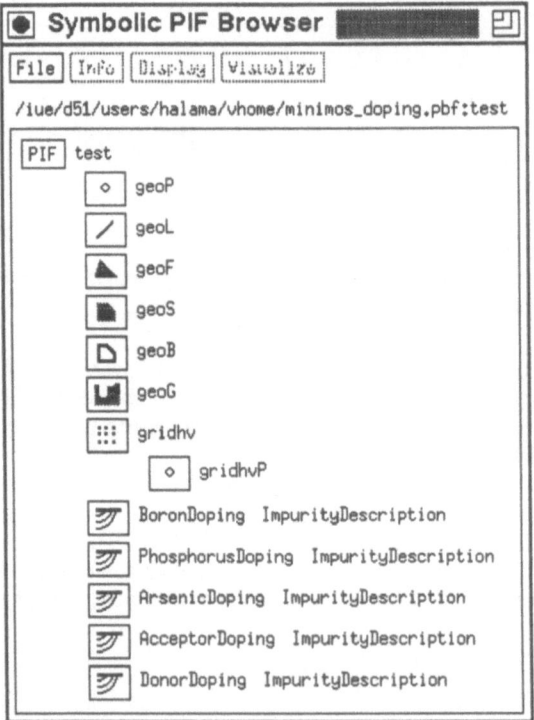

Figure 27: The symbolic PIF browser bouquet is created according to the data contained in a (binary) logical PIF file and reflects its hierachical structure.

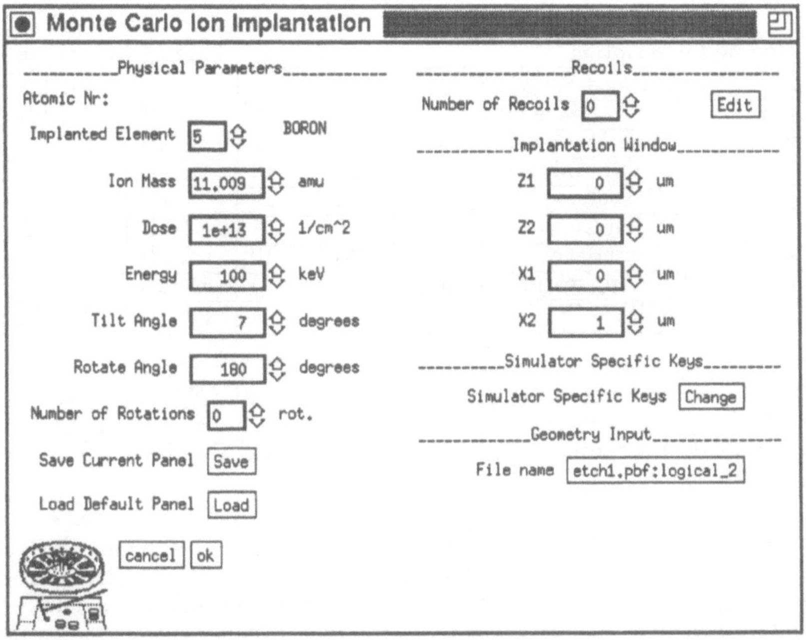

Figure 28: This widget bouquet is the user interface for the PROMIS Monte Carlo ion implantation module.

The *XLISP* interpreter which is used for the task level implementation of VISTA has also bindings to the *VUI-library*, the *X Toolkit*, and the widget sets. This integration of the user interface into the task level interpreter (which can be viewed as the "main program" of the TCAD system) is indispensable for supporting the casual user with a comfortable point-and-click interface and for providing the experienced user and programmer with a high degree of customizability and extendability.

Furthermore, some parts of the user interface have been implemented as *LISP macros* and are loaded at run-time. The XLISP integration has also proven to be a valuable tool for the rapid prototyping of user-interface concepts and for tasks like the automatic generation of widget bouquets, like the tool panels in Fig. 28 or Fig. 30.

The X Window interface is not a feature of the original XLISP interpreter, although OSF/Motif bindings[22] are available, which, unfortunately, besides the disadvantages mentioned earlier, make heavy use of the "non-standard" object system of XLISP.

To preserve the consistency and simplicity of XLISP and in order to provide a homogeneous procedural interface and programming environment, we had to implement the X Window interface (*VISTA UI Bindings* in Fig. 21) for XLISP from scratch. As there are other C-coded parts of the framework which need to be accessible on the extension language level, a generic, automatic method, for linking given functions with the XLISP interpreter has been implemented.

The size of the original XLISP interpreter (Version 2.1) is 270kB. Some specific extensions have been made to the interpreter, mainly for event handling and for introducing miscellaneous operating system interfaces, like for process control (for running simulators), and other features which are needed for TCAD purposes. These 200kB of extensions thus increase the size of the task level interpreter to 470kB.

Module	description	N_f	N_c	binding	code size
xvw	Extended Widget Set	105	19	115 kB	531 kB
vui	User Interface Library	83	3	96 kB	254 kB
ve	Global Error System	14	37	20 kB	165 kB
svg	Graphics Library	22	10	21 kB	73 kB
ptb	PIF Toolbox	8	29	26 kB	190 kB
ver	Version Control	0	7	1 kB	0 kB
vos	OS Interface	(63)	12	(2) kB	214 kB
pai	PIF Application Interface	(20)	(117)	(54) kB	2256 kB
	total	315	234	335 kB	3683 kB

Table 1: Number of functions (N_f), number of constants (N_c), size of code for the XLISP interface, and module size of every module that is linked with the XLISP interpreter. Numbers in parentheses indicate manual binding, all other code is generated automatically.

The code required to implement LISP interfaces for framework modules which are relevant for the task level is currently 335kb (see 1), but is steadily increasing. There are only a few functions (indicated by parentheses in Table 1) which are manually bound to the interpreter. The vast majority of the interface code is generated automatically during the build phase of VISTA.

In order to integrate the X11 event handling mechanism in a consistent way, the standard read-eval-print loop was extended to handle events coming from the X11 system: Events coming from the user interface (like an expose request for a window) are passed to the interpreter and processed simultaneously with keyboard input or events coming from other (network) streams connected to the XLISP interpreter or from signals which indicate events like child process termination. During the read phase, the interpreter first checks whether the evaluation queue of pending expressions needs to be emptied and optionally evaluates these "callback" expressions. Then it waits for any event coming from streams, terminal input, signals (e.g. from terminating simulator runs), or the X Window system. When any of these events occurs, the interpreter puts the associated callback expression into the abovementioned evaluation queue — or in the case of terminal input — waits for completion of the expression and then evaluates it.

6.1   The Callback Concept

The object-oriented callback concept of the X Toolkit may be generalized in a very straightforward manner and successfully applied to those parts of the TCAD framework where a proper decoupling and high flexibility of the control flow is desirable. It is obvious that this is of special value for a flexible task level implementation.

Events coming from the X Window system are passed to the XLISP interpreter. If a LISP expression was associated with the activated widget at creation time, this expression is evaluated by the interpreter and can be used to change parameter values, trigger other events like the execution of a simulator or start the evaluation of a LISP program or any other tool.

The same callback concept is also used for the control of simulator execution. If a simulation tool terminates, it signals the termination to the parent process, which again causes an associated callback expression to be evaluated. Callbacks can be triggered by the user interface, error handler, network layer, timer, or by termination of child processes.

By agreeing upon a standard prototype for callback functions (which is already specified by the X Toolkit):

```
void callBackFunc(object_identifier, client_data, call_data);
```

it is possible to use a unified consistent method in various places throughout the framework, thereby gaining in simplicity and flexibility.

6.2   Task Level Tool Integration

As already pointed out earlier, there are different methods to create a task level interface for a given tool. Linking the tool with the interpreter is only feasible for small-sized applications, which have library-like properties. For larger tools (like simulators) a LISP interface must be coded to be able to run them from within the TCAD shell.

Fig. 29 shows a simple example of the task level interface of a conversion tool, which converts SUPREM 3 output files to PIF. The corresponding tool control panel, which is automatically created at run-time, is shown in Fig. 30.

7.   Conclusion and Future Aspects

It is the scope of methods and unpredictable requirements that makes technology computer-aided design a challenging discipline. It is yet this property that dictates the rigorosity of a future-oriented TCAD framework. The need for comprehensibility, on the other hand, prohibits the (otherwise feasible) creation of a framework by combining existing solutions.

In VISTA, we have achieved a homogeneous and comprehensible architecture by favoring the generalization of existing concepts over the introduction of new (maybe even better suited) partial solutions. A bottom-up *design* has been used, wherever possible, to be prepared for unforseeable future requirements.

Using PIF as the interchange format of the data level was the initial choice, which was motivated by the sole existence of the PIF standard at that time, and by its intrinsic flexibility and open-endedness. However, the *crucial* part of the data level is an efficient application interface like the VISTA PAI.

One of the major reasons for using the X Toolkit to implement the specialized user interface functionality as widgets is that it provides a *clear concept* for re-use and for future extensions. Finally, the well-known advantages[6][7][22] of an interpretive language for composing a user interface from widget-level building blocks have verified the choice of XLISP as extension language interpreter. The generalized use as extension language interpreter, "main" program of the user interface, and central facility for CASE-related tasks contributes significantly to the consistency and maintainability of the system. Automatic code generation (also using XLISP) helps to raise the level of abstraction on which problems like language binding are solved. Because of its homogeneity, the combination of LISP as the task level extension language and PIF as the data level interchange format has proven to be a flexible basis for a comprehensible and powerful TCAD framework.

```
; This function actually runs the sup2pif executable
;
(defun sup2pif (asc pbf pif)
  (vos::run
    (vos::os2vospec "sup2pif")
    (format NIL "~A ~A ~A" asc pbf pif)
  )
)

; This function is called when the confirm button is pressed
;
(defun sup2pif-confirm (widget client-data call-data)
  (sup2pif
    (vos::get-vospec  (vuu::get-value 'ASC client-data))
    (vos::get-vospec  (vuu::get-value 'PLB client-data))
    (vos::get-logical (vuu::get-value 'PLB client-data))
  )
)

; This defines the control panel for the suprem 3 wrapper
;
(setq sup2pif-widget
  (vui::generate-tool-panel
    (setq sup2pif-description
      '(
        (NAME     "SUPREM3 to PIF")
        (ICON     "topif")
        (CONFIRM  #'sup2pif-confirm)
        (FORM
          (ASC  VOSFILE  "SUPREM3 .sav file: "  "*.sav" NIL)
          (PLB  VOSPLB   "         output PLB: "  "*.pbf" "sup3prof.pbf")
        )
      )
    )
  )
)

; This creates an entry in the pulldown menu, which will pop up
; the tool control panel
;
(vui::add-to-menu vui::current-menu-button
      "suprem3->PBF"
      #'(lambda (widget client-data call-data)
          (vui::popup-tool-panel sup2pif-widget)
        )
      NIL
)
```

Figure 29: Task level LISP code which is required to integrate the SUPREM 3 output wrapper.

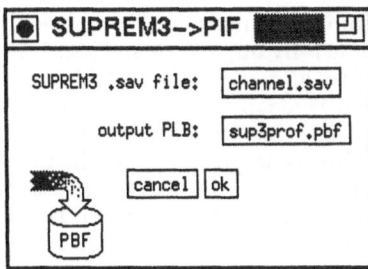

Figure 30: The tool control panel for the SUPREM 3 to PIF converter.

7.1 Tool Abstraction

The use of high-level tool abstraction methods for CAD tool management has been demonstrated through the Cadwell design framework[29]. We believe that a generalized and unified concept for the abstract characterization of tools (even in a low-level fashion, down to single functions) is highly desirable as it can be employed for many different purposes such as the automatic creation of a user interface or the generation of different language bindings. From an abstract tool description in LISP syntax, several pieces of interface code (generating a main program which takes care of argument-passing for a given function, thereby forming a stand-alone executable, for instance) can be generated automatically. The same tool abstraction can also be used to describe a tool for optimizers or other sequencing or analysis tools.

Another — already existing — application for a unified tool abstraction concept is the automatic integration of given C functions into the XLISP task level interpreter.

7.2 Visual Programming Interface

Visual programming capabilities are very valuable for the efficient support of any user. Almost every non-trivial task in TCAD is to a considerable extent data-flow oriented. The whole task is thus fairly well defined by the arrangement of modules and the flow of data between them, which also implies the sequence of tool execution. Again, both the callback concept (for module activation) and the tool abstraction concept (for module description) can be used for the implementation of a visual programming widget, thereby not increasing the system complexity from the programmer's point of view.

Visual programming is especially well-suited for building specific applications through the assembly of several generic functions or modules. We have tried to implement all new parts of VISTA (especially the visualization and the PIF ToolBox functions) using generic approaches. Visual programming puts the process/device engineer in a programmers role without the need to do programming work. It is expected that a visual programming interface will significantly contribute to the ease of use of the TCAD framework.

7.3 Object-Oriented Design Representation

In a TCAD environment, it would be convenient to represent devices to be simulated as objects belonging to a device class hierarchy and with methods attached to them. Thus a device would "know" how to simulate itself, i.e. its class would have methods attached which call the appropriate simulator. To achieve this, the design representation of the data level has to be fully object-oriented, and the procedural interface has to provide means to build class hierarchies and attach methods to classes. Since PIF provides a LISP-like syntax it is ideally suited to extend it with such object-oriented features. A C++ language interface would present those features to applications. Methods attached to PIF objects would be coded in C++ and made available to the extension language through the Tool Abstraction Concept. However, since only a minority of today's TCAD applications are written in C++, there is presently no strong need for such an interface.

Acknowledgements

The VISTA project has been sponsored by the research laboratories of AUSTRIAN INDUSTRIES - AMS at Unterpremstätten, Austria; DIGITAL EQUIPMENT at Hudson, USA; SIEMENS at Munich, FRG; and SONY at Atsugi, Japan, and by the "Forschungsförderungsfonds für die gewerbliche Wirtschaft", project 2/285 and project 2/299, as part of ADEQUAT (JESSI project BT1B), ESPRIT project 7236.

We are very grateful to

A. Gabara
University of California, Berkeley, California

N. Khalil
Digital Equipment Corporation, Hudson, Massachusetts

P. Lindorfer
National Semiconductors, Santa Clara, California

E. Masahiko, M. Mukai, and P. Oldiges
SONY Corp, Atsugi, Japan

H. Masuda
Hitachi Device Development Center, Tokyo, Japan

L. Milanovic, G. Nanz, C. Schiebl, R. Strasser, and M. Thurner
Campusbased Engineering Center, Digital Equipment Corporation G.m.b.H, Vienna, Austria

M. Noell
Motorola APRDL, Austin, Texas

H. Read
Carnegie Mellon University, Pittsburgh, Pennsylvania

K. Traar and G. Punz
SIEMENS AG, Vienna, Austria

for their patience and support, for their efforts in installing and testing VISTA, and for their contributions and criticisms.

References

[1] F. Fasching, W. Tuppa, and S. Selberherr. VISTA — The Data Level. *IEEE Trans.Computer-Aided Design*, 1993 (in publication).

[2] S. Halama, F. Fasching, H. Pimingstorfer, W. Tuppa, and S. Selberherr. Consistent User Interface and Task Level Architecture of a TCAD System. In *Proc.NUPAD IV*, pp 237–242, 1992.

[3] M.R. Simpson. PRIDE: An Integrated Design Environment for Semiconductor Device Simulation. *IEEE Trans.Computer-Aided Design*, 10(9):1163–1174, 1991.

[4] E.W. Scheckler, A.S. Wong, R.H. Wang, G. Chin, J.R. Camanga, A.R. Neureuther, and R.W. Dutton. A Utility-Based Integrated System for Process Simulation. *IEEE Trans.Computer-Aided Design*, 11(7):911–920, 1992.

[5] A.S. Wong. *Technology Computer-Aided Design Frameworks and the PROSE Implementation*. PhD thesis, University of California, Berkeley, 1992.

[6] J.K. Ousterhout. Tcl: an Embeddable Command Language. In *1990 Winter USENIX Conference Proceedings*, pp 133–146, 1990.

[7] J.K. Ousterhout. An X11 Toolkit Based on the Tcl Language. In *1991 Winter USENIX Conference Proceedings*, pp 105–115, 1991.

[8] H. Matsuo, H. Masuda, S. Yamamoto, and T. Toyabe. A Supervised Process and Device Simulation for Statistical VLSI Design. In *Proc.NUPAD III*, pp 59–60, 1990.

[9] P. Lloyd, H.K. Dirks, E.J. Prendergast, and K. Singhal. Technology CAD for Competitive Products. *IEEE Trans.Computer-Aided Design*, 9(11):1209–1216, 1990.

[10] H. Jacobs, W. Hänsch, F. Hofmann, W. Jacobs, M Paffrath, E. Rank, K. Steger, and U. Weinert. SATURN - A Device Engineer's Tool for Optimizing MOSFET Performance and Lifetime. In *Proc.NUPAD III*, pp 55–56, 1990.

[11] C.H. Corbex, A.F. Gerodolle, S.P. Martin, and A.R. Poncet. Data structuring for process and device simulations. *IEEE Trans.Computer-Aided Design*, CAD-7:489–500, 1988.

[12] D.M.H. Walker, C.S. Kellen, D.M. Svoboda, and A.J. Strojwas. The CDB/HCDB semiconductor wafer representation server. *IEEE Trans.Computer-Aided Design*, CAD-12:283–295, 1993.

[13] A.S. Wong and A.R. Neureuther. The Intertool Profile Interchange Format: A Technology CAD Environment Approach. *IEEE Trans.Computer-Aided Design*, 10(9):1157–1162, 1991.

[14] D.S. Boning, M.L. Heytens, and A.S. Wong. The Intertool Profile Interchange Format: An Object-Oriented Approach. *IEEE Trans.Computer-Aided Design*, 10(9):1150–1156, 1991.

[15] SWR Working Group of the CFI/TCAD TSC. *Semiconductor Wafer Representation Architecture*. CAD Framework Initiative, Austin, Texas, USA, 1.0 edition, 1992.

[16] D. Boning, G. Chin, R. Cottle, W. Dietrich, S. Duvall, M. Giles, R. Harris, M. Karasick, N. Khalil, M. Law, M.J.McLennan, P.K. Mozumder, L. Nackman, S. Nassif, V.T. Rajan, D. Schröder, R. Tremain, D.M.H. Walker, R. Wang, and A. Wong. Developing and Integrating TCAD Applications with the Semiconductor Wafer Representation. In *Proc.NUPAD IV*, pp 199–204, 1992.

[17] S.G. Duvall. An Interchange Format for Process and Device Simulation. *IEEE Trans.Computer-Aided Design*, CAD-7(7):741–754, 1988.

[18] P.J. Asente and R.R. Swick. *X Window System Toolkit, The Complete Programmer's Guide and Specification*. Digital Press, 1990.

[19] D.M. Betz. *XLISP: An Object-Oriented Lisp, Version 2.1*, 1989.

[20] R. Stallman. *GNU Emacs Manual*, 1986.

[21] *AUTOCAD Release 11 Reference Manual*, 1990. Publication AC11RM.E1.

[22] N. Mayer. WINTERP: An object-oriented rapid prototyping, development and delivery environment for building user-customizable applications with the OSF/Motif UI Toolkit. Technical report, Hewlett-Packard Laboratories, Palo Alto, 1991.

[23] F. Fasching, C. Fischer, S. Selberherr, H. Stippel, W. Tuppa, and H. Read. A PIF Implementation for TCAD Purposes. In Fichtner and Aemmer [30], pp 477–482.

[24] J. L. Hennessy and D. A. Patterson. *Computer Architecture: A Quantitative Approach*. Morgan Kaufmann Publisher Inc., 1990.

[25] P.H. Winston and B.K.P. Horn. *Lisp*. Addison Wesley, 1989.

[26] Institute for Microelectronics, Technical University Vienna, Gußhausstraße 27–29, 1040 Wien, AUSTRIA. *PAI Release*, 1.0 edition, 1992.

[27] *OSF/Motif Programmer's Guide, Release 1.1*, 1991.

[28] H. Pimingstorfer, S. Halama, S. Selberherr, K. Wimmer, and P. Verhas. A Technology CAD Shell. In Fichtner and Aemmer [30], pp 409–416.

[29] J. Daniell and S.W. Director. An Object Oriented Approach to CAD Tool Control. *IEEE Trans. Computer-Aided Design*, 10(6):698–713, 1991.

[30] W. Fichtner and D. Aemmer, editors. *Simulation of Semiconductor Devices and Processes*, volume 4, Konstanz, 1991. Hartung-Gorre.

Technology CAD at NEC

N. Tanabe

NEC Corporation, ULSI Device Development Laboratories,
1120 Shimokuzawa, Sagamihara, Kanagawa 229, JAPAN

Abstract

An overview of the TCAD systems at NEC is presented on the types of problems encountered in the LSI technology developments and the electrical designs. Most of the simulation tools described here are ones that have been developed over the past 10 years and exclusively used in the actual LSI design cycles. The framework tools, that support the practical use of TCAD systems, are presented first. Then, focusing on our contribution to the related fields as well as on our industrial activity, the development and the use of process, device, circuit and interconnect simulations are presented. Some concerns that may enforce the TCAD regime are also discussed.

1. Introduction

In this paper, we survey the development and the use of simulation in semiconductor technology developments and electrical designs at NEC. Here, we do not emphasize the details of any of physical models as well as numerical computation methods usually involved in the simulation. Rather, we intend to outline the individual tools in the relationship with an integrated TCAD environment accomplished at NEC. Nevertheless, by reviewing our works done over the past 10 years, we briefly refer to the topics involved in or configured to the systems that, we believe, have more or less contributed to the TCAD evolutions. In addition to those topics, we also refer to some of our activities that attempt to remove the gap between the simulations and the experimental results, either by the engineering methodology or by the task-force. Although process and device simulations are the key issues in this field, we include, in the category of TCAD, circuit and interconnect simulations. The reason of this is that the device technology has to be fed to the circuit design properly, likewise electrical characteristics and parameters obtained through process and device simulations or by the direct measurement have to be represented correctly by the compact model. Because there exists an inherent relationship between device technologies and/or TCAD and computer resources, we especially note some of our systems as a driving force to the research of advanced computer architecture in NEC. The tools described hereafter have been developed by many of colleagues listed in the references as well as the collaborations with University of Aachen, Carnegie Mellon University and University of Bologna.

2. Framework Tools

Process and device simulations usually impose a heavy burden on the designers in preparing input data and in compiling simulation results. An integrated simulation system enables an excellent user interface if Engineering Workstation (EWS) is adopted as a front-end processor. Likewise, quick turn around time is achieved if a high performance processor is attached to the EWS as a back-end processor, especially in executing CPU intensive jobs such as multi-job in a parametric space and Monte Carlo simulation. Here, the back-end processor can be EWSs as a distributed network computing environment, main-frame computers and high performance (vector or parallel) machines. The P&D Workbench (Process and Device simulation Workbench)[1] realizes an integrated process and device simulation environment on EWS in conjunction with several framework tools and server machines, thus facilitating the use of simulation in device technology developments. Fig. 1 shows the system configuration of P&D Workbench. Besides the simulation programs, the system is composed of five framework tools; a data and job manager MEDLEY+ESCORT[2], an interactive input data capture DAIJOBDA, a data-base for standard process recipes and simulation results, a graphic output processor VDP-CORE and a job/data transfer for server machines. Fig. 2 shows an example of DAIJOBDA windows. Each of them represents; process editor (A), parameter table (B), data-base reference (C), utilities for device construction (D) and simulation (E), graphic output menu (F) and job sequence menu including selection of server machine (G). The system is installed on EWS4800 and is distributed to many of device technology development divisions as well as fabrication lines.

A similar facility is also realized in the use of electrical circuit simulation. The system is configured in either multi-tasking or encapsulation to the front-end tools[2]. The former enables a bustle simulation typically being required in analog cell designs. To the contrary, the latter is mainly used for large scale DRAM circuit simulations.

As a framework tool from another viewpoint, XPLAN is a tentative, however, a promising tool that adopts the experimental design method into the simulation[3]. The system is composed of experimental simulation, analysis of variance (ANOVA) and optimization. Here, the sequence of experimental simulations is determined by the Taguchi-method. With this method, the simulation is needed only in the sets of factor levels in terms of orthogonal array, therefore, the number of experimental simulation runs is extremely reduced. For example, four simulation runs are enough for three factors each of which has two level conditions. Fig. 3 shows the system flow of XPLAN. In the experimental simulation domain, the controllable factors X and their fluctuations dX are assigned to the inner and the outer orthogonal array respectively, then the elements of response matrix Y are simulated. The signal level S and the signal to noise ratio S/N are obtained by taking the mean value and the standard deviation in each row of Y. Here, S/N corresponds to the robustness in terms of fluctuation and S regards to the target of response. A factor effect graph visualizes the sensitivity and the dominance over the factor levels. In order to attain the composite target optimization in multiple factor and response space, the factor effects are transformed to orthogonal polynomials, then a computational optimization technique is applied. Our belief is that the system described above provides not only the design optimization but also a better engineering method in tunning model parameters usually involved in attaining the practical use of simulations.

3. Process Modeling and Simulation

The origin of process simulation in NEC is traced to a prototype, however, a novel thermal oxidation simulation[4]. In this program, the oxidant diffusion through the growing oxide and the change in shape in the total system were modeled concurrently. The viscoelasticity was taken into account that formulates the oxidation-induced deformation based on the balance-of-force viewpoint. Thus both compressive/tensile and shear components of stress vectors were calculated. The use of this model in LOCOS oxidation simulation demonstrated good agreement in comparison with experiment(Fig. 4) and proved the feasibility of using this model in general 2-D oxidation simulation.

For ion implantation, besides conventional analytical distribution models such as Gaussian, joined half-Gaussian and Peason distribution functions, numerical Monte Carlo simulation[5] has been applied. In this method, ion behavior is determined by directly referring to 3-D crystal structure data. Both channeling and dechanneling are well simulated by dealing with many-body scattering with lattice atoms as well as atom thermal vibration effects. Therefore, tail spreading of boron distribution due to subchanneling is calculated even if the ion beam is tilted or the crystal surface is covered with amorphous layers. The defect distribution due to implantation damage is also calculated and is used in the subsequent annealing process simulation. The Monte Carlo simulator is a versatile, however, rather CPU intensive tool, typically requiring more than 10 hours on a RISC EWS. Such a capability is then exploited to generate new and improved parameters for analytical models, or to extend their range to higher implantation energy.

Advanced dopant diffusion models have been also developed. A partial differential equation solver ZOMBIE[6] has been intensively used in order to compose and verify the models. One of them is a boron diffusion model in short time annealing process subsequent to ion implantation[7]. It formulates the non-equilibrium diffusion phenomena and the reaction of point defects as well as of defect-dopant pairs, by taking into account their charge states and the dopant inactivation due to clustering. Fig. 5 shows a simulation of boron distribution profiles that are reproduced well in comparison with experiments. As for the dopant diffusions in polysilicon, a formulation of dynamic clustering and grain growth kinetics, tightly coupled with dopant diffusion and segregation, has been proposed[8].

In our integrated TCAD environment, a system of 2-D process simulation SPARTAN2[2] serves for the practical use in process development. The system embodies fractions of SUPREM3 simulation models, empirical topography models and some of the advanced proprietary models described above in the finite element discretization method or by plugging in the case of Monte Carlo simulation. Although advanced process modeling will continue toward finer technologies, experiments are always needed to confirm accuracy and calibrate parameters no matter what capability of the resulting model is. Therefore, as some of CAD vendors advocating, it seems essential to construct experimental process data-base and incorporate it into the process calibration routines.

4. Device Modeling and Simulation

An early work in the field of device modeling in NEC is 1-carrier, 2-D MOS device simulation[9]. In this work, mechanisms for normalized drain current reduction and saturation voltage shift down to the $1\,\mu m$ channel length region had been studied and led to a formulation of analytical short channel MOS model. With experiences obtained thereby, a standard simulation package BIUNAP[10] was developed. The program features 2-carrier, 2-D drift-diffusion model incorporating with physical models such as bulk and channel mobilities, carrier recombination and generation, and band-gap narrowing effect. The program has been widely used in designing both MOS and BJT devices. The variants of the program, for the use in compound devices, were also produced. In order to extend the simulation ability into 3-D space, where the shape descriptivity is one of the key issues to realize an applicable tool, the "triangular prism" as an element for representing 3-D device structures was proposed[11]. The triangular prism itself is represented by face, edge and vertex primitives and the program data structures describing their connections are then generated. Therefore, the formation and deformation of shape is manipulated arbitrarily by modifying the data sequence in accordance with fabrication process steps. Fig. 6 shows a typical example that has a snapped interface commonly appearing in trench structures. In reality, however, the use of 3-D simulation is quite restricted, typically simulating alpha particle effects in the bulk. To bridge the gap between the simulation capability and the users demand, better discretization method for non-planer device structures, that is crucial even in the case of 2-D simulation, is required. The simulation turn around time, that also prevents the use of 3-D simulation, has to be improved as well.

Optimizing spatial discretization is still left to users that also reduces the practicality of device simulation. The adaptive mesh technique, that refines the mesh posteriorly according to the discretization error, is expected to resolve this problem. Simple error indicators have been incorporated into an adaptive mesh simulator and their capabilities have been examined[12]. The following are selected for error indicators:

$$\beta = \frac{\psi''}{2\sqrt{1 + \psi'^2}} dx^2 \tag{1}$$

$$\gamma = \frac{|J_1 - J_2|}{\max(|J_1|, |J_2|)} \tag{2}$$

$$\eta_i = \left(\int_{E_i} |F - F*|^2 dV \right)^{\frac{1}{2}} \tag{3}$$

To limit β is almost equivalent to limiting the spatial change of $\nabla\psi$, and reduces the discretization error in the Poisson's equation. γ is, on the other hand, applied to the current evaluation for the rectangular box, where J_1 and J_2 denote the current densities along two parallel edges of the box. Thus, γ essentially examines the uniformity of current density normal to the Voronoi boundary typically being required in the MOS channel region. η_i is a more general error indicator applicable to both Poisson's and current continuity equations, and is not restricted to the rectangular box structure. Here, F and $F*$ denote the calculated flux density and its true one both to be defined as the uniform quantity in each mesh element E_i. Fig. 7 shows an experimental result, where starting from an initial coarse mesh, meaningful mesh refinement is achieved.

For the analysis of hot carrier transport phenomena, a population transaction method has been proposed[13]. In this method, momentum space electron distributions are calculated by the direct integration of Boltzmann transport equation. Therefore, it outperforms the Monte Carlo simulation in determining the distribution evolution and enables the non-steady-state analysis with a smaller CPU time. Further compact model, that is still required for the analysis of hot carrier transport within a limited computer resource, was derived by means of relaxation time approximation[14]. Following the resulting 2-D energy transport simulation, we have undertaken the generalized hydrodynamic model[15] in our standard device simulation environment.

5. Circuit Modeling and Simulation

The use of circuit simulation in electrical design has still been broadening. Ensuring electrical characteristics of devices, therefore, is of importance in order to feed device technologies to circuit designs. For the period of three years from 80s to 90s, a working group, gathered from several design divisions, was formed to accomplish the standerizations for test chip fabrications, measurements and parameter extractions[16],[17]. Subsequently, it led to the development of new analytical MOS model for circuit simulation[18] that attempts to remove the difference between the electrical characteristics obtained by the conventional model and the measurement of advanced device. The new model represents the effective channel carrier mobility and the difference between the effective and the mask gate lengths as a function of external voltages. The threshold voltage shift including the reverse short channel effect is also derived as a function of channel length. The resulting model is a fairly compact one. However, with a single set of model parameters, it covers a wide range of biases and channel lengths down to the sub-half micron region. Fig. 8 shows an example being applied to a $0.4\mu m$ channel length process.

A system of circuit aging simulation CASSY[19] predicts the circuit behavior following a period of stress due to hot carrier. The system is configured in a pre- and post-processor configuration to NECSPICE. Similar to the well known CAS system[20], the degradation factor is defined as a function of drain and substrate currents with impact ionization parameters. Different to the previous system, however, in which device parameters such as the flat-band voltage and the mobility are assumed to be "aged", only the drain resistance is selected for representing the degradation of electrical characteristics. It is experimentally found that the drain resistance extracted from a series of stressed devices has a good linear relationship with the age factor. A use of this system in a SRAM circuit design clarifies that the dynamic stress decreases by the factor of 7 in comparison with DC stress.

A system of parallel circuit simulation CENJU is used in designing large scale DRAM circuits. The hardware is a distributed memory architecture multiprocessor(Fig. 9), currently operating under 8, 16 and 64 processor configurations. Together with a novel circuit partitioning method, node tearing enables parallel circuit simulation[21]. The system has achieved a processing speed up of 28 with 64 processors(Fig. 10). Although the CENJU hardware was designed mainly by studying the above algorithmic features, due to the resulting versatile architecture, the system is also used to develop the parallelisms for other types of numerical computation problem such as partial differential equation and Monte Carlo simulation.

6. Interconnect Simulation and Parameter Extraction

Besides the standard finite difference simulations, new approaches mainly focusing on link to the circuit simulation have been developed. In the program 3D-AWE[22], an interconnect structure is represented by the equivalent 3-D RC mesh networks. The stimulus/response transfer function of the networks is then approximated by using a partial fraction series of a few poles in the frequency domain[23]. The time response is, therefore, directly calculated by means of domain transformation. Fig. 11 shows a comparison between the proposed method and a standard 3-D simulation with respect to the time response of crossing wire structure. The program has been applied to the analysis of interconnect in a SRAM cell as well as the Silicon well in the substrate, and has produced transient responses one order of magnitude faster than conventional finite difference simulations.

In the case of package simulation, spatially large aspect ratio has to be dealt with.

A 3-D LCR extraction program JUKAI resolves this problem by using mixed analytical and numerical integration method[24]. Incorporating with a potential-adaptive
grid scheme, the program extracts capacitances about two order of magnitude faster
than the standard 3-D simulation in a range of 5% relative error tolerance. Fig. 12
shows a comparison between this method and a standard finite difference simulation
with respect to the extracted capacitance value and the simulation speed. A part of
off-diagonal L and C matrices negligibly small for the design purpose is removed and
SPICE LCR nets are generated.

7. Conclusion

We presented an overview of our TCAD systems and the use of them in LSI device technology development and design cycles in NEC. The framework tools that
facilitate the use of simulation were successfully developed and have served substantially for broadening simulation in this field . We also reviewed individual tools for
process, device, circuit and interconnect simulations with an intention to mark our
contributions to the related fields. It is true that the cost savings through the use of
TCAD systems dominantly depend on the user's skill. However, it is also true that
the goal of computer-aided design is to release the quality of design from the degree
of designer's experience. Several limitations exist in the TCAD regime, namely ease
to use, predictivity and processing speed. They influence and complement each other.
The evolution of simulation models in the balance of computing power seems to be
a key issue. Nevertheless, we always need to struggle for better simulation results in
comparison with experiments no matter what modeling capability is. In this respect,
we also noted our in-house activities, thereby bridging the gap between simulations
and experiments by means of either engineering methodology or task-force.

Acknowledgement

The author wishes to express his thanks to M. Nakamae and M. Fukuma for
their encouragement, S. Asada, H. Matsumoto, H. Kato, S. Hasegawa, T. Ohta, T.
Nakata, H. Furuta, K. Soejima, S. Kumashiro, T. Saito, Y. Tamegaya, T. Iizuka, K.
Tanaka, M. Hane, Y. Akiyama, H. Onozuka, Y. Nakatani and W. Specks for valuable
discussions and advice in preparing this paper.

References

[1] Y. Tamegaya, H. Ikeuchi, H. Kuge, Y. Akiyama, Y. Hatanaka and M. Asou,IEICE
    Trans. Electron., Vol. E75-C, No.2, p. 234(1992)

[2] NEC Res. & Develop., No.96, p. 339(1990)

[3] K. Soejima and K. Kitagawta, Proc. SEMI Technology Symposium, p. 87(1992)

[4] H. Matsumoto and M. Fukuma, IEEE J. Solid-State Circuits, vol. SC-20, p.
    52(1985)

[5] M. Hane and M. Fukuma, IEEE Trans. Electron Devices, vol. 37, p.1959(1990)

[6] W. Jüngling, et al., IEEE Trans. Electron Devices, vol. 32, p. 156(1985)

[7] M. Hane and H. Matsumoto, IEDM Tech. Dig., p. 701(1991)

[8] M. Hane and S. Hasegawa, Proc. VPAD 93, p. 52(1993)

[9] M. Fukuma and Y. Okuto, IEEE Trans. Electron Devices, vol. 27, p. 2109(1980)

[10] S. Kumashiro and M. Sakurai, Proc. NASECODE 4 p. 365(1985)

[11] K. Tanaka, S. Kumashiro, H. Katoh, N. Tanabe, T. Kurobe and M. Fukuma, Proc. NASECODE 6, P. 317(1989)

[12] K. Tanaka, P. Ciampolini, A. Pierantoni and G. Baccarani, Proc. VPAD 93, p. 118(1993)

[13] T. Iizuka and M. Fukuma, Solid-State Electronics, vol. 33, p. 27(1990)

[14] M. Fukuma and R. H. Übbing, IEDM Tech. Dig, p. 621(1984)

[15] R. Thoma, A. Emunds, B. Meinerzhagen, H. Peifer and W. L. Engl, IEEE Trans. Electron Devices, vol. 38, p. 1343(1991)

[16] Technical Report no. JEL-0065, NEC Co. (1991)

[17] Technical Report no. JEL-0067, NEC Co. (1991)

[18] Technical Report no. JMR-CB-92097, NEC Co. (1993)

[19] Technical Report no. JEL-0097, NEC Co. (1992)

[20] P. M. Lee, M. M. Kou, K. Seki, P. K. Ko and C. Hu, IEDM Tech. Dig., p. 134(1988)

[21] H. Onozuka, M. Kanoh, C. Mizuta, T. Nakata and N. Tanabe, Proc. EDAC, p. 12(1993)

[22] S. Kumashiro, R. A. Rohrer and A. J. Strojwas, IEDM Tech. Dig., p. 193(1990)

[23] L.T. Pillage and R. A. Rohrer, IEEE Trans. Computer-Aided Design, vol. 9, p. 352(1990)

[24] W. Specks, Proc. The 6th Karuizawa Workshop on CAS, p. 453(1993)

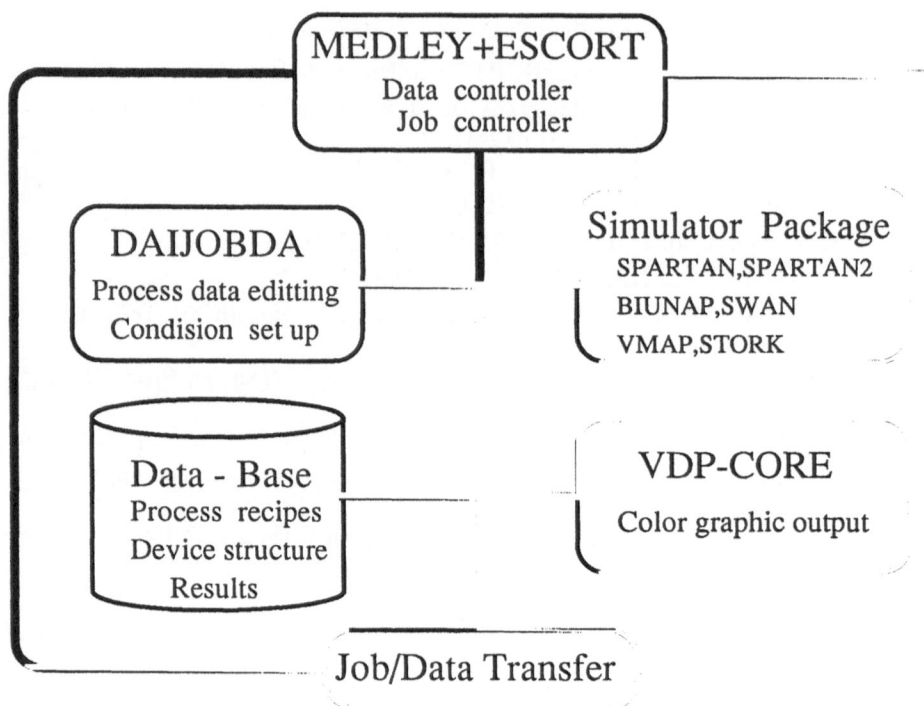

Figure 1: System configuration of P&D Workbench

(A)

(C)

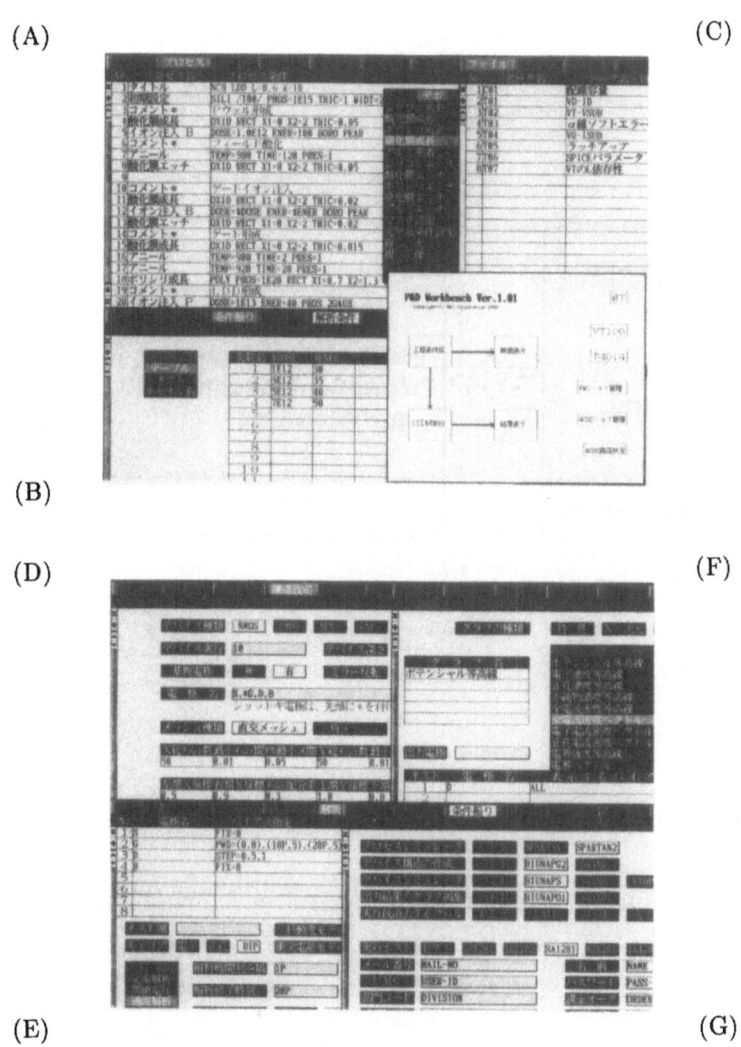

(B)

(D)

(F)

(E)

(G)

Figure 2: Example of DAJOBDA windows.

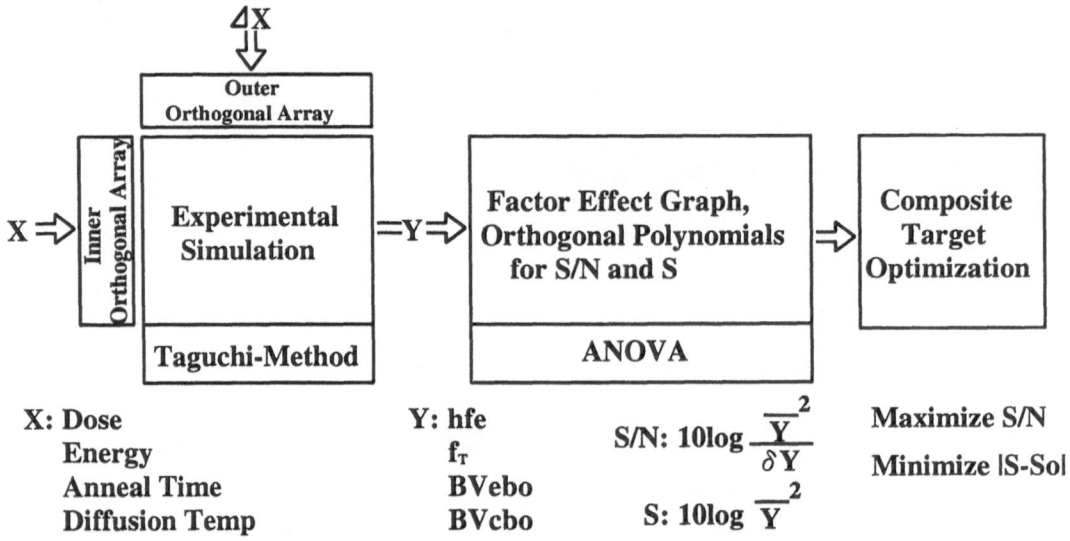

Figure3: System flow of experimental simulation system XPLAN

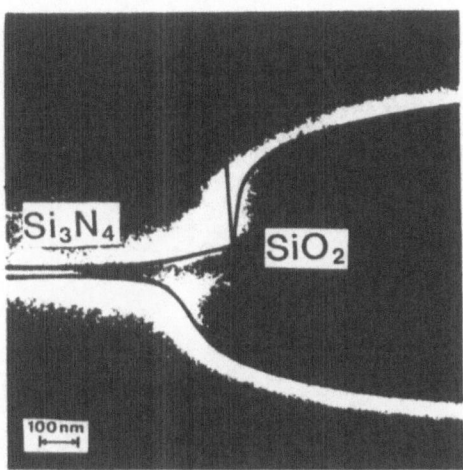

T_{pad}: 25 nm
T_{SiN}: 210 nm
1000°C, 4 hrs

wet

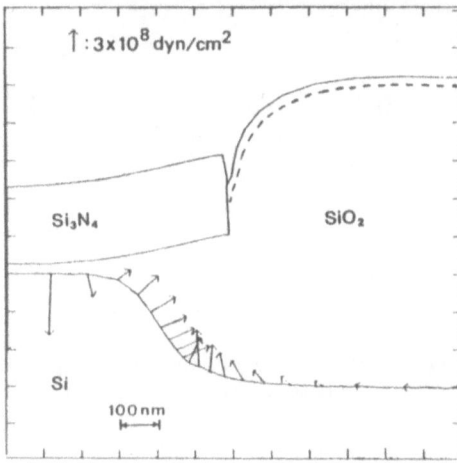

Figure 4: Comparison between thermal oxidation shapes obtained by experiment and simulation. Simulated stress vector is also shown.

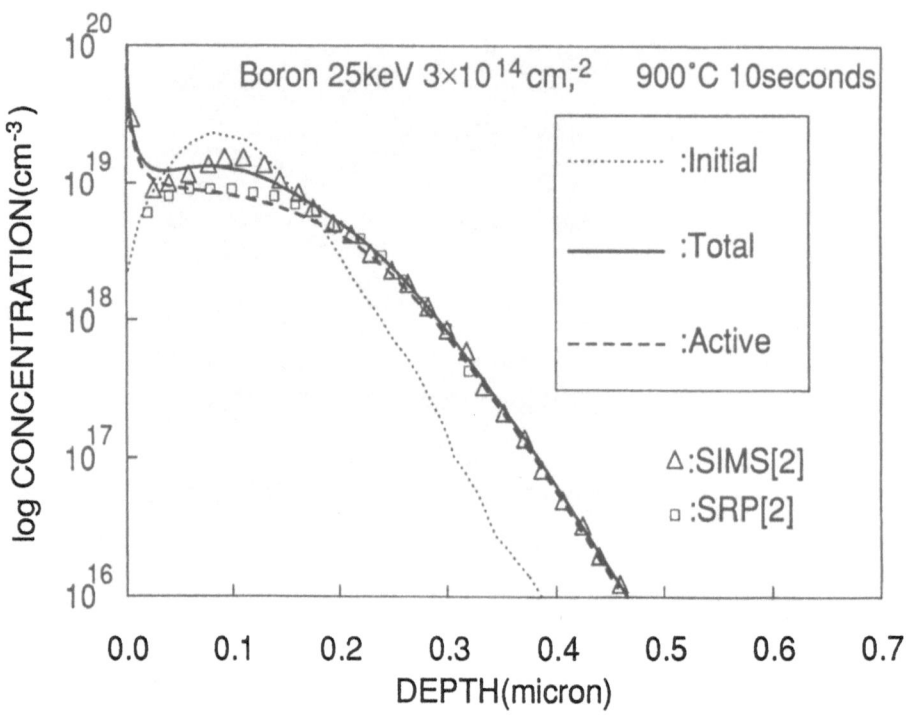

Figure 5: Boron initial distribution profile and profile after diffusion for 900°C 20 second anneal, subsequent to a 25KeV $5 \times 10^{15} cm^{-2}$ implantation.

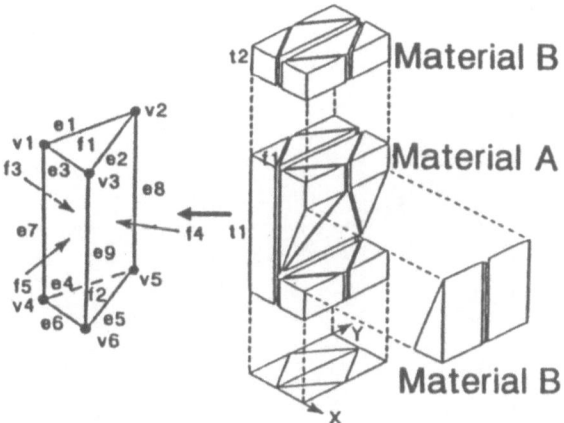

Figure 6: Shape division into triangular prisms for 3-D device simulation.

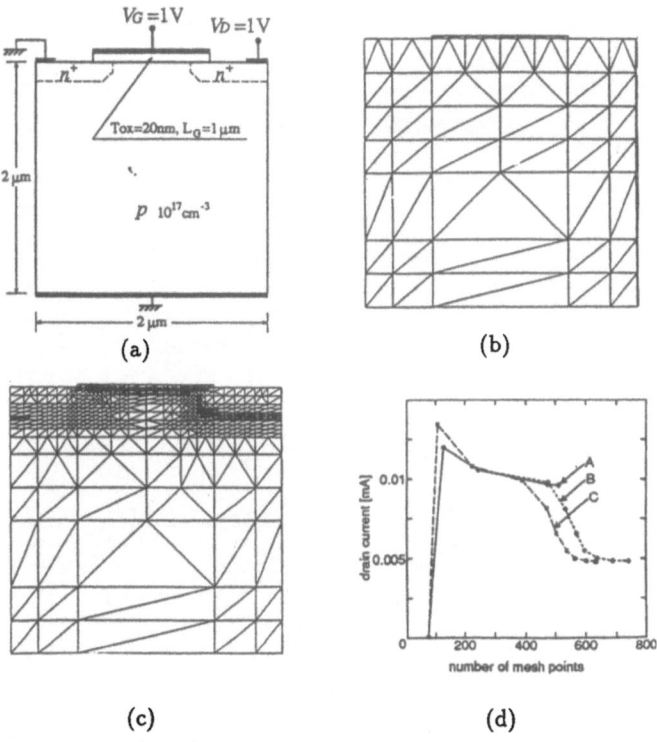

Figure 7: Example of adaptive mesh generation. (a) Simulated MOS structure. (b) Initial coarse mesh. (c) Optimized mesh according to error indicator η. (d) Drain current as a function of # of mesh points. Error indicators are A: β, B: β and γ, and C: η.

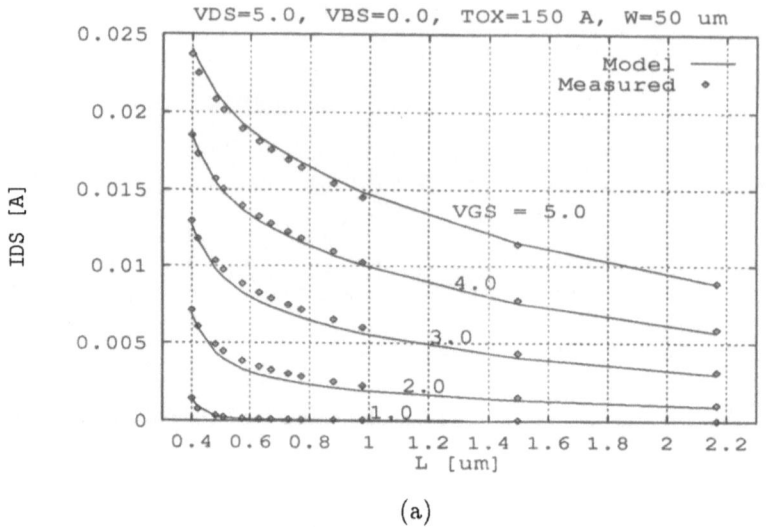

(a)

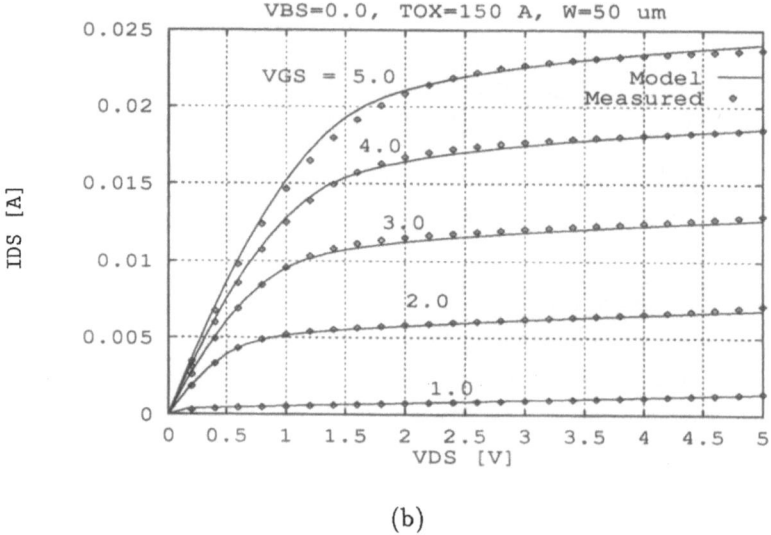

(b)

Figure 8: Comparisons between electrical characteristics obtained by new analytical model and measurements. (a) Drain current as a function of gate lengths. (b) Drain current as a function of biases.

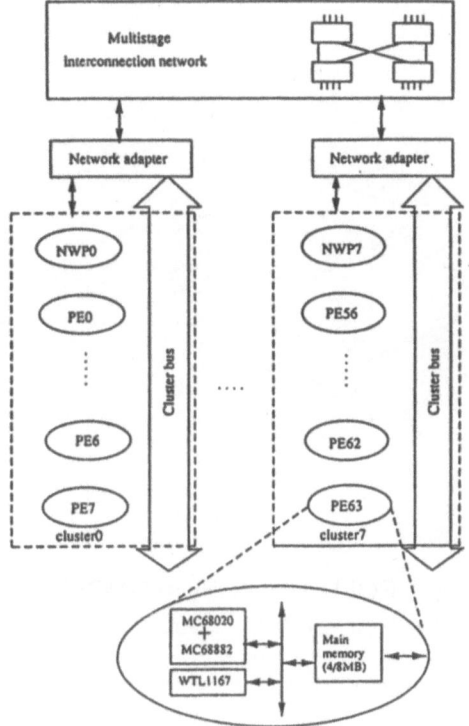

Figure 9: Hardware configuration of multiprocessor system CENJU.

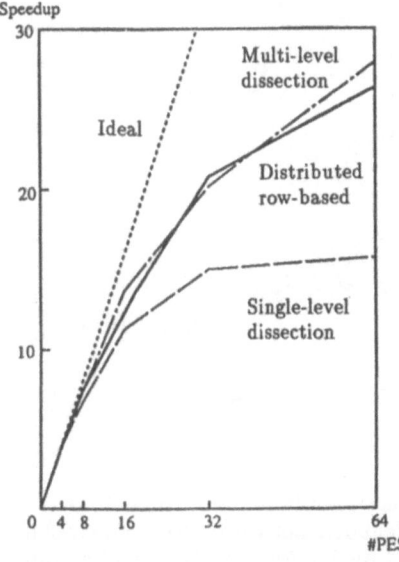

Figure 10: Parallel performance evaluations of 64Mb DRAM circuit simulation.

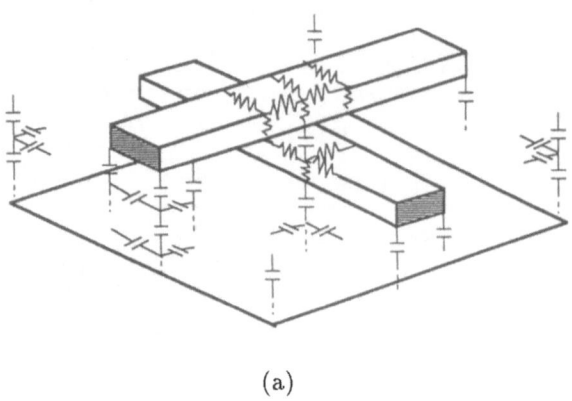

(a)

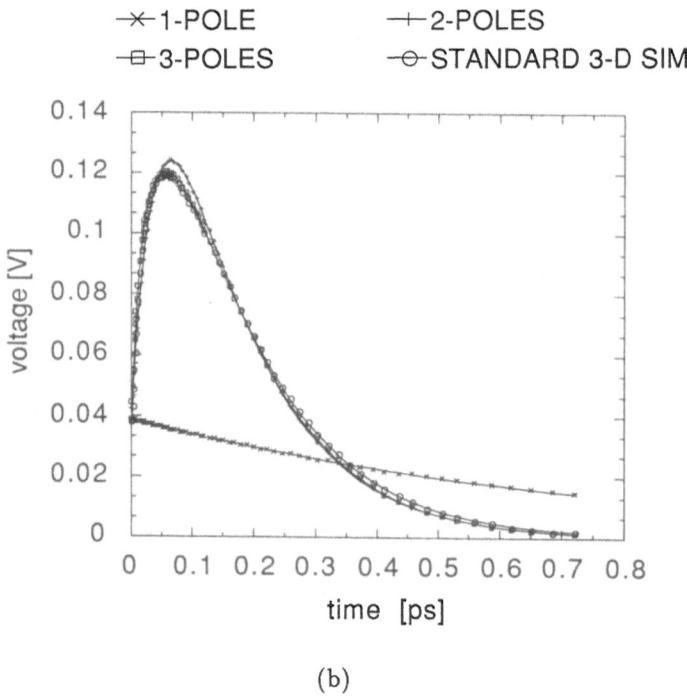

(b)

Figure 11: Interconnect simulation by using 3D-AWE method. (a) Crossing wire structure and equivalent RC network. (b) Comparison between voltage responses of crossing wire obtained by 3D-AWE method and a standard 3-D simulation.

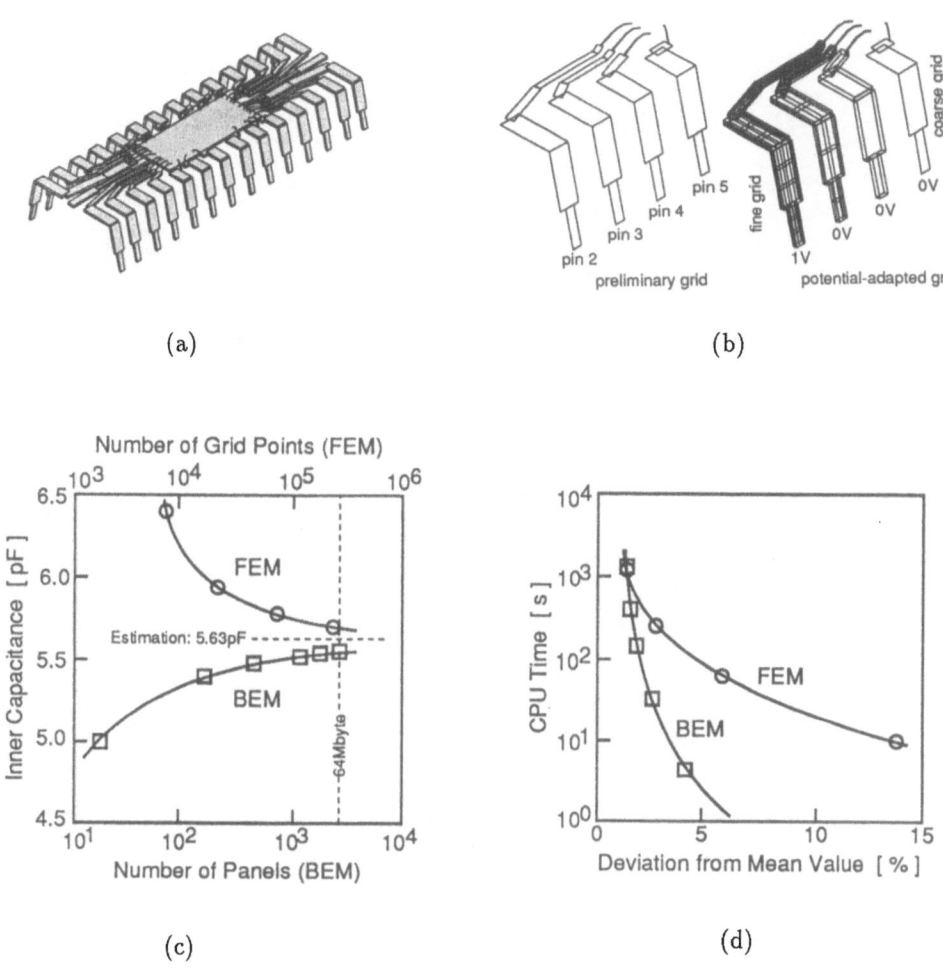

(a)

(b)

(c)

(d)

Figure 12: Package parameter extraction by JUKAI. (a) 28-pin lead frame structure of a memory chip. (b) Capacitances extracted from a preliminary grid are used to define the potential-adaptive grid. (c) Capacitance extracted by BEM (JUKAI) and FDM (a standard 3-D simulation) methods as a function of the grid resolution. (d) CPU time of BEM and FDM methods on a RISC EWS as a function of the derivation from the estimated real capacitance value.

TECHNOLOGY CAD SYSTEMS   255
Edited by F. Fasching, S. Halama, S. Selberherr – September 1993

Technology CAD at OKI

K. Nishi and J. Ueda

OKI Electric Industry Corporation,
550–1 Higashiasakawa, Hachioji, Tokyo 193, JAPAN

ABSTRACT

In this article, the outline of Technology CAD at OKI is described.   TCAD system at OKI, or UNISAS, is a unified
process/device/circuit simulation system, and offers a user-
friendly simulation environment for OKI's process, device and
circuit engineers. Multi-dimensional process and device simulators, OPUS and ODESA constitutes the core of UNISAS and can be
used for versatile device structures for various purposes.
Physics-based simulators constitute another part of UNISAS, and
are used mainly for modeling purpose. UNISAS has already been
used quite extensively by engineers for actual process and device
development and is now an indispensable tool for low-cost, fast
TAT VLSI development.

1.Introduction
     With the first introduction of a process simulator,
SUPREM[1], to analyze a full LSI manufacturing process, the idea
to fully describe circuit performance on a computer coupled with
already-existing device simulators and circuit simulators, or
TCAD in a current term, evolved as an effective tool for VLSI
development in future. OKI recognized the importance of TCAD at
the early stage, and began R&D work from the beginning of 80's.
The first target was to develop a multilayer process simulator,
which was first announced as early as 81's at the Electrochemical
Society meeting[2]. Then, our efforts were drawn more to develop
general-purpose process/device simulators for versatile devices
and various analysis purposes. The unified simulation system,
UNISAS (UNIversal Semiconductor Analysis System), was developed,
first with a 1D process simulator, ASPREM and a 2D device simulator ODESA[3], and   later, with a 2D general-purpose process
simulator OPUS[4]. Recently, ODESA was linked with a circuit
simulator, OCAP, with a table look-up model, and now, UNISAS is
established as a user-friendly effective tool for coupled multi-
dimensional process/device simulation system for general purposes.
     While major efforts were made to develop general-purpose
simulators, research works were oriented toward precise modeling
of processes and devices. Several simulators were developed to
understand physical phenomena during specific processes or in the
devices under operation. These simulators are linked with general-purpose simulators and constitute parts of UNISAS.
     Modeling work based on experiments was also done to develop
such process models as diffusion, and oxidation, and also carrier
mobilities. The models have effectively implemented in OPUS and
ODESA, and contributed to more accurate process/device simulations.
     In this paper, we show details of UNISAS system, and de-
scribe outlines of each simulator. Finally, some of the applications are described.

2.UNISAS system
2.1 General features
     The criteria for developing UNISAS are summarized in Table
1.
     The first criterion is an applicability to versatile de-
vices. Application-specific simulators like a MOSFET simulator
are much faster to develop. In a long term, however, we believe
that to maintain many simulators each for specific devices are
not effective. For the model enhancement, we have to implement

Table 1:Criteria for UNISAS development

Applicability to versatile devices
Flexibility for future enhancement
Flexibility for various simulation purposes
User-friendliness
Fast turn-around-time
Maintenability of the system

them into each simulator.   Also pre- and post-processing work is much more time-consuming with many simulators.

The second criterion is a flexibility for future model enhancement.   This include well-ordered file structures as well as well-ordered module organization and programing itself.

A flexibility for various simulation purposes is also a concern.   Some engineers need fast simulations for evaluating only device sensitivity on some process or device parameters, and other engineers may need more precise simulations to understand device operations.

The fourth criterion is a user-friendliness.   From the beginning of the system development, we intended UNISAS not only for research purposes but also for a practical tool used by actual device designers.   Graphics was also one of a major target at the beginning.

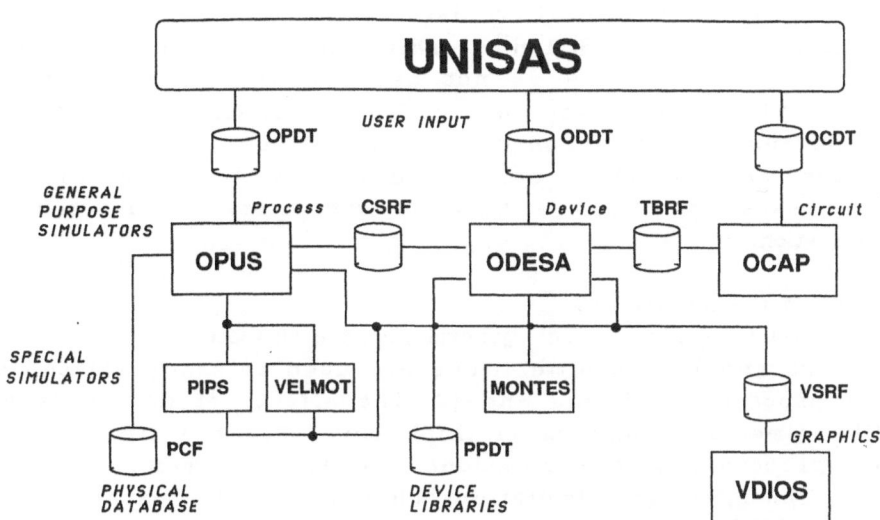

Fig.1:System structure of UNISAS

Other criteria include a fast turn-around-time (TAT) and a maintainability of the system.

In order to meet these criteria, UNISAS is organized as shown in Fig.1. UNISAS can be divided into three parts: 1)General-purpose simulator group, 2)Special simulator group, and 3)Graphical utilities.

General-purpose simulator group consists of OPUS for process simulation, ODESA for device simulation and OCAP for circuit simulation. In order to achieve a flexibility for simulation purposes, OPUS and ODESA themselves have capabilities to simulate in 1D to 3D spaces. An applicability to versatile devices is not easy to achieve, but upto now, they are successful in simulating MOSFET, bipolar, power devices, SOI devices, CCDs, compound semiconductors and even wiring capacitances.

Special simulator group now consists of VELMOT for stress-oriented oxidation simulation, PIPS for Monte Carlo ion implantation simulation, and MONTES for Monte Carlo device simulation. They are linked with OPUS and ODESA. Right now, they are more in research-oriented applications including modeling, but also help to achieve more accurate simulation of device characteristic. We note that other than these in-house simulators, a lithography simulator and a device simulator based on a hydrodynamic model are also used for R&D.

Results files of these simulators have the same unified format and can be visualized by an in-house graphic system VDIOS.

All the simulators have the same input language called UNICOL. UNICOL is an in-house-developed natural language, and endures a full or a simplified description of process conditions and applied voltage conditions. This language is easy to learn, and also device engineers do not have to worry about different grammar. This constitutes one of the easinesses for maintainability because we need only one UNICOL compiler. This also constitutes one of the user-friendlinesses together with the practicalnesses of linkage between simulators as described in the next section.

2.2 Linkage of the simulators

Care should be taken to link general-purpose simulators.

Process and device simulations are used to study (1) the effects of process conditions and (2) the effects of device parameters like dimensions and oxide thickness. For the first purpose, coupled process/device simulations are the necessity. For the second purpose, device simulations changing device parameters are sufficient, even though dopant profiles by process simulations are preferred in order to correlate the simulation results with

experiments. It is plausible that users just change input parameters like implantation condition or gate oxide, and simulations
will run automatically.

Thus, UNISAS users can have two choices for a coupled-
process/device simulation. The first is to use directly the
results of OPUS. Accurate device shape and dopant distribution
are reflected to the device simulations, and this is suitable for
the simulations where complex device shapes are crucial for the
device characteristics. The other choice is to construct more
simplified standard device structures. Since not the details but
the essences of the device structure are reflected, these standard device structure libraries are suitable to understand the
basics of the device characteristics depending on the device
parameters. Note that in the device simulation, dopant profiles
can always be taken from OPUS results.

In either case, linkage of 3D process and device simulations
are quite complicated. We already developed GCOS[5] for this
purpose. The system structure is shown in Fig.2. GCOS has capabilities not only to handle device structure by process simulation but also to generate 3D device shape by interpreting
pseudo-process. The pseudo-process is used to generate standard
device structures. Boundary conditions and mesh generations are
done interactively using multi-windows, graphics and a mouse.
Resulting standard device structures are symbolized using UNICOL
language, which makes the modification of the device structure
easier.

Another pre-eminence of GCOS is that GCOS has the capability

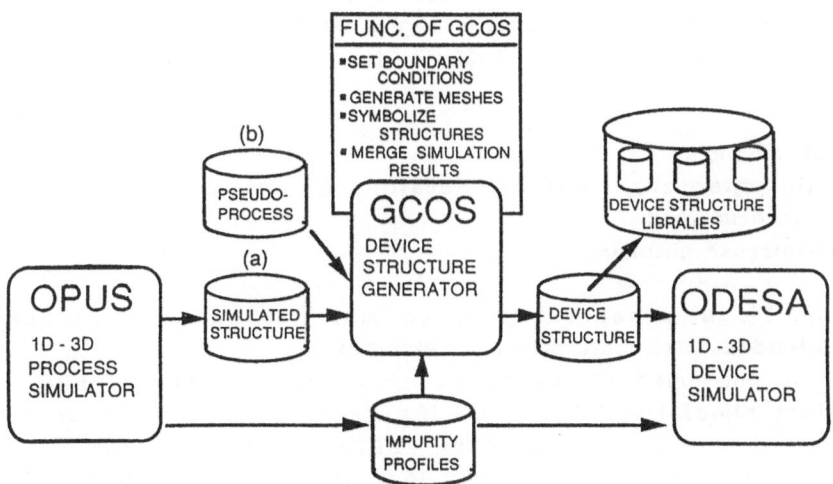

Fig.2:System structure of GCOS

to merge two or more process simulation results. For symmetrical devices like conventional MOSFET, process simulation of only half of the device is sufficient. For CMOS simulation, nMOS and pMOS processes are independently simulated by OPUS and can be merged to form one CMOS device for the device simulation. This lessens the burden on the process simulation and could save CPU time and memories drastically.

Currently, we already have a standard device structure library including 2D- and 3D-MOSFETs, 2D- and 3D-exotic bipolar devices, CCDs, power devices and various 3D wiring capacitance.

For a linkage between ODESA and OCAP, the first choice was to have a table look-up model. For a hand-free linkage of ODESA and OCAP, analytical models may not be appropriate because of difficulties in automatic parameter extraction. The number of current/voltage data points for the input of OCAP is minimized through error evaluation of currents and circuit delay time (tpd). In addition to current/voltage table data, junction capacitance is also linked by tables simulated by ODESA. Other device-oriented parameters like flatband voltage and gate-drain overlap length are also automatically calculated from the device structure for device simulation.

Thus, what are needed for engineers to run unified simulations from process to circuit are only the upper-level operation to control each simulation, in addition to process flow, bias points and netlist for process simulation, device simulation, and circuit simulation, respectively.

UNISAS is currently running on the computer network including a general-purpose machine with a vector facility and various work stations. UNISAS is being used extensively by many engineers at OKI for a cost-effective, fast development of various VLSIs.

3.Description of Tools
In this chapter, outlines details of our major in-house simulators are described.
3.1 General-purpose simulators
A)OPUS
OPUS[4,6] is a general-purpose multi-dimensional process simulator intended to simulate arbitrary structures.

OPUS is composed of two programs: (a)pre-processor PPPS, and (b)numerical simulator PSS. The inputs of PPPS are process flow by users and physical constants. PPPS interprets these inputs, checks the grammatical and also logical errors in these inputs, extracts physical data pertinent only to this specific simulation, and transforms them to numerical files. PSS runs with these numerical files.

General features of OPUS are described in the followings.

First, OPUS can be applied to arbitrary structures. In order to achieve this, OPUS has orthogonal finite-difference discretization scheme with a pre-determined grids. In order to reduce numerical errors for diffusion calculation due to curved boundaries, special grids are added to pre-existing grids only at the boundary. For the boundary movement, OPUS employs a string

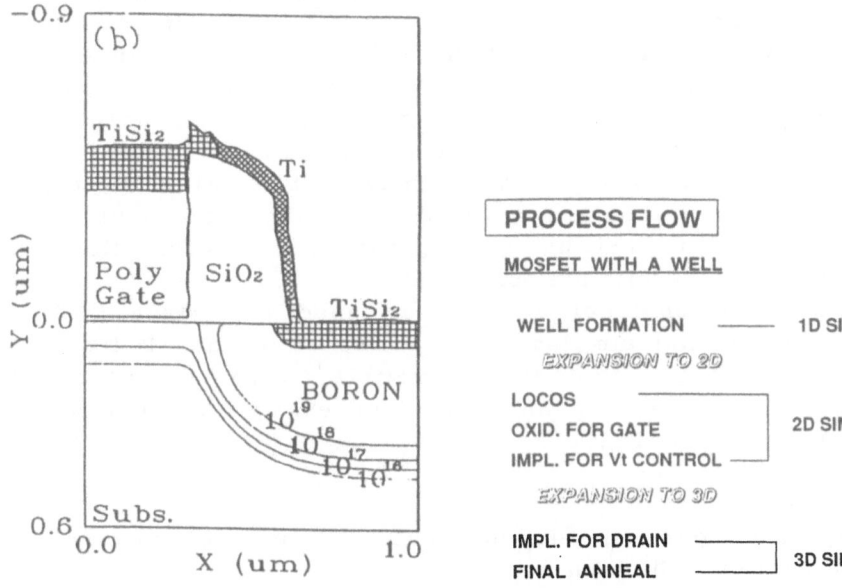

Fig.3:Salicide process   Fig.4:Method to execute 3D
simulation by OPUS   process simulations of MOSFET

Table 2:Classification of physical constants in OPUS

NAME OF CONSTANTS	MODIFIERS	EXAMPLE
UNIVERSAL	— — —	BOLTZMAN CONST.
MATERIAL	MATERIAL	NUMBER DENSITY
MATERIAL TRANSFORMATION	MATERIAL.MATERIAL	GROWTH RATES FOR OXIDATION
IMPURITY	MATERIAL.IMPURITY	DIFFUSIVITY
ION	MATERIAL.IMPURITY.ION	MOMENTS FOR IMPL.
INTERFACE	MATERIAL.MATERIAL.IMPURITY	SEGREGATION CONST.

model in 2D simulation, and has a capability of a so-called topographical simulation.

Second, program enhancement for new processes and new physical models is easy. In order to achieve this, PSS has modular structures throughout the program, or otherwise, well-disciplined program structure. The recent enhancement to salicide process[7] proves this as shown in Fig.3.

Third, arbitrary materials and impurities are added to OPUS quite easily. In order to achieve this, all the physical models and constants in OPUS are classified and modified by combination of materials, impurities, and ions. An example of physical constants is shown in Table 2. When new materials or impurities are added, what we have to do is only to register a new name, specify a model to be used, and enter model parameter values to be used. These can be done only with a physical constant file, and the change in the program is not necessary. At this stage, we note that OPUS can simulate upto 10 materials and 8 impurities at the same time with maximum grid points of 200,000, although species of materials or impurities are not limited.

Fourth, multi-dimensional analysis is used to reduce CPU time in high-dimensional simulations. A method to do 3D MOSFET simulations is shown in Fig.4 as an example. The method could reduce CPU time by more than an order. An example of 3D simulations of MOSFET is shown in Fig.5.

B)ODESA

ODESA[3] is a general-purpose multi-dimensional device simulator based on a drift diffusion model.

ODESA is composed of two programs: (a)pre-processor PPDS, and (b)numerical simulator DSS. Inputs to PPDS include process

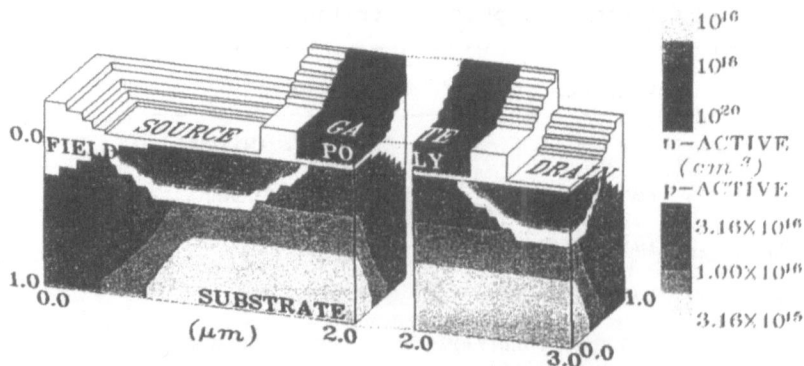

Fig.5:3D process simulation of MOSFET

simulation results, and either stored or user-specified device structure for the device simulation in addition to bias conditions. PPDS interprets user inputs, checks the grammatical and also logical errors in these inputs, finds the optimal solution strategy, and transforms them to numerical files. DSS runs with these numerical files.

General features of ODESA are described in the following.

First, ODESA employs a modified step design method to solve Poisson and carrier continuity equations in a Gummel iteration scheme. Thus, fast CPU time is achieved by an initial guess method independent from bias conditions, and by a step design optimizing the degree of convergence in each step.

Second, ODESA is intended to be used for different purposes. Potential alone can be solved for capacitance calculation. Users also has a choice for solving either one carrier or two carriers for different purposes. Transient analysis can be used to analyze latch-up and funneling phenomena induced by arufa-ray.

Third, ODESA is designed for simulation of versatile devices with arbitrary structures. In addition to silicon devices, the model is enhanced to silicon-oriented hetero-devices and also to compound semiconductor devices only by using a different data base. ODESA can simulate such standard devices like MOSFETs, bipolar devices, SOI structure, CCDs, and also user-specified structures often by using GCOS system as described in the previous section.

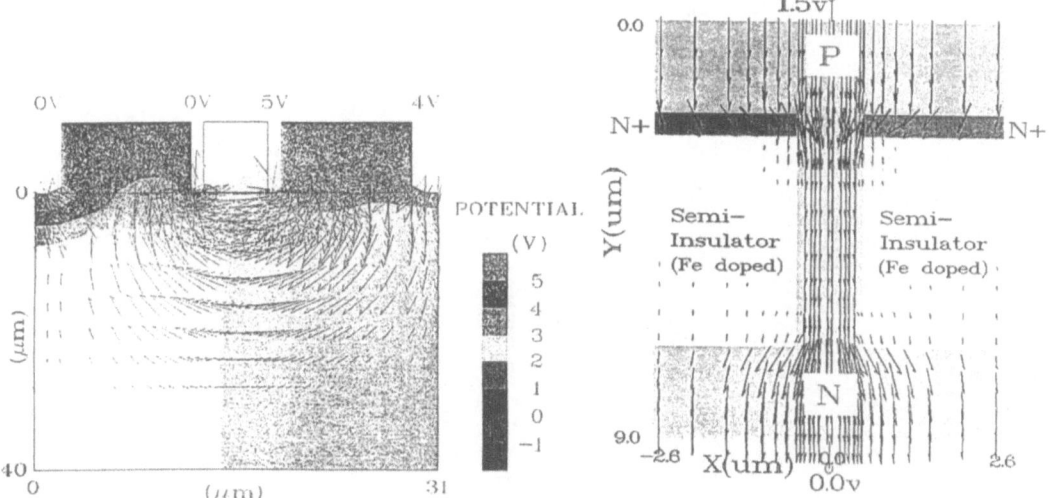

Fig.6:Transient analysis of CMOS

Fig.7:Current flow in
an InP laser diode

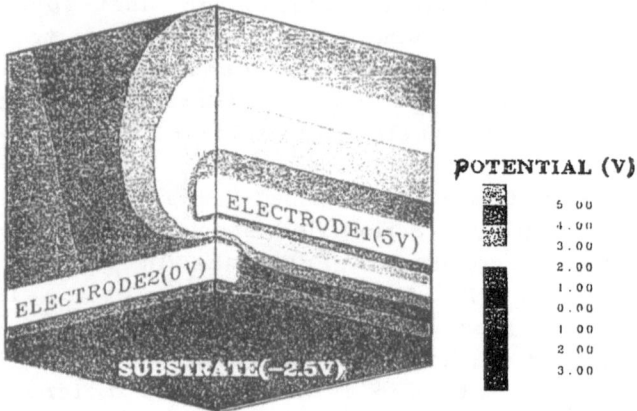

Fig.8:Simulation of Wiring capacitance.   Shade shows the equi-
potential region.

Fig.6-8 show some examples of device simulations.   Fig.6 is
a transient analysis of CMOS device and shows a potential distribution and current flow at latch-up.   Fig.7 is an application to
current flow in an InP laser diode.   Finally, Fig.8 is an application to 3D wiring capacitance and shows a potential distribution.

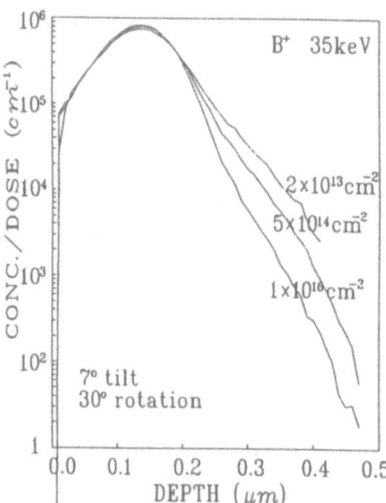

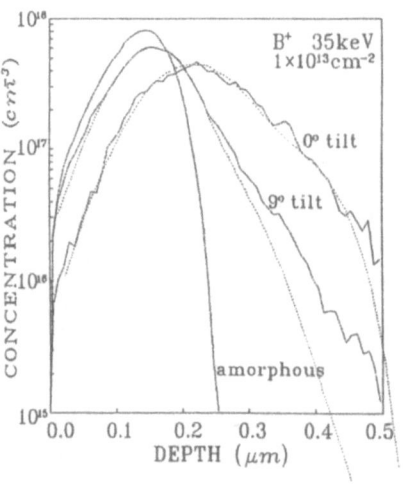

Fig.9:B+ implantation to different crystalline orientation
of Si

Fig.10:BF2 ion implantation
with different doses

3.2 Special simulators
A)PIPS

PIPS is a 3D ion implantation simulator based on Monte Carlo method, and employs vector algorithm for the fast calculation. PIPS has capabilities to analyze trajectories of knocked-on atoms, implantation to crystalline material, generation of point defects, and creation of an amorphous layer. Physical models are usually gathered from published papers. As for channeling, atomic area density in any of the crystalline orientation is calculated. Mean free path of implanted ions varies accordingly. Fig.9 shows comparison between simulations and experiments to show the channeling effects. We see good agreements which indicate the effectiveness of the channeling model.

This channeling model is expanded to account for generated point defect concentration. The basic idea comes from that atomic area density used for calculation of ion channeling should decrease with generated self-interstitials. Fig.10 shows simulation results for different BF2 ion doses. During implantation, penetration depth of boron decreases due to generated point-

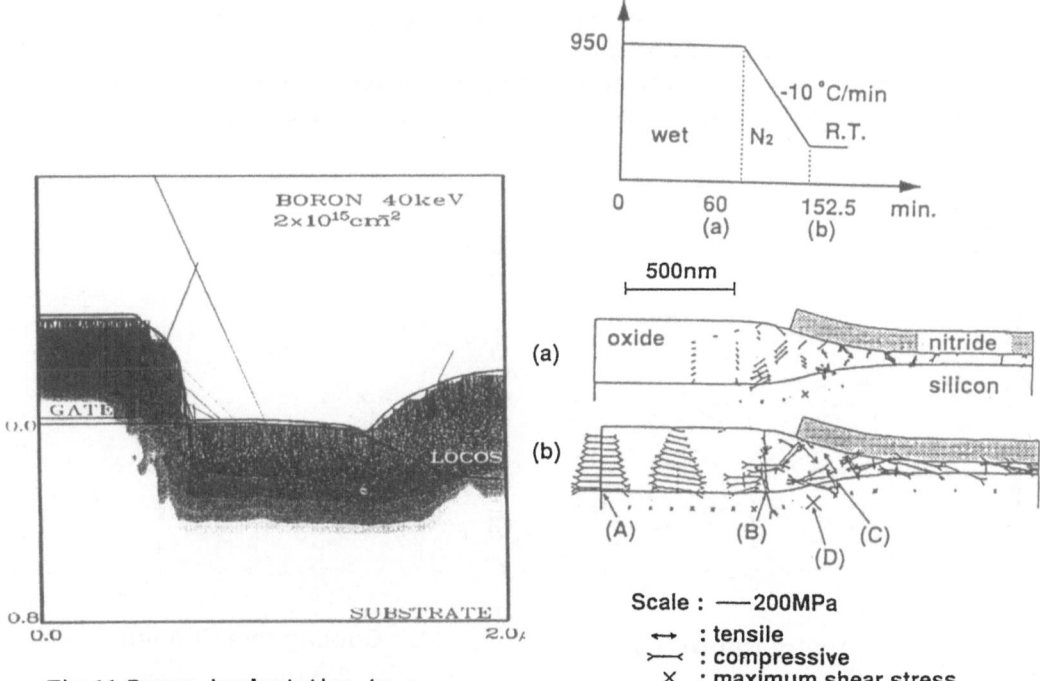

Fig.11:Boron implantation to a MOSFET drain. Shade shows equi-concentration region.

Fig.12:Stress distribution after LOCOS(a) and cooling down(b)

defects.   Thus, boron distribution normalized to the dose becomes shallower for higher doses.

PIPS is linked with OPUS and users can use PIPS for more accurate analysis of ion implantation at specific process during the whole manufacturing flow.   Fig.11 shows the boron implantation to a realistic MOSFET drain region.

PIPS is also used for modeling purpose.   2D parameters for the use by analytical models were extracted with multilayer structures depending on implant energy and oxide thickness.   Also PIPS was used to generate parameters for implantation into a TiSi2 layer.

B)VELMOT

VELMOT is a 2D stress-oriented oxidation simulator[8] based on a viscoelastic model described by a spring and dashpot. Stress-dependent models of oxidant-diffusion, oxidant reaction rate and oxide viscosity are incorporated in VELMOT.   The stresses by viscoelastic deformation and thermal mismatch among the materials are taken into account.   Noted is that the all the materials under consideration including a nitride film is treated as a viscoelastic material.   This is not only for more broader modeling but also for applications to more thermal stress-oriented problems other than oxidation.

An example of oxidation simulations with thermal mismatch is shown in Fig.12[8].   Fig.12 shows the oxide shape and stress

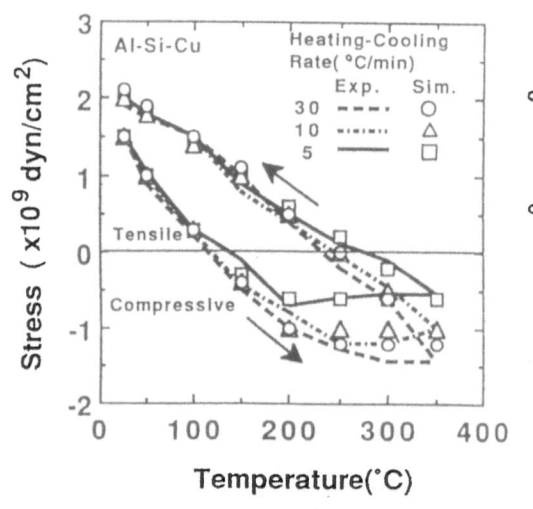

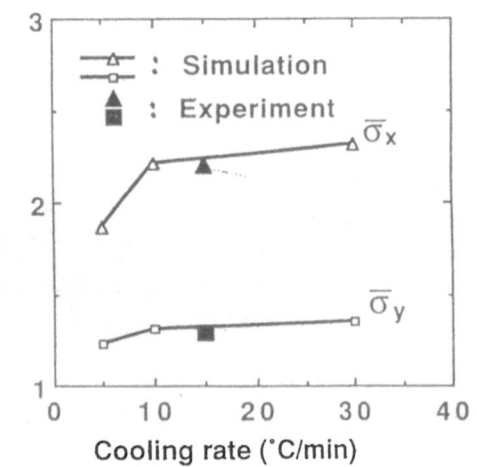

Fig.13:Stress of an Al film on a silicon substrate during heating and cooling cycles[9].

Fig.14:Stress of a sub-micron Al interconnect.

distribution after LOCOS(a) and cooling down to room temperature(b). The stress after cooling down is much larger than that of right after LOCOS due to thermal mismatch between oxide and nitride. We have found that these stress values depend on the cooling temperature. These kind of analyses are possible only by viscoelastic modeling.

Another interesting application of viscoelastic modeling is to the stress redistribution of. metal interconnect during temperature cycles. Fig.13 shows the simulation and experimental stress values of a plane aluminum on a oxide film[9] during temperature cycles with a parameter of different heating (cooling) rate. Different hysteresis curves were simulated for the first time, which shows excellent fit to the experiments. Simulation of a submicron aluminum interconnect was also carried out and is shown in Fig.14. The stress values are in good agreement with experiments, suggesting that viscoelastic modeling can be used to stress migration of a aluminum interconnect.

VELMOT can run with an output file of OPUS. Thus, it is easy to do oxidation simulation and stress calculation with a realistic structure. A future challenge is the stress dependent diffusion modeling using stress distribution calculated by VELMOT.

C)MONTES

MONTES is a 2D MOSFET device simulator based on Monte Carlo method. MONTES employs a modified window Monte Carlo method in which a conventional window is enlarged for the calculation of Poisson equation. This reduces the CPU time drastically without losing accuracy. Thus, typical CPU time to 1 hour and enables a simulation of realistic MOSFETs with a high drain concentration.

Physical models include non-parabolic band for both electrons and holes, a Brooks-Herring scattering model modified to account for high-doping region, carrier-carrier scattering and surface roughness model. Impact ionization is also included in the lucky-electron model with a soft-threshold energy. Fig.15 shows a simulation example of MOSFET. Here, impact ionization is calculated and carriers generate at the region where carriers surpass the electric field maximum point.

Fig.16 shows the drain current versus effective channel length of MOSFETs. Here, solid line indicates experimental data, squares indicates the results by drift diffusion model, and circles indicates the results by MONTES. For longer channel-devices, we see excellent agreements among the three, which indicates the effectiveness of the Montes. However, the results by a drift-diffusion model are smaller than experiments with

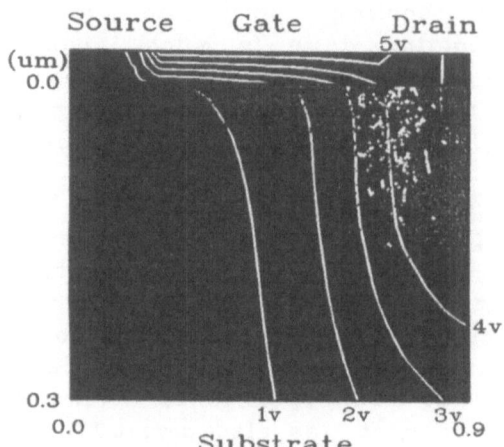

Fig.15:Simulation example of
MOSFET by MONTES

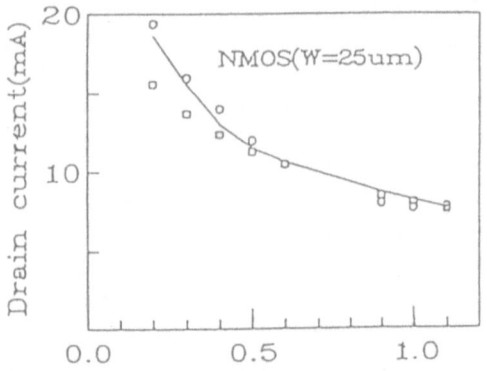

Fig.16:Drain current vs. effective channel length of MOSFETs.
Circles and squares show simula
tions by MONTES and ODESA.
Solid line shows experimental
results.

shorter effective channel length as expected. On the other hand,
the results by MONTES show excellent agreements with experiments.
This indicates that MONTES can be used for simulation of realistic MOSFETs as well as modeling of higher energy carrier transport.

4.Applications
    We show here three examples of UNISAS applications.

4.1 Reverse short channel effect in threshold voltage of PMOSFET
There have been several reports on reverse short channel (RSC)
effects in threshold voltage (VT) which result in higher VT
values with shorter gate length. Many explanations were already
proposed, which include gate bird's beak, and enhanced diffusion
due to point defect injection into channel region generated by
oxidation, silicidation or implantation damage to drain [11,12].
    We found that another different case causes much higher RSC
effects in pMOSFETs. Fig.17 shows VT versus gate length of pMOS-
FETs with a parameter of spacer length. It is seen that for
spacer width of 0.25 um, RSC effects are clearly seen.
    2D process and device simulations are used to investigate
the causes of these RSC effects. We comment that this technology
avoids boron implantation to form LDD, and utilizes an oxidation
process prior to boron implantation to form high-concentration

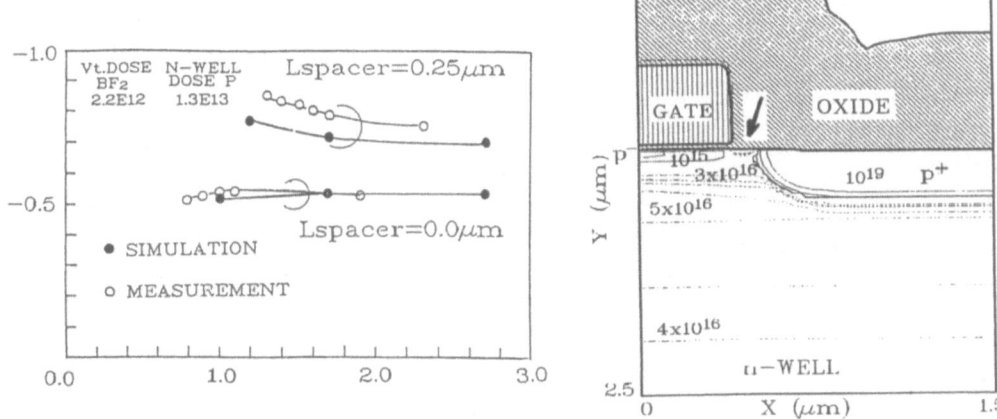

Fig.17:VT vs. Leff of pMOSFETs   Fig.18:Simulated carrier profile
with or without spacer.

drain region.   Fig.18 shows the process simulation results by
OPUS.   It is seen that under the spacer, p-type layer is gone.
This is explained as follows.   During oxidation process, not only
high-concentration drain region but the region under the spacer
is oxidized.   Since segregation coefficient of boron is less than
unity, most of the boron segregates into the growing oxide,
reducing the boron concentration at the silicon side.   This
offset region accounts for large RSC effects in this case.   Vt's
using coupled process/device simulations are also plotted in
Fig.17, showing excellent agreement between simulations and
experiments.

4.2 3D simulations of bipolar device
     As an example of coupled 3D process/device simulations with
a complex device structure, applications to the ultra-high speed
bipolar transistor named SATURN[13] are described.   Fabrication
process of SATURN extensively utilizes polysilicon.   In 3D simulations, OPUS utilizes 1D and 2D simulation efficiently to save
CPU time not reducing the accuracy.   Fig.19 shows the simulated
structure of SATURN by OPUS.   In polysilicon, we can see curved
arsenic-concentration contours, which are not symmetrical in the
two cross sections due to 3D diffusion effects.   By using GCOS,
device structure for the device simulation is constructed.
Fig.20 shows the current distribution simulated by ODESA.   A
cross section along the horizontal plane around the original
Poly/Si interface is also shown.   Current distribution is completely different from 2D simulation and base resistance is much

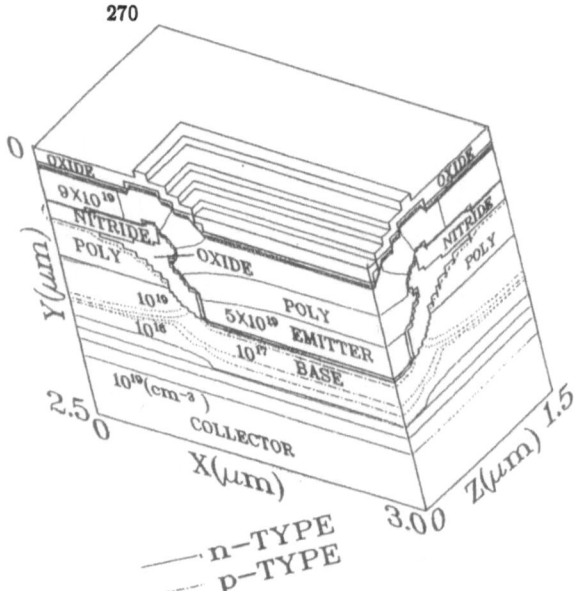

Fig.19:Simulated structure
of SATURN by OPUS

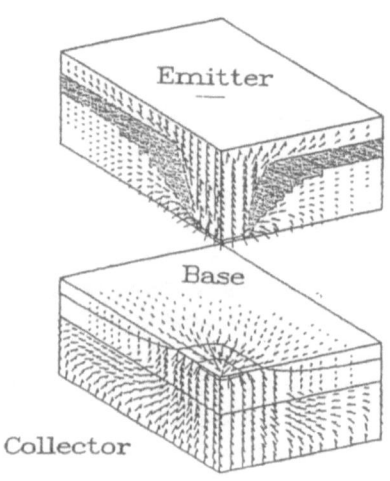

Fig.20:Current distribution
simulated by ODESA.

lower in 3D simulation.   These kind of 3D simulation is possible
only by a coupled 3D process/device simulation.

4.3 Unified simulations of deep-submicron CMOS circuits
      As a final example, we will describe application of UNISAS
to the design of half-micron CMOS technology.   Main purpose was
to investigate the effects of narrower spacer[14] to tpd, and
also to get an optimal process conditions with respect to gate
oxide thickness(t_{OX}) and V_T.   Table 3 summarizes the simulated
samples.   First, process conditions are roughly adjusted to
obtain desired V_T.   Both shallow and deep implantation are done
for V_T control.   Care is taken so as to reduce V_T implantation
conditions.   Simulated and measured V_T are also shown in Table 3,
and show good agreements.   Fig.21 shows simulated and measured
saturation currents (I_{DS}).   Significant improvements in I_{DS} are
observed for narrower spacers compared to a wider spacer.   This
is due to smaller resistance in LDD.

Finally, tpd's of an 2-NAND gate are shown in Fig.22. tpd
is usually deeply correlated with I_{DS}. For different t_{OX}, however, higher gate and junction capacitance may cause t_{pd} reduction
in thinner t_{OX}, which appear in sample B and E. In any case, we
see an excellent agreement between simulations and experiments.

CPU time for the simulation of one sample on 30-MIPS machine
is 2 hours for process simulation and 2.5 hours for device simulations. CPU time for circuit simulation of a simple 2-NAND
circuits is negligible. Noted is that these simulations can be
done without any interruption by users. Thus, only one day was
necessary in order to get the thorough results. Excellent agreement between experiments and simulations and reasonable CPU time
indicate the practicability of the system, and can be the key
technology to minimize the development cost of future VLSI's.

Table 3:Simulated samples.

SAMPLE	SPACER	t_{OX} (A)	VT(sim.) nMOS	VT(sim.) pMOS	VT(experi.) nMOS	VT(experi.) pMOS
A	WIDE	120	0.59	0.72	0.58	0.77
B	NARROW	120	0.53	0.53	0.52	0.53
C	NARROW	120	0.60	0.68	0.60	0.70
D	NARROW	120	0.65	0.81	0.70	0.84
E	NARROW	100	0.52	0.65	0.51	0.66
F	NARROW	100	0.57	0.82	0.60	0.82

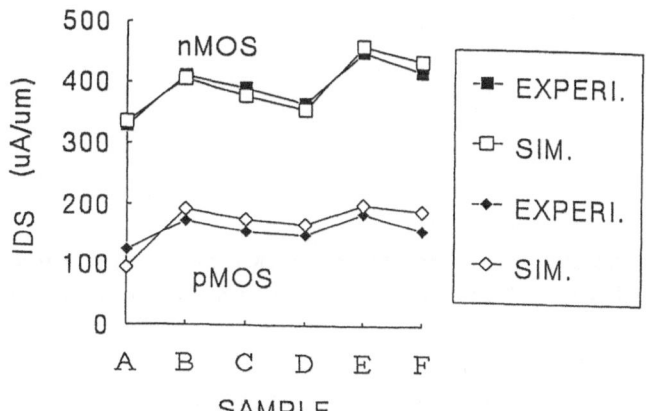

Fig.21:Simulated and measured saturation currents (I_{DS})

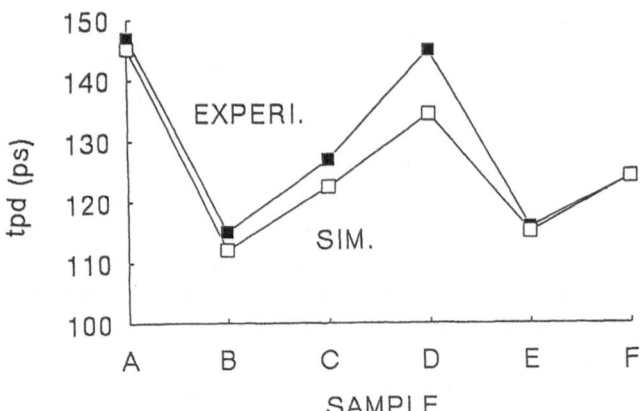

Fig.22:tpd's of an 2-NAND gate.

5.Summary

     TCAD system at OKI, UNISAS, was described. UNISAS is a uni-
fied process/device/circuit simulation system, and offers a
user-friendly simulation environment. Multi-dimensional process
and device simulators, OPUS and ODESA constitutes the core of
UNISAS and can be used for versatile device structures for various purposes. Physics-based simulators constitute another part
of UNISAS, and are used for modeling purpose. Some applications
are described which show good examples of the effectiveness of
UNISAS. UNISAS is already used quite extensively by engineers
for actual device development and is now an indispensable tool
for low-cost, fast TAT VLSI development.

Acknowledgements
     The authors would like to thank S.Kuroda, K.Fukuda, K.Kai
for their simulations. The authors also would like to J.Ida for
his experiments.

References

[1]:D.A.Antoniadis and R.W.Dutton, IEEE Trans. Electron Devices, ED-26, 490 (1979)

[2]:K.Sakamoto, K.Nishi, T.Yamaji, T.Miyoshi and S.Ushio, J.Electrochem. Soc. 132, 2457 (1985). First report appeard in The Electrochem. Soc. Extended abstracts, Vol.81-2, 411, 1981 by K.Nishi, K.Sakamoto, and S.Ushio.

[3]:J.Ueda, Y.Namba, K.Sakamoto, T.Miyoshi and S.Ushio, J.Inst. Elec.Commun.Eng.Japan, J67-C, 825, (1984)

[4]:K.Nishi, K.Sakamoto, S.Kuroda, J.Ueda, T.Miyoshi and S.Ushio, IEEE Trans. Computer-Aided Des., CAD-8, 23, 1989

[5]:H.Kajitani N.Shimizu, K.Fukuda, S.Baba, K.Nishi and J.Ueda, NUPAD IV, p.245, 1992

[6]:S.Ushio, K.Nishi, S.Kuroda, K.Kai and J.Ueda, IEEE Trans. Computer-Aided Des., CAD-9, 745 (1990)

[7]:K.Kai, S.Kuroda and K.Nishi, VPAD'93, p.66, 1993

[8]:S.Kuroda and K.Nishi, IEICE Trans. Electron. E75-C, 145, 1992

[9]:S.Kuroda, Y.Kawai, H.Onoda and and K.Nishi, IEDM Tech.Dig., p.713, 1991

[10]:F.Fukuda, S.Baba, T.Miyoshi and J.Ueda, NASECODE VI, p.422, 1989

[11]:C-Y Lu and J.M.Sung, IEEE Electron dev. Lett., 10, 446, 1989

[12]:C.Mazure and M.Orlowski, IEEE Electron dev. Lett., 10, 556,

[13]:Y.Okita, M.Shinozawa, A.Kawakatsu, A.Umemura, K.Yamaguchi and K.Akahane, IEEE Proc. CICC, p.22.4, 1988

[14]:J.Ida, S.Ishii, Y.Kajita, T.Yokoyama, and M.Ino, IEICE Trans. Electron. E76-C, 525, 1993

The MASTER Framework

P.J. Hopper and P.A. Blakey

Silvaco International,
4701 Patrick Henry Drive, Bldg. 3, Santa Clara, CA 95054, USA

Abstract

The MASTER Framework has a two level architecture. The first level consists of a standard structure format and a set of 'MASTER Tools'. This provides a high quality utility based framework that supports highly interactive use. The second level is a 'Virtual Wafer Fab' that adds capabilities for large-scale simulation-based design and experimentation. Powerful semiconductor technology CAD systems are constructed by populating The MASTER Framework with conforming process and device simulators.

1. Introduction

Many simulation tasks, especially those performed as part of semiconductor technology development, require multiple simulators to be used in combination. For example, process simulators are used with device simulators to predict the influence of process parameters on electrical behavior. Semiconductor technology CAD (S-TCAD) frameworks and systems make it possible, and even convenient, to use multiple tools to perform simulation tasks.

The papers in these proceedings demonstrate that the philosophies, architectures, features, and uses of different S-TCAD systems differ considerably. The MASTER Framework which is described in this paper is oriented towards the needs of industrial users. It can be configured to meet the needs of a wide range of users, it supports highly interactive use, and it supports exceptionally high levels of task integration.

The organization of the paper is as follows. The first section provides some background material. The second section reviews strategic issues that are of concern to both implementors and users of S-TCAD frameworks. The general architecture of The MASTER Framework is reviewed, and the component parts are then described in detail. A brief description of the presently available range of MASTER-conforming simulators is provided in an Appendix.

1.2. The Evolution of S-TCAD Use

In the early days of S-TCAD it was very difficult to use multiple simulators to perform a task. The first problem was portability: implementors wrote code for the operating system and language dialects of their local mainframe computer, and other users were expected to port the code to their own system. Significant impediments remained after tools were ported. Each simulator used a different input syntax and a different data representation; and each simulator required users to perform generic operations (such as visualization) in different ways. Users were responsible for data transfer between tools, and learning to use multiple simulators was unnecessarily complicated. Tool quality suffered in various ways as implementors used scarce resources to replicate generic functionality.

The situation improved greatly during the 1980's. Workstations became widely available and very affordable, thereby providing engineers with powerful, convenient, interactive computing environments. Standards for operating systems (Unix), programming languages (C and C++), and graphical display across networks (X-Windows) emerged. Competition among commercial developers of S-TCAD software led to improved tools, greater affordability, and multi-platform support.

During the mid to late 1980's it became clear that the remaining impediments to the more widespread use of simulation would be eliminated if data representations were standardized, and if special software were written to perform generic operations required by many simulators. This led to the present interest in S-TCAD frameworks and systems.

1.3. Frameworks and Systems

A *framework* is a software environment that supports the use of multiple simulators, while working independently of any particular simulator. Frameworks normally provide convenient data transfer between simulators, a uniform user interface, and well defined procedures for adding new tools. A *system* is a collection of simulators and interfaces that allows users to perform tasks that involve the use of multiple simulators. It is not necessarily flexible or extensible. A *framework-based system* is obtained when a framework is populated with simulators.

Virtually all frameworks provide the benefits that result from the standardization of data and generic operations. Beyond this base level of functionality frameworks differ markedly with respect to the additional capabilities they provide. A basic choice is between adding features that are useful to developers, or features that are useful to users. 'Developers frameworks' are designed to make it easy for developers to add new features. They supply capabilities that allow developers' subtasks to be performed in a uniform, consistent manner. 'Users frameworks' focus on making it convenient for users to perform engineering tasks. The MASTER Framework is a users framework.

2. Strategic Design Issues

The strategic issues in the design (or assessment) of a semiconductor technology CAD system center on defining: the intended users; the system architects/designers; the system implementors; maintenance and support responsibilities; and how the system is financed. These questions can be investigated before any specific technical details are considered. The answers will often give a lot of insight into the nature of the system under consideration, and its probable strengths and weaknesses.

2.1. Organizational Strengths and Weaknesses

Several types of organization are involved in the design, implementation, financing and use of S-TCAD systems and frameworks. These organizations include: companies that manufacture semiconductor products; companies that develop S-TCAD software; university based research groups; government funding agenices; and centralized research consortia. In many cases organizations drawn from one or more of the above categories co-operate in a decentralized manner.

Each category of organization has characteristic strengths and weaknesses with respect to financial resources, personnel resources (number, experience, skill mix, and motivation level), continuity, and accountability. Understanding these strengths and weaknesses is helpful when assessing the nature and long term viability of a particular development effort.

Companies that produce semiconductor products generally have significant financial resources. The issues they face are: is it appropriate to start a peripheral development activity; can an internal development group be staffed adequately; compared to what is available commercially, can an internal group develop unique capabilities or provide significant cost advantages; and are internal developers willing to provide adequate levels of support to users?

Companies that develop S-TCAD software products can provide continuity and customer support. They are faced with the issues of recruiting personnel with the appropriate mix of skills necessary to design and implement products; funding development prior to the occurrence of sales; and making enough sales to ensure long term commercial viability.

Universities have a culture of innovation, and access to graduate student labor that is typically cheap, flexible, and enthusiastic. The issues faced by universities are: can inexperienced graduate students implement capabilities with adequate speed and quality; are routine development tasks an appropriate part of graduate student research; can continuity of effort be maintained after students graduate; who is responsible for support and maintenance; can funding be obtained for work that is not in the mainstream; and can funding be retained for work that competes directly with commercial organizations?

For government funding agencies the issues are: how can certain organizations be selected for subsidy without undermining unsubsidized organizations; how can self-selection by commercially unviable organizations be avoided; how can progress be monitored and evaluated effectively; and will the work continue after government funding ceases?

Centralized research consortia can spread development costs over their membership. The issues faced by centralized research consortia include: developing an initial concensus; acquiring and retaining funding; recruiting staff; setting detailed technical directions; monitoring progress; establishing technology transfer mechanisms; and providing maintenance and support.

Co-operating groups that work in a decentralized fashion may appear to have the resources and skills necessary to perform a task. However, they usually find it difficult to overcome the problems of coordination and information transfer. The associated inefficiencies are usually much greater than anticipated, and the diffusion of accountability often means that few worthwhile results are produced.

2.2. The Rationale For Commercial Frameworks

Only a few of the very largest microelectronics companies can now consider developing and supporting S-TCAD frameworks and systems internally. A decision to support a major in-house development activity can no longer be justified on the grounds of cost, and only very seldom can it be justified on the grounds of capabilities. However some organizations continue to place a significant value on ownership and control of what they perceive as a strategic resource.

The vast majority of the microelectronics industry prefers to purchase affordable, high quality, well supported S-TCAD software from a commercial developer. The MASTER Framework was developed to meet the needs of this market. Silvaco management realized several years ago that there were several factors operating in favor of starting to implement a commercial framework. Silvaco was not locked in to a traditional batch-oriented way of using process and device simulation. Silvaco was also in a position to hire experienced development engineers and programmers in the right mix, and in sufficient numbers, to complete the project in a timely fashion. The company has a support and applications organization that is highly regarded by customers. Finally, it was anticipated that a framework product could be made available at an affordable price, because the development costs could be distributed over a large customer base.

The problems of implementing and maintaining large software systems are not amenable to solution by relatively small academic groups. Such work has a relatively low academic content, and requires greater resources and more continuity than most academic groups can provide. There is still a place for academic activity in the area of S-TCAD frameworks and systems. Such work can be used to evaluate concepts, and to educate students in the associated issues and techniques. However academic work is not likely to be of direct significance to industry because of concerns about quality, support, and continuity of effort.

From a slightly broader perspective, the problems that are encountered in S-TCAD can be divided into two categories, 'isolatable' problems, and 'systemic' problems. The isolatable problems are aspects of the associated physics, chemistry, and mathematics that can be identified and solved using a divide-and-conquer approach, with particular subproblems becoming the focus of the efforts of a small group. The 'systemic' problems of developing and maintaining a software infrastructure has to be addressed by a different type of organization. Silvaco management saw the opportunity to be the organization that supplies the infrastructure for S-TCAD systems.

Silvaco's model for the future of S-TCAD development is that Silvaco will attack the systemic problems. The infrastructure that is developed will enable Silvaco's academic and industrial collaborators to focus on research into the 'isolatable subproblems'. The solutions that are developed will be fed back to enhance the capabilities of future S-TCAD systems.

3. The General Architecture of The MASTER Framework

The constraint of economic viability has had significant impact on the design and implementation of The MASTER Framework. The need for commercial acceptance means that the design is oriented towards the needs of users, not the needs of developers. An incremental development plan was designed so that revenues from the first phase of implementation could fund a second phase; and performance, simplicity, and maintainability were given precedence over academic elegance and conceptual originality, whenever such choices needed to be made.

The MASTER Framework has two levels (see Figure 1). The first part consists of a standard structure format (SSF) and a set of interactive, graphically-oriented 'MASTER Tools'. The SSF and the MASTER Tools provide the *tool integration level* of the MASTER Framework. This is a high quality, file- and

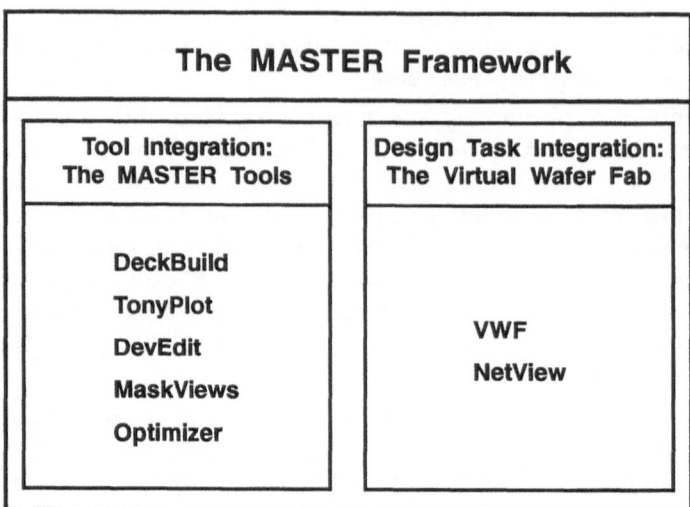

Figure 1. The Two-Part Architecture of the MASTER Framework.

utility-based framework with extensive capabilities built in for the tasks of input deck debugging and calibration.

The second level of the architecture builds on the first level, providing additional capabilities that support large-scale simulation-based design and experimentation. This level is referred to as the Virtual Wafer Fab (VWF). The VWF provides the computational analog of the 'split-lot' methodology that forms the basis of empirically-based technology development. The VWF provides the *design task integration level* of the MASTER Framework.

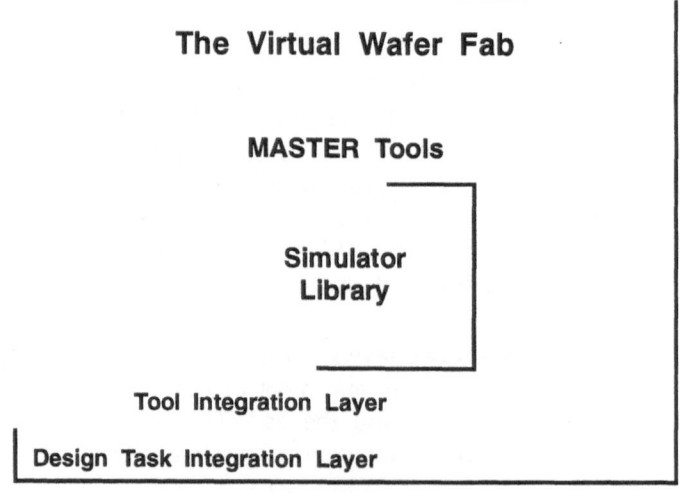

Figure 2. S-TCAD System Structure.

A conceptual view of S-TCAD systems that are based on The MASTER Framework is shown in Figure 2. The heart of each system is the MASTER-conforming process and device simulators that are contained in the system's simulator library. A brief summary of the simulators that are presently available is provided in the appendix.

The architecture of The MASTER Framework allows exceptional flexibility in the construction of S-TCAD systems. Systems can be configured to meet the needs, experience levels, and budgets of a wide range of customers, and systems can be extended as users' needs evolve. It is normally recommended that The Virtual Wafer Fab is added only after a customer has developed familiarity with the simulators and the MASTER Tools.

4. The MASTER Tools

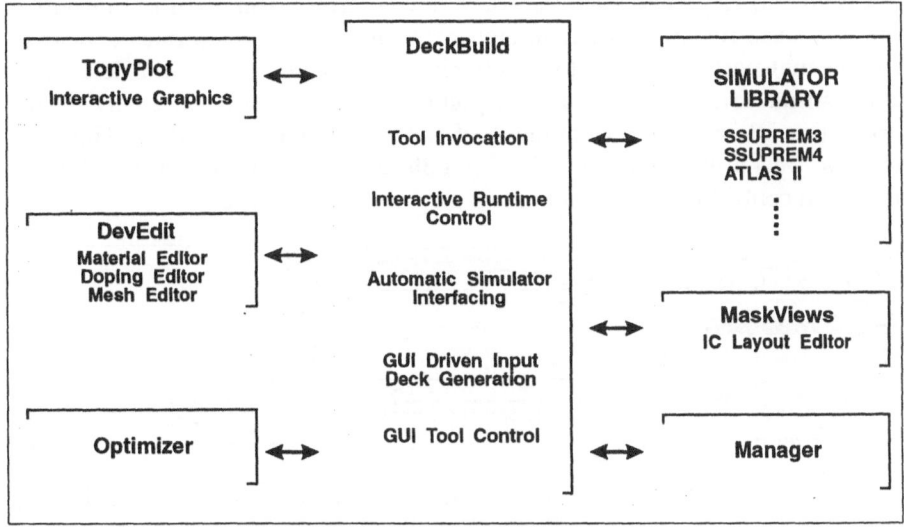

Figure 3. The MASTER Tools.

The MASTER Tools are DeckBuild, TonyPlot, DevEdit, MaskViews, Optimizer, and Manager. The functionality of the MASTER Tools, and the connections between them, are shown in Figure 3. DeckBuild is the central tool. It provides an interactive run time environment, and can invoke and control simulators and other MASTER Tools. The remaining MASTER Tools provide capabilities for: scientific visualization; structure, doping, and mesh specification editing and meshing; specification of layout information; black-box optimization; and click-and-drop use of MASTER Tools and SSF files. The MASTER tools were written in C and C++ by professional programmers using extremely high standards of software engineering.

4.1. DeckBuild

DeckBuild supplies a flexible environment for generating, editing, and running input decks. It provides a graphical user interface that allows users to produce input decks without knowing simulator-specific input syntax. Information is typed into a series of pop-up windows. When specification is complete DeckBuild can automatically produce a syntactically correct input deck. Existing decks can be read in to DeckBuild, and decks can be edited directly by the user. Multiple simulators can be called from within a single input deck. Simulator invocations and information transfer between simulators are taken care of automatically. Input parameters in decks can be optimized to match known target results. A sophisticated extraction capability allows engineering parameters to be obtained from calculated results.

DeckBuild allows precise user control of how an input deck is run. It provides stop at, pause, restart and single step capabilities. It also supplies a history function that permits a user to backtrack to a previous point in a deck, and then continue computation from this point. This capability is extremely useful during the interactive development of a simulated process flow. Figure 4 shows a DeckBuild window with an edit window, an output window, a simulation control panel, and a process step definition pop-up.

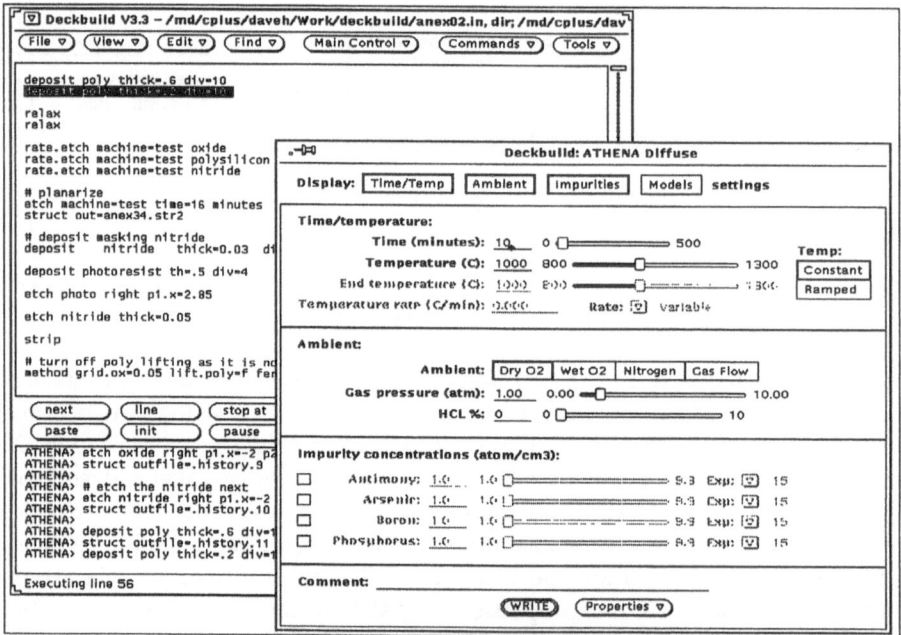

Figure 4. A DeckBuild window showing edit and output windows, with a process step definition pop-up in the foreground.

DeckBuild not only makes it easier to run simulators in traditional ways. It also provides an environment that supports the development of input decks that give results that are calibrated to experimental results. Such input decks provide a good baseline from which to experiment.

4.2. Other MASTER Tools

TonyPlot is an interactive scientific visualization tool that provides comprehensive capabilities for viewing the results produced by process and device simulators. All of the usual ways of displaying scientific data are supported. These include x-y plots with linear and logarithmic axes, contour plots, and surface plots. Virtually every characteristic of the plots, including the text and position of labels may be specified by the user. TonyPlot also includes an animation feature that provides a way of viewing a sequence of plots in a manner that shows solutions as a function of some parameter. The parameter can be varied under push button control, or the frames can be looped continuously. This animation feature is very helpful for developing physical insight. TonyPlot supports the production of hard copy plots on a wide range of printers.

DevEdit is an interactive tool for specifying and editing structures, doping profiles, and meshes. The meshing capabilities include the production of a high quality initial mesh based on geometrical considerations, and refinement and unrefinement based on user-specified criteria.

DevEdit can be used stand-alone, or it can be invoked by DeckBuild. Large devices with many grid points may be specified completely using DevEdit, making this tool useful as a pre-processor for 2-D device simulators. A special mode of DevEdit supports the meshing of 3D structures. Figure 5 is a view of a DevEdit window showing a structure that was meshed using DevEdit.

MaskViews is a versatile IC layout editor used to specify layout information to process simulators. Simula-

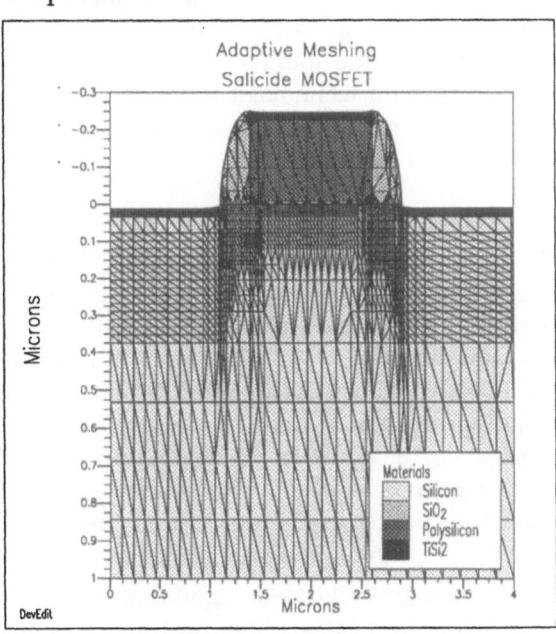

Figure 5. A MOSFET meshed using DevEdit.

tion based experimentation was previously restricted to the varying of process flow parameters only. MaskViews supports simulation-based experimentation with layout variations. MaskViews enables users to investigate critical dimensions, polygon reshaping, misalignment tolerances, global shrinks, and phase shift mask parameters. It is fully interfaced to GDS2 Stream formats so that complete IC layouts may be imported and exported, and small subregions can be selected for detailed analysis.

The Optimizer provides black-box optimization, calibration, and tuning capabilities. Control of the Optimizer is integrated into DeckBuild. The Optimizer can be used across multiple simulators, i.e. it is possible to tune parameters of process simulators to obtain specified electrical characteristics from a device simulator. Optimization targets may include structural dimensions, device parameters after a complicated electrical test, and any intermediate outputs.

Manager is a simple application manager that supports interactive point-and-shoot and drag-and-drop use of a variet of files and MASTER tools. The use of this tool is very intuitve. For example, a structure file can be selected using the mouse, and dragged to to the TonyPlot icon. It will then be plotted on the screen.

5. The Virtual Wafer Fab

The VWF automates large-scale, simulation-based, design, calibration and optimization. The underlying paradigm of the VWF is the computational split lot. This mirrors the split lot methodology used for experimental development of semiconductor technologies. The differences are that simulated experiments substitute for real experiments, and workstations and software substitute for operators and equipment.

5.1. Maximizing The Economic Benefits of Simulation

Simulation provides several benefits. These include: predictive 'what-if' capabilities; physical insight; and knowledge encapsulation and reuse. Predictive capabilities allow the substitution of simulation for experiments. Eliminating some of the experiments that would otherwise be performed during technology development lowers costs and saves time. The physical insight provided by simulation helps engineers develop experience more quickly, i.e. it improves the productivity of individuals by giving them additional knowledge. The knowledge encapsulation and reuse provided by simulators improves the productivity of individuals by making it easy for them to access the knowledge of others.

All of these benefits have significant economic value, but the value is not always easy to measure. The most directly quantifiable benefits are the 'costs of experiments avoided' when the use of simulation reduces the number of experiments required for technology development. This suggests targeting the capabilities of S-TCAD systems at maximizing the extent to which simulation is substituted for experiment. The VWF is designed to achieve this goal.

The VWF mirrors experimental split-lot methodology very closely. There are several reasons for this. First, the approach is probably very close to optimum: if more effective methodologies existed, they would already be used for experimental development. Second, the simulation methodology can be learned easily by engineers, since the underlying paradigm is already familiar. Third, when simulation mirrors experimental procedures very closely there is an increased probability that simulation will be used as a direct replacement of some experiments.

It may seem surprising that the use of simulation has not previously mirrored experimental procedures. There are several reasons for this. The most important reason is that using simulation in this way has been too time consuming and tedious. The user has been responsible for designing the 'computational experiment', generating all the associated input decks, submitting each individual simulation run, transferring data between simulators, handling storage for the generated data, and extracting useful information from the results. Some of these activities (e.g. automated deck generation, simulator invocation, and inter-simulator information transfer) are automated by DeckBuild. The VWF automates the other activities to a very high degree.

5.2. VWF Capabilities

The VWF employs a database, rather than files, for storing information; and it uses a worksheet as the primary representation of a computational experiment. The database and worksheets are central concepts of the VWF and are reviewed first in this section. The other features of the VWF will be described in the context of describing how a computational experiment is performed within the VWF.

Information storage and handling in the VWF is database-oriented rather than file-oriented. All the information associated with computational experiments is stored in an object oriented database and is accessed through an intuitive icon-oriented and menu-driven interface. The MASTER database supports libraries and workspaces. Libraries are used to store persistent objects that are useful across a project or organization. For example, standardized process recipes and electrical tests can be stored in libraries. Other possible objects are mask layouts and cross-sections. Workspaces typically contain shorter lived information used by an individual engineer, including computational-experiments-in-progress.

Vt adjust implant dose	Vt adjust implant energy	P-LDD implant dose	P-LDD implant energy	Nvt	Nbeta	Ntheta	Pvt	Pbeta
1.5e+12	25	1e+14	40	0.728865	7.65048e-05	0.0859445		
1.5e+12	28.5	1e+14	40	0.661566	7.6585e-05	0.0841255		
1.5e+12	32	1e+14	40	0.610565	7.65616e-05	0.0818454		
1.5e+12	25	1e+14	40	0.799401	7.22224e-05	0.0825794		
1.5e+12	28.5	1e+14	40	0.732001	7.30774e-05	0.0835462		
1.5e+12	32	1e+14	40	0.679009	7.31596e-05	0.0821723		
1.5e+12	25	1e+14	40	0.851855	6.6642e-05	0.0741935		
1.5e+12	28.5	1e+14	40	0.791352	6.85354e-05	0.0804045		
1.5e+12	32	1e+14	40	0.738936	6.92576e-05	0.081021		
1.5e+12	25	1e+14	40	0.91138	6.17652e-05	0.0599253		
1.5e+12	28.5	1e+14	40	0.842814	6.35572e-05	0.0653263		
1.5e+12	32	1e+14	40	0.793337	6.51546e-05	0.0706666		
1.5e+12	25	1e+14	40				1.33229	2.06681e
1.5e+12	28.5	1e+14	40				1.39092	1.9531e-
1.5e+12	32	1e+14	40				1.42539	1.79319e
1.5e+12	25	1e+14	40				1.38226	1.90277e
1.5e+12	28.5	1e+14	40				1.42195	1.75042e
1.5e+12	32	1e+14	40				1.45908	1.63788e
1.5e+12	25	1e+14	40				1.41337	1.72536e
1.5e+12	28.5	1e+14	40				1.4467	1.59016e
1.5e+12	32	1e+14	40				1.50668	1.58461e
1.5e+12	25	1e+14	40				1.42364	1.58143e

Figure 6. A portion of an interactive VWF worksheet.

A computer generated worksheet is a major focus of the computational split-lot methodology implemented within the VWF. The worksheet allows users to view input parameters and output results. The worksheet provides a way of simultaneously presenting the definition of a computational experiment, and the results that have been obtained. This representation may be manipulated very naturally through a graphical user interface. The worksheet of a computational experiment can be edited by the user at any time. The worksheet supports filtering of values, and allows columns to be specified as functions of other columns. Experiments can be added or deleted at any time. Results are logged and appear automatically as simulations are completed. A portion of a VWF worksheet is shown in Figure 6.

A computational experiment starts from an input deck developed for a baseline process flow. This deck will often have been tuned to match experimental measurements. Splits may be defined at three levels: IC layout cross-sections; processing flow split points; and device tests. This computational procedure mirrors the specification of test structures, the fabrication of various process flow wafer splits, and the testing of individual devices.

If desired, a sensitivity analysis on specified output values with respect to selected input parameters may be performed prior to the definition of the computational split-lot experiment. The preliminary sensitivity analysis is used to identify which parameters should be varied in a subsequent split-lot experiment. Parameters are ranked in order of their influence on the output, and the effect each parameter has on the final results is listed. A special feature on the split point editor allows a sensitivity design tree to be created automatically without any further user interaction.

Computational split-lot experiments can be defined by the user, or they can be generated automatically through the use of built-in capabilities for the design of experiments (DOE). An arbitrary number of different values can be specified for each selected input parameter, resulting in a "tree" of experiments. The associated worksheet is generated automatically. The user can modify an experiment by directly editing the associated worksheet. The built-in capabilities for experimental design include full and partial factorial, Box-Behnken, random, Latin Hypercube, and Composite.

The VWF automatically generates the input decks associated with a computational experiment, and automatically submits jobs for execution in a networked computing environment, on a MIMD parallel machine, or both. The experimental tree structure is exploited to minimize the total required computation. Intermediate results are reused for child nodes, thereby eliminating unnecessary duplication, and filter limits can be specified so that branches of an experimental tree are eliminated if intermediate extracted parameter values go outside acceptable ranges. Jobs are routed to each of the

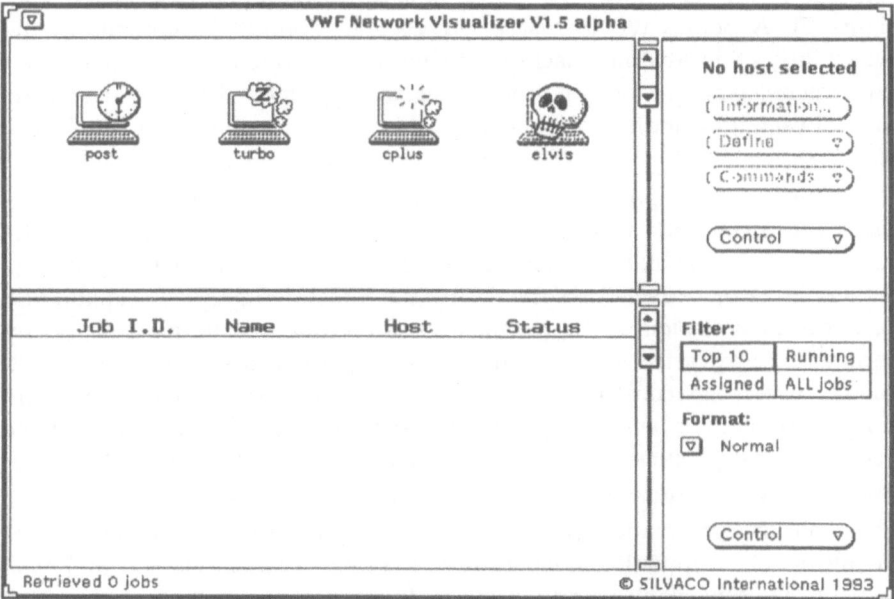

Figure 7. The VWF network control screen.

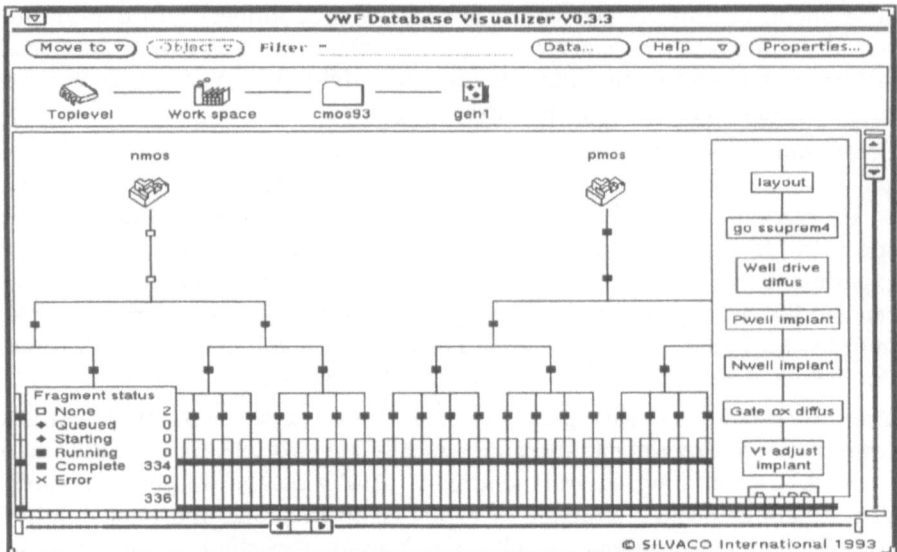

Figure 8. A VWF experimental tree screen.

computers in a network, subject to specified conditions that can include run priorities, and the times and days for job submission to each computer. A network visualizer allows the status of a network to be determined (see Figure 7). A popup window accessed via the 'Control' button allows the availability of individual machines to be set or edited. The status of an experiment in progress can be viewed as an experimental tree in which the color of each node indicates the status of the associated simulation (see Figure 8).

The results of a computational experiment can be analyzed in various ways. Users can inspect the results on the graphical worksheet, which shows the values of all the variable parameters associated with each simulation run, plus all simulated or extracted results. Using the integrated graphics capabilities, users can look at plots of the raw data, or scatterplots. Scatterplots show any parameter or extracted value versus any other parameter or simulated result. Finally, users can construct models of the results using either regression analysis or neural nets. Design and/or optimization can subsequently be performed very efficiently using the modeled data. Figure 9 shows an example of a response surface obtained from modeled data.

The VWF implementation also takes into account the need to communicate the results of computational experiments to others. Worksheets and graphical displays may be conveniently printed. Pagination, formatting, titling and labeling are performed in a way that allows printouts to be used directly in reports and presentations.

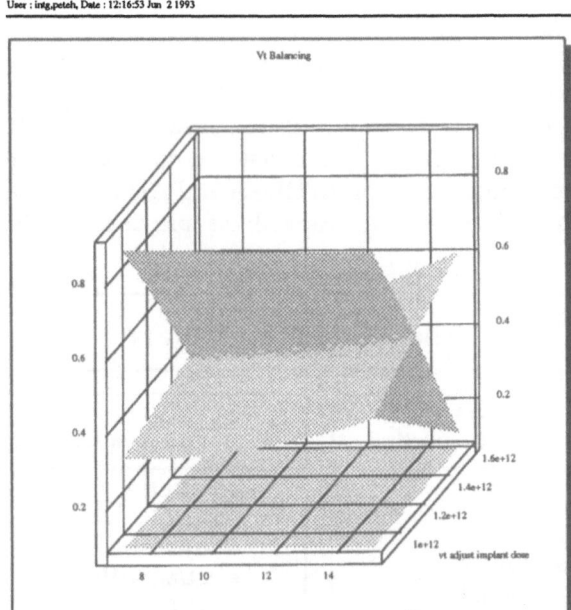

Regression graphics from database entry 'cmos93/expt1'
User : intg.peteh, Date : 12:16:53 Jun 2 1993

SILVACO International, Virtual Wafer Fab V0.3.5

Figure 9. A response surface graph showing threshold voltage balance.

6. Conclusions

The MASTER Framework is a professionally implemented, commercially supported semiconductor technology CAD framework. It provides two levels of functionality, a tool integration level and a design task integration level. Users populate The MASTER Framework with a set of process and device simulators to produce flexible yet sophisticated S-TCAD systems.

7. Acknowledgements

Many customers, collaborators, and Silvaco developers and programmers have contributed ideas that are used in The MASTER Framework. The authors of this article, who had no coding responsibilities, thank the implementation team for working the long hours and late nights necessary to bring the product to fruition.

Appendix: The Simulator Library

Silvaco's process and device simulators can be classified into four groups: a 2-D process simulation system called ATHENA; a 2-D device simulation system called ATLAS; a 3-D device simulation system called THUNDER; and 'Other Tools' (see Figure A1). The industrial use of S-TCAD focuses on 2-D process and device simulation using ATHENA and ATLAS. These systems are designed to make full use of the capabilities provided by The MASTER Framework.

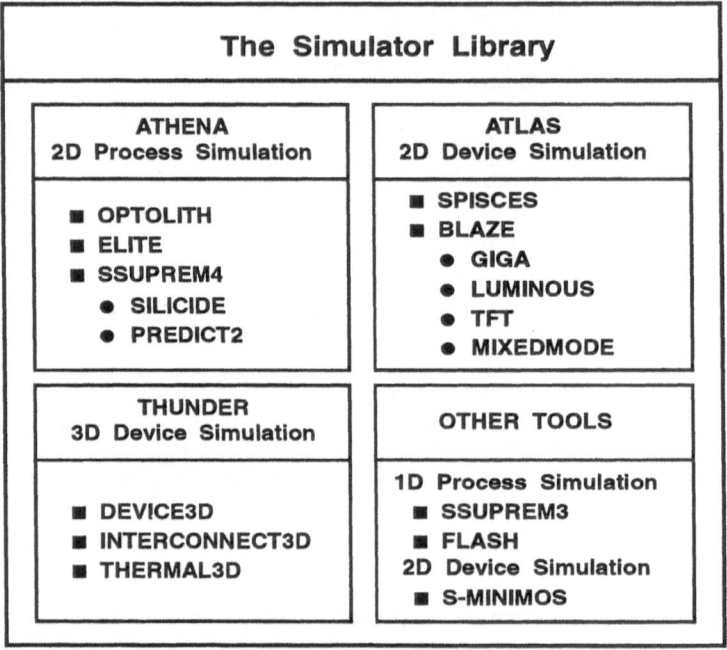

Figure A1.

A1.1. The ATHENA 2-D Process Simulation System

ATHENA is a framework based system for 2-D process simulation. It integrates a set of process simulation tools that provide comprehensive capabilities for 2-D process simulation. The individual tools are OPTOLITH, ELITE, and SSUPREM4.

OPTOLITH is a 2-D lithography simulator. It simulates aerial image formation and photoresist exposure and development. OPTOLITH is available in two versions, one planar and one nonplanar. ELITE is a general purpose 2-D topography simulator that simulates a wide range of deposition and etch processes used in modern IC technologies. These processes include dry etching, wet etching, APCVD, LPCVD, ion milling, metalization and reflow. SSUPREM4 incorporates a range of advanced physical models for diffusion, implantation, and oxidation. SILICIDE is an add-on module for SSUPREM4 that provides unique capabilities for modeling silicide

A1.2. The ATLAS 2-D Device Simulation System

ATLAS is a framework based system for 2-D device simulation. The ATLAS framework can contain either or both of two primary device simulators, SPISCES and BLAZE, and any combination of four products, GIGA, LUMINOUS, TFT and MIXEDMODE that work in conjunction with either SPISCES or BLAZE.

SPISCES simulates structures encountered in silicon technologies. It maintains backwards compatibility with earlier stand-alone versions of SPISCES. BLAZE handles arbitrary semiconductors, and graded and abrupt heterojunctions. GIGA adds the ability to simulate lattice heating and heatsinks. LUMINOUS adds optoelectronic interactions and very sophisticated ray tracing. TFT adds the features required to simulate devices based on charge conduction in polycystalline and amorphous materials. MIXEDMODE is a circuit simulator that allows numerically-simulated ATLAS devices to be used instead of conventional circuit models for some devices.

A1.3. The THUNDER 3-D Device Simulation System

THUNDER is a framework based system for 3-D device simulation. The THUNDER framework is populated by any combination of three products, DEVICE3D, INTERCONNECT3D, and THERMAL3D. DEVICE3D provides capabilities for the full bipolar simulation of silicon devices. INTERCON-NECT3D calculates the capacitances and resistances associated with interconnect structures. THERMAL3D calculates the temperature distribution in 3D thermal environments that can include regions of semiconductor, insulator and heatsink, with user-defined thermal sources in specified regions.

A1.4. Other Tools

The other simulators offered by Silvaco are two 1-D process simulators, SSUPREM3 and FLASH, and a 2-D device simulator, S-MINIMOS.

SSUPREM3 is a mature, well-calibrated tool for simulating ion implantation, oxidation and diffusion in silicon structures. FLASH is a process simulator for general materials. It has sophisticated models for ion implantation, and incorporates a dial-an-operator mechanism that is very convenient for developing sophisticated kinteic diffusion-reaction models.

S-MINIMOS is an enhanced version of MINIMOS 4.0, which was developed at the Technical University of Vienna. S-MINIMOS is integrated with

DeckBuild and TonyPlot for stand-alone operation. More importantly, it is very tightly integrated into the Virtual Wafer Fab. It can therefore play an important role in simulation-based design and experimentation of MOS based technologies.

CAESAR: The Virtual IC Factory as an Integrated TCAD User Environment

V. Axelrad, Y. Granik, and R. Jewell

Technology Modeling Associates, 3950 Fabian Way, Palo Alto, CA 94303, USA

Abstract

TMA's Virtual IC Factory (CAESAR) is seen as a new enabling technology. CAESAR provides a higher level of user-system interaction than previously possible, especially effective in managing large-scale simulated experiments. CAESAR offers a hierarchic approach to specifying a simulation flow. Its open architecture design allows the utilization of heterogeneous simulators, standard or user-defined post-processing tools and extractors. CAESAR supports large-scale simulations by creating and managing intermediate simulation results. This includes automatic and transparent maintenance of data dependencies, incremental simulation support and visual access to simulation results via icons representing wafer data.

Rather than an incremental improvement over previously chosen approaches, CAESAR represents a qualitatively different complete environment to use TCAD with benefits for both the novice and the experienced so-called "power user". The novice user benefits from the ease-of-use and encapsulation features of the system, allowing to conceal complex physics, numerics, simulator syntax and data interfacing details in pre-tested library modules. The experienced user can concentrate on the engineering goal of the simulation study and run large simulated splits without worrying about massive amounts of simulator input files, data files and complex dependencies between them.

Several major semiconductor companies have been involved in the specifications for CAESAR and its daily development. This assures the practical relevance of the tool, its robustness and suitability to fulfill real engineering needs.

Synopsis

1. Introduction

Traditionally, Technology Computer-Aided Design (TCAD) has been centered around complex numerical simulation tools. These tools are highly sophisticated with respect to the physical phenomena being described and the numerical methods used to this purpose. Consequently, the use of such simulation tools requires a high level of user expertise and experience. Characteristic to applications of TCAD tools was thus the use of input languages to control the complex multitude of physical models and numerical methods available to the user. Since the main emphasis of development work in TCAD was on the simulators themselves, tool integration was weak. Inter-tool communication as well as storage and maintenance of simulation results required a significant amount of user involvement and responsibility. This approach was indeed justified at the time, as it offered maximum flexibility for the user and did not impede experimentation with numerical methods and physical models.

As process and device simulation is moving out of predominantly research and development environments into more production-oriented engineering groups, the way engineers use TCAD is changing rapidly. Cost pressures from semiconductor manufacturing dictate the use of simulation as an at least partial replacement for actual fab experiments – split runs. Instead of single process steps and/or device simulation problems, the semiconductor engineer needs to model an entire process flow as well as subsequent electrical test procedures for a set of control parameter values, resulting in massive amounts of inter-dependent individual simulations and data. Use of previously developed process/device simulation modules becomes mandatory, requiring support of process module libraries and electrical test procedure libraries for use within an organization across its R&D and manufacturing departments. All this necessitates an unprecedented level of tool integration as well as automated data handling and storage.

With process and device simulation tools rapidly advancing towards maturity, it became feasible to satisfy the needs of both the traditional R&D-oriented TCAD users and newer requirements of manufacturing engineers. The new integrated user environment CAESAR models the relevant aspects of a real semiconductor factory in a tightly integrated and highly intuitive graphical tool set.

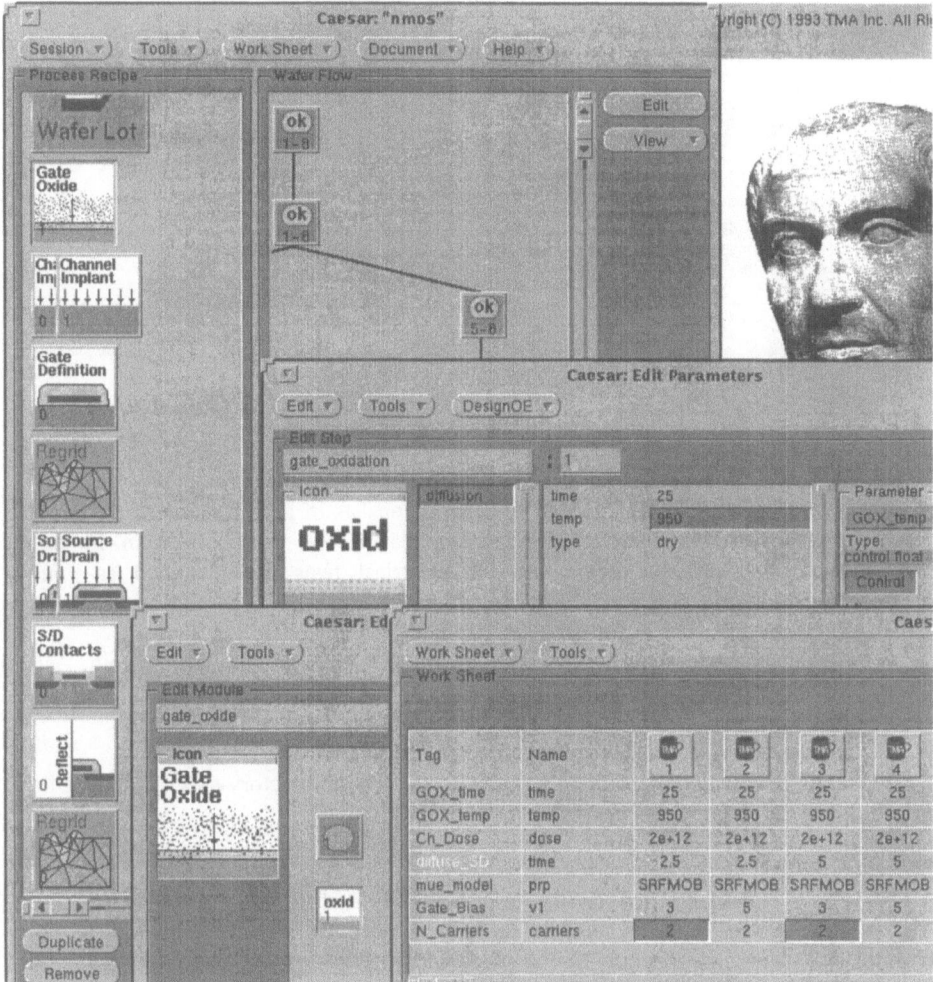

Figure 1. Full View of CAESAR main window, module and step editors and the Work Sheet Dialog.

2.   CAESAR: Scope of the Tool

The basic idea behind the virtual factory concept such as CAESAR's is to emulate the workings of an actual semiconductor fab as closely as possible and necessary. The goal is to obtain relevant data such as electrical product performance, processing time, cost, etc. under varying processing conditions. As a result the number of actual physical experiments can be reduced, resulting in drastic cost savings and improved time-to-market. The "classical" TCAD part of this task is process and device simulation, which is currently the scope of CAESAR:

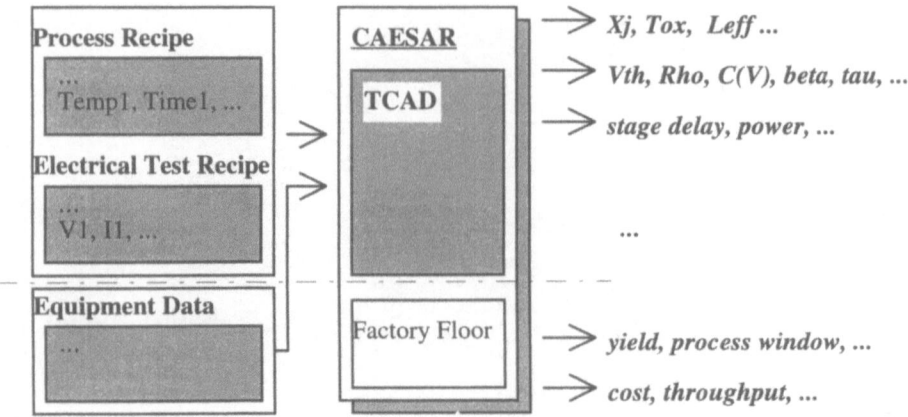

Figure 2. The Virtual Factory Concept and Scope of CAESAR.

As shown in Figure 2, the input to CAESAR consists of a description of the process recipe and the electrical test procedures applied to the device. Certain important parameters in the process recipe and the electrical tests are identified as control parameters. The results of a Virtual Factory simulation can be:

- structural data such as junction depths, oxide thicknesses,
- electrical device data such as threshold voltages, capacitances, current gain,
- circuit parameters such as gate delay time, rise- and fall-times,
- global process criteria such as yield, process window,
- manufacturing parameters such as cost and throughput.

Modeling of cost and throughput requires equipment and factory-specific data as input in addition to process and test recipes.

3. Hierarchy of the Process Recipe

A complex simulation flow can be described in several different ways, which may be equivalent with respect to the information they carry, but quite different from the user's point of view. This situation may be compared with multiple projection images of the same three-dimensional object: they all represent different aspects of the same physical entity. CAESAR offers a variety of viewing and editing modes to accomodate different phases of work with the system as shown in Figure 1 using a simple MOS process as an example.

The simulation flow is represented in CAESAR using the following hierarchy:

- Modules, technically distinct portions of the process recipe (Figure 3, left).
- Steps, a linear sequence of steps comprises a module (Figure 3, middle).
- Commands, each steps maps to a well-defined sequence of simulator commands complete with parameters (Figure 3, right).

Modules are encapsulation objects for steps, providing an additional hierarhic degree of freedom for the specification of the process. Generally, modules can be used to represent *what to do* in a relatively technology-independent way. As an example, a module could be "N-Well Creation" or "Poly-Gate Definition". Steps, on the other hand, are technological steps needed to actually do it. Examples here could be "Implant", "Anneal", "Mask", "Etch", etc. Thus steps are technology-specific instructions of *how to do it*.

Finally, commands are the simulator-specific decomposition of steps into simulator language.

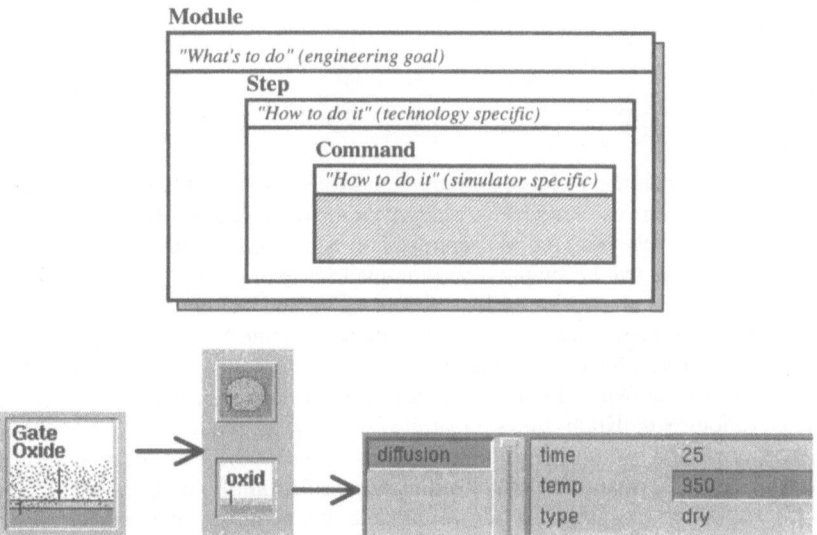

Figure 3. Module - Step - Command Hierarchy of CAESAR.

The user creates a process recipe using pre-tested modules from a vendor-supplied or customer site-specific libraries. Physical model parameters as well as numerical parameters (parameters are attributes of steps) can be adjusted using the "module editor" and "step editor" shown in Figure 1. The simulated experiment is set up as a directed graph visualized in the "Process Recipe" section of the main window (Figure 1). Simulation splits can be specified on module level.

Simulation modules serve mainly as encapsulation objects for simulation steps. Steps are simulator-specific sets of well-defined simulator instructions (commands). The simulator type is an attribute of the step. Simulator commands in a step are represented as a dynamic list of named commands each again with a dynamic list of typed arguments. Possible types are: float, character, logical, lists, etc. In addition, minimum and maximum values can be specified for numeric parameters, parameters can be identified as control parameters to be used in splits or as fixed value parameters, which cannot be modified by the user:

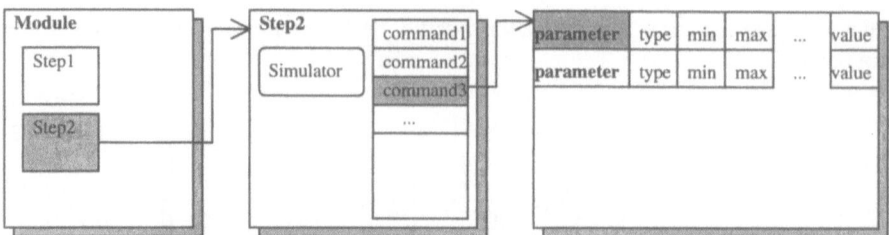

Figure 3a. Schematic Representation of the Module - Step - Command Hierarchy.

At execution time each module is mapped to a simulator input file. The simulator type is determined by the value of the simulator attribute in the first step in the module. Therefore a module can be simply reprogrammed for execution with another simulator by replacing simulation steps. For each simulator we define the methods to load and save a description of the semiconductor wafer, so that a module becomes a portable and self-contained operator, which transforms an input wafer representation into an output wafer representation as shown in Figure 4. As a result, modules can be stored in libraries and used in various simulation flows in arbitrary order as required by the technological specification as long as the used simulators conform to the same format to describe a wafer state. An important side-effect of this encapsulation of simulation steps in modules is that natural check points occur at module boundaries. Intermediate results of a simulation are defined as results of simulation modules and are stored automatically.

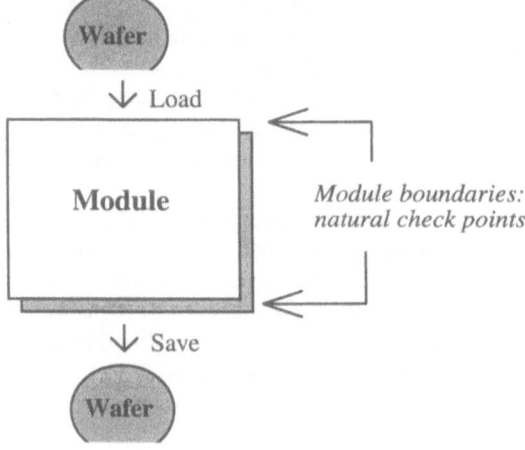

Figure 4. Module as a Wafer State Operator.

4.   Wafer Flow as a Result of the Simulation Flow

A consequence of the encapsulation of simulation steps in modules with intermediate wafer states saved at each module boundary is that a wafer flow can be identified, which corresponds to the specified process recipe. While the process recipe is a directed graph with splits being represented by multiple modules at one level (Figure 5, left), the wafer flow is a tree as shown in Figure 5, right, where splits results in branching of the tree. In the wafer tree each icon represents a wafer, the result of a simulaton module. The simulation status of the wafer (done, queued, aborted, running, etc.) is identified by appropriate symbols.

As modules are executed during the simulation of the split, the status of the wafers changes accordingly. Thus the wafer tree is a natural representation of the simulated split during its execution, allowing visual monitoring of the simulation status. Error recovery, which is neccessary in cases such as simulator divergence due to physical or numerical reasons, system failure, etc. is much easier in a visual environment. Changing a parameter value in one of the simulation modules to resolve the problem will result in an automatic re-simulation of all dependent parts of the tree (incremental simulation, see below). This can be especially important when independent parts of the tree are executed on a network. Automatic saving of intermediate wafer states guarantees high robustness and stability of the simulated experiment. In the event of a simulator failure or even a computer system or network failure, the experiment can be re-started from the failed module, thus avoiding re-running the whole simulation and minimizing the error-recovery effort.

In addition, the wafer tree serves as a graphical data viewer, representing all intermediate and final results of a simulated split. The user can access the data at any time during the experiment by selecting a wafer icon in the wafer tree and applying one from a selection of available tools. These tools can perform post-processing, visualization, extraction or a user-defined set of operations on the wafer representation, such as for instance extracting a peak electric field value and its location from a device simulation result.

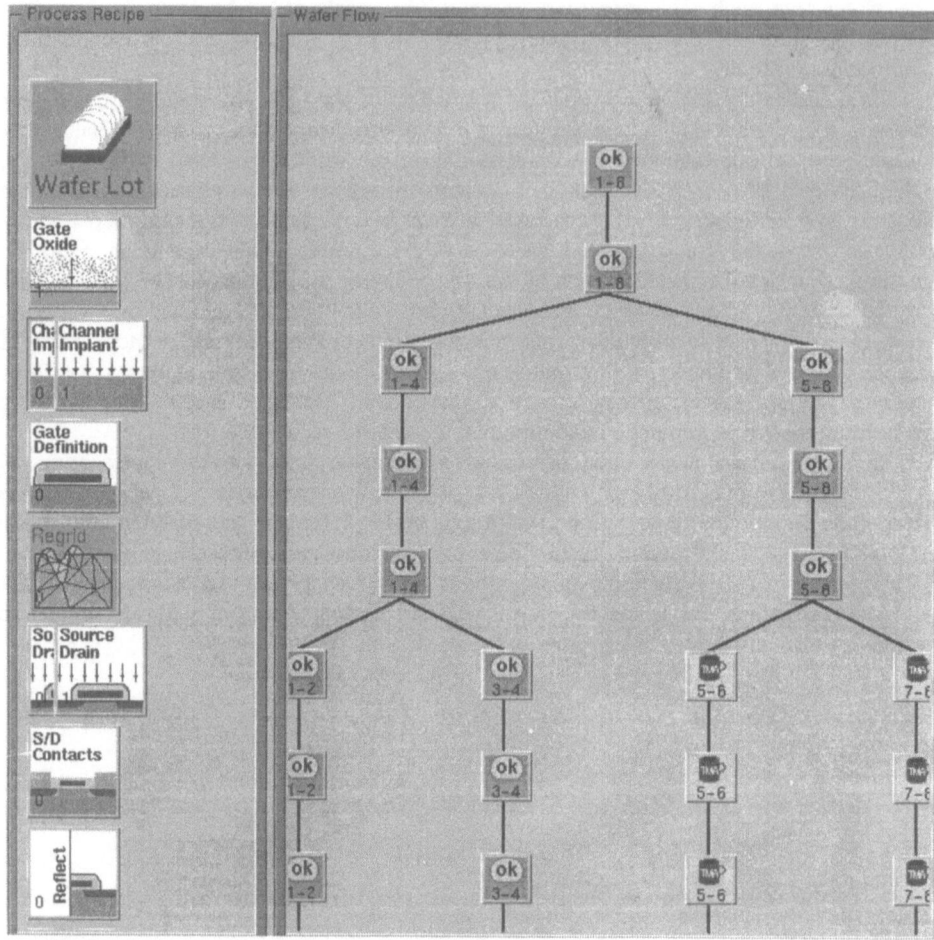

Figure 5. Process Recipe (graph) and Wafer Flow (tree).

5. Running Splits, Incremental Simulation, Data Management

To create splits in a process/device simulation flow the user can simply duplicate simulation modules at any level in the simulation graph by selecting the desired module and applying the duplicate operation to it. As shown in the example in Figure 5, the simulation graph can be split up in an arbitrary number of branches at any of its levels. Once branching is created "Module Editor" and "Step Editor" dialog windows shown in Figure 6 can be used to specify simulation parameters for each branch.

In general, all parameters in parallel modules at a split can be changed for different branches. However, only a few of them will usually sufficiently indentify the physical situation. These are *control parameters* of the split. As an example, control parameters of a MOS process could be channel implant dose and energy. Module/Step parameters identified as control parameters can be used for automated annotation of simulation results, controlling the simulation from a spread-sheet like interface described below, etc.

Control parameters for the simulation can be specified using the "Step Editor" dialog window as shown in Figure 6. Since steps are contained within modules, a module must be selected first from the process flow (See Figures 1 and 5) and a step must be selected in the module. Any parameter of any of the steps contained within a module can serve as a control parameter regardless whether it is going to be used to create a split a not. Thus the user can identify physical process parameters as well as physical and numerical model coefficients as control parameters to study their influence on relevant process or device results or simply to annotate the simulation.

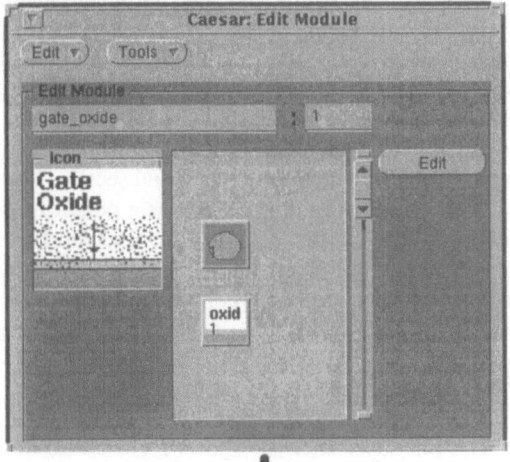

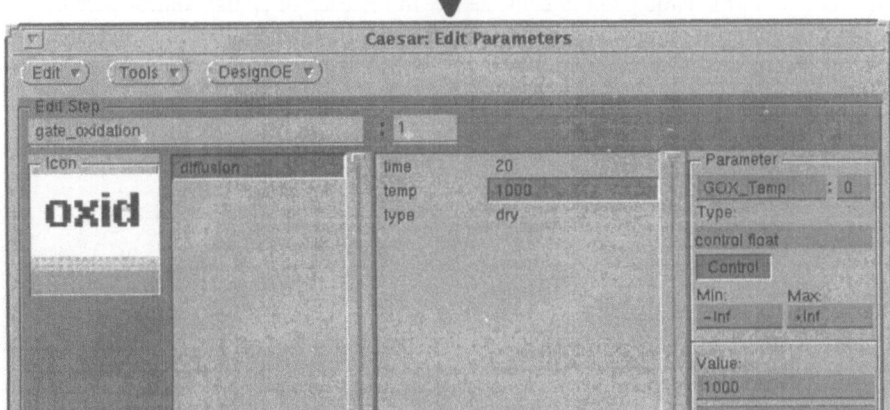

Figure 6. Specifying Control Parameters for Splits on the "Step Editor" dialog window. The edited step is selected on the "Module Editor" dialog window (above).

Once all module/step parameters for the split are specified (control parameters can be adjusted and visualied using the run-sheet interface described in the next section), a wafer tree is constructed and is ready for execution. A make-like algorithm is used by CAESAR to determine which modules in the split need to be executed. This means that changes in either the input wafer state or module/step parameter values result in the re-simulation of the module and all modules whose input depends on the output of this module.

This functionality facilitates so-called *incremental simulation*, meaning that if a part of a complex simulation flow is modified only what is neccessary is re-simulated. This feature is obviously especially important for performing large-scale simulations with complex dependencies. However, even for a relatively simple linear simulation flow it relieves the user from the tedious and error-prone task of creating intermediate simulation files and making sure that changes made in the middle of the process get propagated.

Data management features of CAESAR include automatic naming of files associated with wafers in the wafer tree, removing wafer data whenever the part of the simulation flow is removed which created that data and supplying wafer data to post-processing tools. If a user wishes to examine a wafer state using for instance a graphical post-processor (structural data) or applying a threshold voltage extractor (IV data), he simply selects the desired wafer in the wafer tree window and chooses an appropriate post-processing tool from the "Tools" menu (see Figure 1 and section 7 on Visualization).

CAESAR's wafer tree window can thus be seen as a graphical data viewer, providing visual information about status of the data and interactive access to it. Completed experiments can be "frozen" for archiving purposes by automatically removing unnecessary intermediate wafer states and compressing the relevant ones to reduce memory requirements.

6. The Run-Sheet Interface

A natural interface for visualizing and controlling a simulation, especially one with splits, is a run-sheet. The run-sheet interface shows for each experiment, i.e. each leaf of the tree, a list of relevant control parameter and their values. Control parameters may or may not be used for splits. They represent important simulation parameters, determining the outcome of the simulation in some sense. In the case of splits, control parameters are the most significant parameters that are different in the branching modules. The choice of control parameters is not restricted by CAESAR. As discussed above (section 5), the user can declare any parameter in any module/step to be a control one.

Figure 7. Run-Sheet Interface showing control parameters and their values for an NMOS process and device simulation.

A number of important details can be seen in Figure 7:

• columns in the table (spread-sheet) represent final results of the split, i.e. wafers at leaves of the tree. The wafer status icons used are identical to those used in the wafer tree window. Rows in the table represent control parameter data: Tag, Name, values for each final result wafer.

• Rows are sorted according to their occurence in the simulation flow, so that the table has a similar layout to the wafer tree. The main difference is in the way how common parts of the simulation are shown: branching in the tree and selecting multiple columns in the table (See Figure 7).

• Tags are user-specified names for control parameters (see Figure 6). This name is usually used in the company-specific documentation of the process module. If the user does not wish to specify a name for the control parameter a default name is provided by the system (name of the module).

• The Name field in the table represent the parameter name as used in the simulator command (see Figure 6).

• Values of control parameters can be edited either in-place or through the "Module/Step Editors" shown in Figure 6.

The run-sheet interface is a transparent and convenient way of keeping track of simulations. It is similar to what is being used in real-world semiconductor R&D fabs and as such is expected to be intuitive to engineers. The run-sheet interface is being extended to show extracted scalar results in additional to simulation control parameters. Extracted results would be quantities shown in Figure 2, for instance threshold voltage, oxide thickness, junction depth, etc. (see also section 8).

This combination of the run-sheet with extracted quantities results in a "split-sheet", an interface to show both simulation controls and simulation results. The split-sheet is a natural interface to both statistical Design of Experiment (DOE) tools as well as software packages for statistical analysis of simulation results. It is also a natural interface for optimization software to perform optimization of for instance electrical properties of devices as a function of process parameters and tool calibration, i.e. adjusting simulator model parameters to achieve agreement between experimental and simulated data.

7.   Visualization with Michelangelo and STUDIO Viz

Data management features of CAESAR allow a tight integration with visualization tools. The main visualization tools currently used from CAESAR are Michelangelo (Structure Editor & Visualizer) and STUDIO Viz - a tool for visualization of IV data.

To visually examine a wafer structure the user simply selects a wafer icon in the wafer tree and chooses "Michelangelo" from the "Tools" menu. The result of this operation may look like Figure 8, showing the electric potential distribution in an NMOS device. In addition to visualizing simulation results such as carrier concentrations, temperatures, doping concentration, etc., Michelangelo [4] possesses structure editing capabilities. The user can modify the geometry of the device or even move individual nodes, create new material regions, change material types of regions, etc. Michelangelo can also create a completely new mesh for the device structure and interpolate all solution values onto this new mesh. Two meshing algorithms are accessible: a general unstructured triangular Delaunay-type gridder, and a quad-tree based gridder Meshbuild from ETH, Zurich [6] including improvements done by Texas Instruments [7].

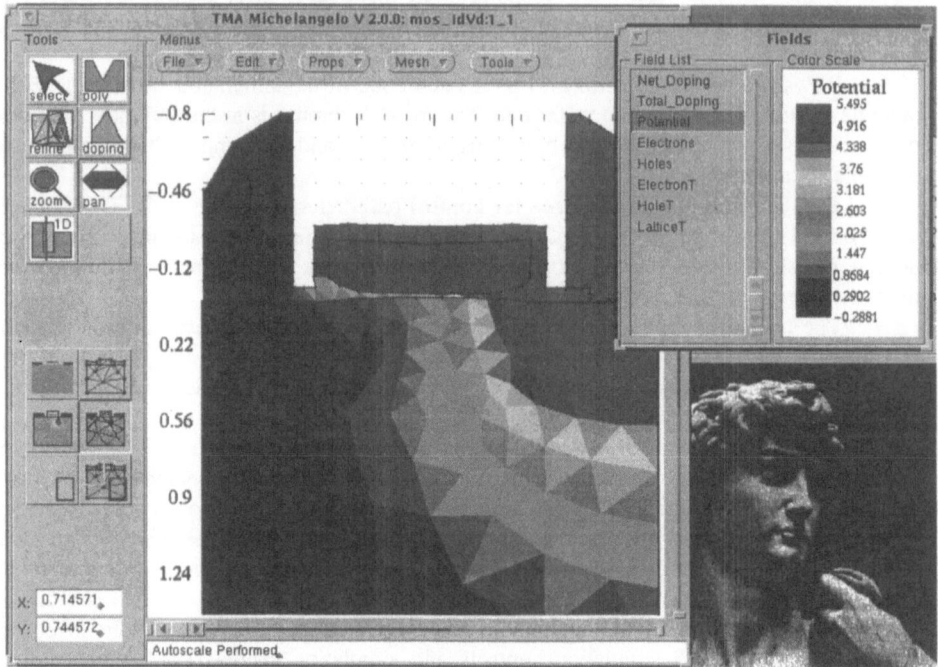

Figure 8. Example for a device structure visualized in Michelangelo (0.35 μm NMOS). The color fill shows the electric potential in the device.

The procedure to examine IV Data is very similar. The user selects the wafer, which in this case must be the result of a device simulation, otherwise IV data will not be present, and chooses "IV Plot" from the "Tools" menu.

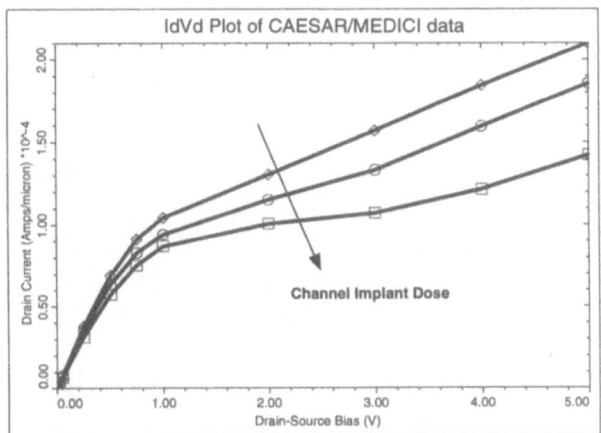

Figure 9. Example for an IV Plot (Drain Current versus Drain-Source Bias for three different channel implant doses) visualized in STUDIO Viz.

8. Post-Processing and Extraction of Results

Post-processing of simulation data is solved in an "open-systems" manner. After a simulation run the user can select a number of wafers in the wafer tree and apply a post-processing tool from the tool palette. The user definable tool driver is then activated. The tool driver may be using a TMA Visualizer such as Michelangelo [4], a parameter extraction tool such as Aurora [5], a device simulator such as MEDICI [1], TSUPREM-4 [2] or an executable created by the user. As a result, a high level of flexibility is avhieved. This is important since it is difficult to predict what the user may need to extract from simulation and not acceptable to restrict the user to a limited selection of extractable quantities.

In many cases the extraction of results will involve multiple wafers. This is typically the case when a comparison between wafers is to be carried out. An example is shown in Figure 9, comparing output characteristics of an NMOS with three different channel implane doses. Another example might be plotting output characteristics with different gate biases or parameter extraction with Aurora using multiple IV Data files (long/short channel, gate/drain characteristics).

Extracted scalar data can be put in the run-sheet user interface, thus showing results of a simulation together with control parameter values. From this interface, the user can access an external software package to perform statistical analysis of the simulation results.

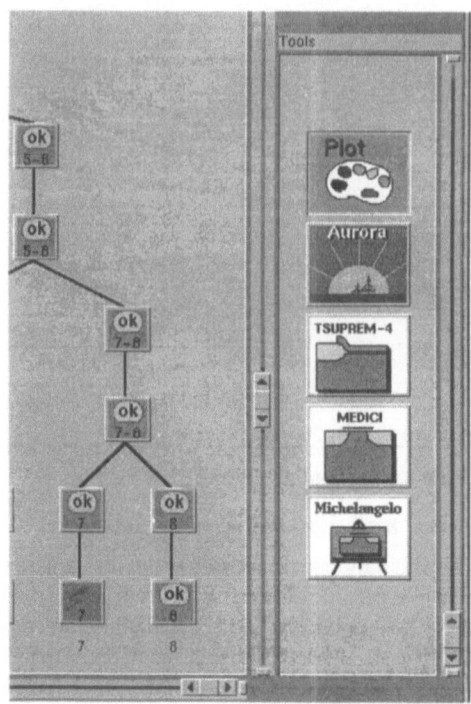

Figure 10. View of the Post-Processing Tools palette. Various tools including parameter extraction, visualizers as well as user-defined post-processing procedures using simulators are visible.

9. Summary

CAESAR, the Virtual IC Factory, is TMA's design of a flexible Open Integrated TCAD system. The tool offers a hierarchical engineering view of the simulation problem, encapsulation of technology- and simulator-specific details and a tight tool integration. A high level of system support is provided for the design and execution of large computer experiments and for the crucial problem of data management. Post-processing and extraction of simulation results are solved in a flexible and customizable manner. User interfaces of CAESAR follow standards and quasi-standards established in the semiconductor industry, thus making sure that semiconductor engineers will not be reluctant to use it. This is achieved by ongoing involvement of major semiconductor companies in the specifications and development of the tool.

References

[1]   Technology Modeling Associates, Inc., Palo Alto, CA., U.S.A., *MEDICI user's manual*, 1993
[2]   Technology Modeling Associates, Inc., Palo Alto, CA., U.S.A., *TSUPREM-4 user's manual*, 1993

[3]   Technology Modeling Associates, Inc., Palo Alto, CA., U.S.A., *TMA SUPREM-3 user's manual*, 1993

[4]   Technology Modeling Associates, Inc., Palo Alto, CA., U.S.A., *Michelangelo user's guide*, 1993

[5]   Technology Modeling Associates, Inc., Palo Alto, CA., U.S.A., *AURORA user's manual*, 1993

[6]   S. Mueller, K. Kells and W. Fichtner, "Automatic rectangle-based adaptive mesh generation without obtuse triangles", IEEE Trans. on CAD/ICAS, 1992

[7]   Texas Instruments, private communication, TI India, Bangalore, 1993

[8]   Technology Modeling Associates, Inc., Palo Alto, CA., U.S.A., *CAESAR user's guide*, 1993

Author Index

Wolfgang Joppich
Slobodan Mijalkovic

Multigrid Methods for Process Simulation

(Computational Microelectronics)
1993. 126 figures. Approx. 340 pages.
Cloth öS 1386,–, DM 198,–
ISBN 3-211-82404-9

Prices are subject to change without notice

This book is the first one that combines both research in multigrid methods and a particular application field here - process simulation. It is the declared intention of this book to convince by practically demonstrating the power of the multigrid principle and to establish an example of fruitful interdisciplinary interaction. The introduction to multigrid is therefore strictly directed towards the goal to provide the algorithmical overview one needs to compose optimal multigrid algorithms for evolution problems of process simulation and similar applications. The necessary explanation how and why multigrid works is derived from the roots. So the book preassumes no advanced familiarity with numerical analysis. Additionally a complete strategy to implement different algorithmical components on an adaptive multilevel grid structure is presented. The outlined principle of grid definement and adaption is based on the control of errors and is reliable as well as general. Last but not least the described strategies are applied to "real life" problems of process simulation.
Consequenly this book is an important contribution to the interdisciplinary challenge of improving numerical techniques for diffusion problems of process simulation.

Springer-Verlag Wien New York

Sachsenplatz 4-6, P.O. Box 89, A-1201 Wien · Heidelberger Platz 3, D-14197 Berlin
175 Fifth Avenue, New York, NY 10010, USA · 37-3, Hongo 3-chome, Bunkyo-ku, Tokyo, 113, Japan

Narain D. Arora

MOSFET Models for VLSI Circuit Simulation
Theory and Practice

(Computational Microelectronics)
1993. Approx. 260 figures. Approx. 600 pages.
Cloth öS 2086,–, DM 298,–, US $ 198.00
ISBN 3-211-82395-6

Prices are subject to change without notice

The book covers the MOS transistor models and their parameters required for VLSI simulation of MOS integrated circuits. It gives the first detailed presentation of model parameter determination for MOS models. Various models are developed ranging from simple to more sophisticated models that take into account new physical effects observed in submicron devices used in today's MOS VLSI technology. The assumptions used to arrive at the models are emphasized so that the accuracy of the model in describing the device characteristics are clearly understood. Understanding these models is essential when designing circuits for the state of the art MOS IC's. Threshold voltage being the single most important MOSFET parameter, a full chapter is devoted to the development of the device threshold voltage model. Due to the importance of designing reliable circuits, the device reliability models as applied for circuit simulations are also covered. Since the device parameters vary due to inherent processing variations, how to arrive at worst case design parameters are covered.
Presentation of the material is such that even an undergraduate student not well familiar with semiconductor device physics can understand the intricacies of MOSFET modeling. The book serves as a technical source in the area of MOSFET modeling for state of the art MOSFET technology for both practicing device and circuit engineers and engineering students interested in the said area.

Springer-Verlag Wien New York

Sachsenplatz 4-6, P.O. Box 89, A-1201 Wien · Heidelberger Platz 3, D-14197 Berlin
175 Fifth Avenue, New York, NY 10010, USA · 37-3, Hongo 3-chome, Bunkyo-ku, Tokyo, 113, Japan